Marine Power Plant

Zongming Yang · Huabing Wen · Xinglin Yang ·
Viktor Gorbov · Vira Mitienkova · Serhiy Serbin

Marine Power Plant

SSTP

Springer

Zongming Yang
Jiangsu University of Science and Technology
Zhenjiang, Jiangsu, China

Xinglin Yang
Jiangsu University of Science and Technology
Zhenjiang, Jiangsu, China

Vira Mitienkova
Admiral Makarov National University of Shipbuilding
Mykolayiv, Ukraine

Huabing Wen
Jiangsu University of Science and Technology
Zhenjiang, Jiangsu, China

Viktor Gorbov
Admiral Makarov National University of Shipbuilding
Mykolayiv, Ukraine

Serhiy Serbin
Admiral Makarov National University of Shipbuilding
Mykolayiv, Ukraine

ISBN 978-981-33-4934-6 ISBN 978-981-33-4935-3 (eBook)
https://doi.org/10.1007/978-981-33-4935-3

Jointly published with Shanghai Scientific and Technical Publishers, China
The print edition is not for sale in China Mainland. Customers from China Mainland please order the print book from: Shanghai Scientific and Technical Publishers.

This Springer imprint is published by the registered company Springer Nature Singapore Pte Ltd.
The registered company address is: 152 Beach Road, #21-01/04 Gateway East, Singapore 189721, Singapore

Preface

The book "Marine Power Plant" provides an insight of the history and development of the whole variety of power plants, the current state of the marine power engineering and its achievements.

For the first time ever in the educational literature, the book presents a detailed retrospective of the appearance and establishment of the main types of heat engines used on ships, ranging from steam engines to gas turbine engines and nuclear power plants.

The issues of energy conversion and transfer in marine power plants are discussed in detail, since they lay the basis for determining the plants' energy efficiency.

When considering characteristics of the fuels employed in marine power plants, a lot of attention is paid to alternative fuels, which have found an extensive application on ships over the last few years.

There is also an elaborate analysis of the sources of air pollution during the operation of marine power plants with heat engines of various types.

The greatest attention has been given to diesel marine power plants, which entirely corresponds to the place they occupy on ships as compared to the plants with other types of heat engines. The analysis of thermal and structural diagrams of diesel power plants reveals the sheer variety of ways to use the products of engine building in marine power engineering. The detailed description of the issues of labeling marine diesels of the world's leading manufacturers makes it possible to be conversant in their types, dimensions, special features, and structures.

The atlas of major devices, components, assemblies, and systems of marine diesel engines is based on the design of modern diesel units; considerable attention is paid to the engines with an electronic control system.

On top of that, there are presented the main methods and patterns for reducing the harmful emissions released by diesel plants into the atmosphere. Of undoubted interest is the example of placing the equipment of a diesel power plant in the modern ship's engine room.

The book also shows the principal and thermal schemes of marine steam turbine units with account to their characteristics and general arrangement of their elements.

Gas turbine plants have found application not only on warships, but also on transport vessels, especially the high-speed ones. Thus, formation of the world and

Preface

The book "Marine Power Plant" provides an insight of the history and development of the whole variety of power plants, the current state of the marine power engineering and its achievements.

For the first time ever in the educational literature, the book presents a detailed retrospective of the appearance and establishment of the main types of heat engines used on ships, ranging from steam engines to gas turbine engines and nuclear power plants.

The issues of energy conversion and transfer in marine power plants are discussed in detail, since they lay the basis for determining the plants' energy efficiency.

When considering characteristics of the fuels employed in marine power plants, a lot of attention is paid to alternative fuels, which have found an extensive application on ships over the last few years.

There is also an elaborate analysis of the sources of air pollution during the operation of marine power plants with heat engines of various types.

The greatest attention has been given to diesel marine power plants, which entirely corresponds to the place they occupy on ships as compared to the plants with other types of heat engines. The analysis of thermal and structural diagrams of diesel power plants reveals the sheer variety of ways to use the products of engine building in marine power engineering. The detailed description of the issues of labeling marine diesels of the world's leading manufacturers makes it possible to be conversant in their types, dimensions, special features, and structures.

The atlas of major devices, components, assemblies, and systems of marine diesel engines is based on the design of modern diesel units; considerable attention is paid to the engines with an electronic control system.

On top of that, there are presented the main methods and patterns for reducing the harmful emissions released by diesel plants into the atmosphere. Of undoubted interest is the example of placing the equipment of a diesel power plant in the modern ship's engine room.

The book also shows the principal and thermal schemes of marine steam turbine units with account to their characteristics and general arrangement of their elements.

Gas turbine plants have found application not only on warships, but also on transport vessels, especially the high-speed ones. Thus, formation of the world and

Ukrainian marine gas turbine construction and application of gas turbine plants on marine transport are considered in detail; the characteristics of advanced gas turbine engines of the world's leading manufacturers are given as well.

The book contains a lot of illustrations, which are mainly published for the first time; they help to get a better understanding of the educational content. The actual designs of modern ship internal combustion engines, gas turbine engines, and steam turbine units, specifications of power equipment and propulsion complexes, thermal and principal schemes of marine power plants given in the book all offer ample opportunities for their application in the readers' individual work.

The book content is systematized in such a way that the reader could navigate through it easily at the self-study of the subject.

Zhenjiang, China Zongming Yang
Zhenjiang, China Huabing Wen
Zhenjiang, China Xinglin Yang
Mykolayiv, Ukraine Viktor Gorbov
Mykolayiv, Ukraine Vira Mitienkova
Mykolayiv, Ukraine Serhiy Serbin

Contents

Chapter 1
History of Marine Power Development

Since ancient times, to provide vessel traffic men persistently searched the way to replace the unpredictable in strength and direction of wind, and to replace oars, which required many workers. In the history of shipbuilding, there are many attempts to build a boat, which is driven by a motor controlled by a man, but only when mankind has learned to use the power of steam, the creation of such an engine has become a reality.

The history of the development of marine power plants is inextricably connected with the history of heat engines creation to solve general energy problems. It is a reflection of the almost four centuries of human development period, when the truth was extracted step by step by the method of trial and error, in conditions of almost complete lack of theoretical basis of ideas, projects, schemes and processes.

Even such a modest excursion into the history of a particular area of technology, which is carried out within this publication, is related to the common challenges that consist of contradictions of dates, facts, parameters, characteristics which are outlined in the materials from different sources and by different authors, with their subjective approach, which is not always possible to check.

At the same time possession of a history of the issue makes it possible not to repeat the mistakes made before, gives the opportunity to use the ideas that were ahead of their time and could not be implemented earlier.

1.1 Pioneers of Steam Era

Patents, books, articles written at different times, in different languages, dedicated to the history of attempts to use steam power to produce useful work, being collected in one place could constitute a considerable library.

After all, the invention of the steam engine, which has provided humanity energy it needs, significantly sped up the movement by not only technical, but also social progress.

Z. Yang et al., *Marine Power Plant*,
https://doi.org/10.1007/978-981-33-4935-3_1

The beginning of this history goes to the hoary antiquity—ancient times. Ancient Greeks skillfully used the power of steam, but as a rule, in religious rituals.

It is impossible not to remember the invention of Heron of Alexandria—famous aeolipile—the ball spinning under the influence of steam that comes out of the two bent tubes. It transforms heat into mechanical work, and the idea of using steam energy by leakage in the circumferential direction of the streams was applied later at creating steam turbines.

More than fifteen centuries after Heron no one made attempts to use steam.

In 1615 in Frankfurt the book "The reasons of driving forces with various useful and funny stories" was published, which belonged to a peer of the French engineer and architect Salomon de Couss. It mentions the possibility of raising water by fire from the ball, in which steam is generated. The pipe passed through the round container cover almost to the bottom. The ball was filled with water, sealed and the fire was spread under it. The resulting steam pressed on water above, and the water rose through the tube, even if it was sufficiently long.

Marquis of Worcester (England) in 1663 published a book which described the invention, which was connected with the original device for lifting water by means of steam [1].

Almost the same way as Worcester, another contender to the title of inventor of steam machine—Samuel Morland—moved further in creating it, as evidenced by the title of his work "The rise of water by all kinds of machines, reduced to the weight measurement". Morland owned the processes of transformation of water into steam.

These inventions were not engines; they could not be used for certain technical use. However, they demonstrated the possibility of obtaining mechanical work using fire and steam. Creating heat engines would not be possible without these inventions.

The appearance of thermal engines is connected with the origination and development of industrial production in the early seventeenth century. The biggest success in this area has been achieved in England, the most developed country by that time. Thermal engines were needed to solve the important practical problem—pumping water out of coal and ore mines.

In 1698, Thomas Savery, a former miner, and then captain of the merchant fleet and coal-owner, received a patent for a steam engine. Patent № 356 testified that it is issued to a device for lifting water and receiving traffic for all types of production by the driving force of the fire. Savery was the first to separate the working body—water steam—from the pumped water and used the condensation of steam through its cooling.

The steam from the boiler (Fig. 1.1) was ejected through tap into a water container and forced the water into the pressure (upper) pipe. Savery managed to implement repeatable cycles unlike all precursors. This was achieved when water was forced out completely and the vessel remained filled with steam and the pressure pipe was overlapped (the suction valve was installed at the bottom of pipe, and the discharge valve—at the top). Next, the vessel was poured over by cold water so that the steam condensed in the vessel and the rarefication appeared, due to that the next portion of water was sucked through the inlet pipe. Savery's machines were used for water supply, drainage of wetlands and meadows; they had little in common with modern

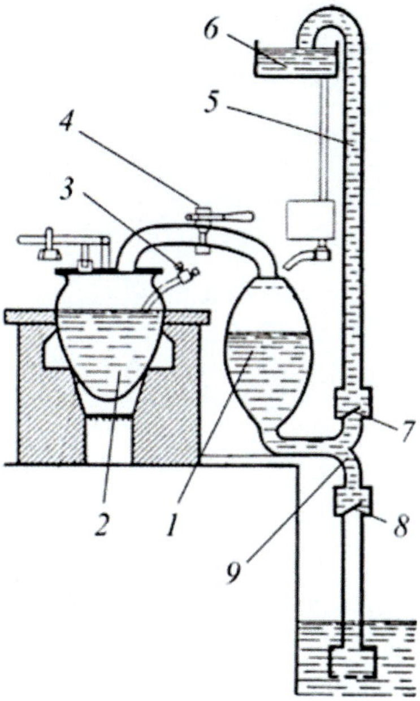

Fig. 1.1 Savery's steam water lift: 1—vessel containing water; 2—boiler; 3—water level control pipe in the boiler; 4—valve; 5, 9—pressure and suction pipes; 6—upper reservoir; 7, 8—discharge and suction valves

concepts of steam engines, because they do not have moving parts (now they are referred to as thermomechanical pumps). Still, it was the first machine capable of operating continuously [2].

A significant influence on the further development of heat engines was a French physicist Denis Papin, known for the invention of the autoclave, the relief valve and powder machine. In 1698, in search of a universal motor, he decided to apply water steam, the condensation of which is accompanied by a decrease in volume by more than 1500 times at atmospheric pressure.

The scheme of D. Papin's steam-atmospheric engine is presented in Fig. 1.2. By heating in the cylinder water turned into steam under the influence of which the piston, overcoming the air pressure, was raised to the upper position in which it was fixed by the stopper. After this, the heating source was removed, the steam was condensed while cooling the cylinder, and the rarefication appeared. If we now remove the stopper, under the influence of atmospheric pressure difference and the rarefication inside the cylinder the piston will move down and lift the load.

Steam engine cycle was determined correctly by Papen. The modern steam-powered installations carry out such cycle in a complex chain of devices: the boiler, engine, condenser, and pump. The Papin's engine combines all processes in a single cylinder. Therefore, it could not do more than one stroke per minute, and it was almost unusable.

The main merit of Savery and Papin is that they applied in practice the main conditions of heat transformation into work: it is not enough to have an elastic

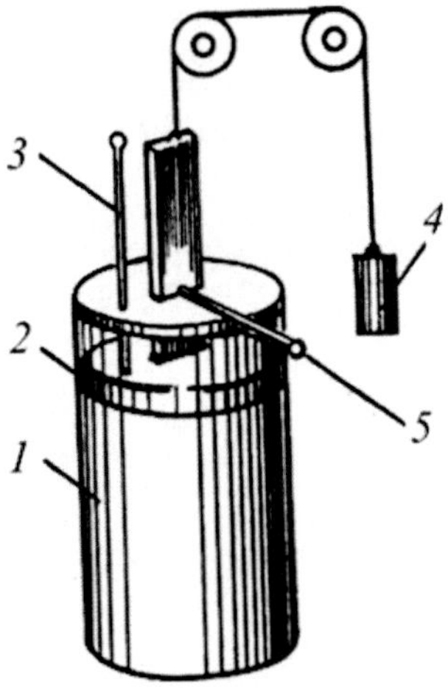

Fig. 1.2 D. Papin's engine scheme: 1—cylinder; 2—piston; 3—air release valve rod; 4—load; 5—stopper

working body and supply the heat to it; this working body should give the part of the heat to the cold source after expansion.

The next important step towards creating a heat engine was carried out by an English inventor, a blacksmith by profession, Thomas Newcomen.

He attempted to combine the ideas of Savery and Papin in his machine, to use the Papin cylinder with a piston, but to carry out the condensation of steam not by removing the fire out of the cylinder, as proposed by Papin, but due to its cooling.

Newcomen's machine (Fig. 1.3) consisted of two vertical working cylinders, the steam came by turns under the pistons of these cylinders from the common pot through a distribution valve. The steam pressure raised the piston, after reaching the highest point the water was injected in the cylinder, the steam condensed and the piston went down under atmospheric pressure implementing stroke. Pumping

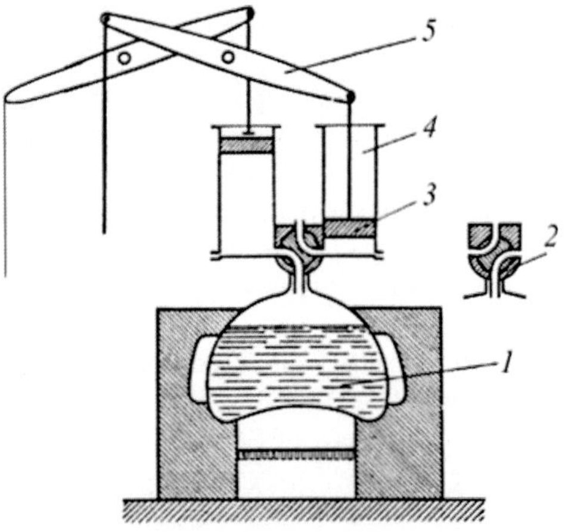

Fig. 1.3 Newcomen's water bilge machine: 1—boiler; 2—distribution valve; 3—piston; 4—working cylinder; 5—rocker

cylinders were arranged parallel to the working ones, their pistons were connected to the pistons of working cylinders through rocking bar—rocker.

In 1705, to implement his ideas Newcomen concluded an agreement with an industrialist—glassmaker John Cowley—and soon Newcomen's machines became widely used.

The alternation of the cylinder cooling by injected water with its following heating with steam brought to a significant loss of heat, and therefore the efficiency was about 1%. There were up to 50 horses kept for some Newcomen machines, which barely had time to bring up fuel. Disadvantages of steam-atmospheric machines made themselves felt, and soon the engines were dismantled.

The attempts to apply the existing steam engines for ships movement ended after the famous Russian scientist Daniel Bernoulli proved that these engines, including Newcomen's machine, are not able to provide vessel movements.

In 1736, Newcomen's compatriot Jonathan Hoole adapted his machine to vessel movements [3].

He set the steam-atmospheric machine on several sailboats as an auxiliary motor. The transfer from the machine to the paddle wheel located at the stern was carried out through a system of belts and blocks.

The first universal heat engine has been created in Russia by a mechanic of Resurrection plants in the Altai I. I. Polzunov. He first realized that one could make a steam engine propel not only the pump, but also bellows. Working bodies of his cars constantly passed the motion to PTO shaft. This feature gave the Polzunov's machine versatility.

The project was described in the representation of the head of the mining office. The scheme of Polzunov's steam engine is shown in Fig. 1.4. Two pistons are connected to the main shaft by means of chains. Water distribution valve alternately supplies cooling water to the subpiston cavities. Steam distribution valve combines the cavity with the boiler. When one of the cavities was connected to the boiler, the piston rose up, the steam distribution valve cranked and cut the subpiston cavity from the boiler. Water was injected through the tube, the steam condensed and vacuum was formed under the piston, and under the influence of atmospheric pressure the piston went down and carried out useful work. In contrast to all previous engine designs, working bodies of I. I. Polzunov and power consumers are not connected one by one.

Operating force is transmitted from the pistons to pulley, from which steam-water distribution device, nutrient piston pump and the working shaft are driven, and the last one in turn actuates the smelters bellows. Proposed summation by Polzunov on the same shaft of several cylinders in the future was the basis for the creation of multi-cylinder engines.

The test of built Polzunov's machine was held in August 1766, after his death. For nearly two months it worked for its intended purpose.

The scales of Polzunov's machine are indicated by its dimensions: height 11 m, for its manufacture 25 tons of iron and copper, more than two tons of lead and tin were expended, machine power was 30 hp.

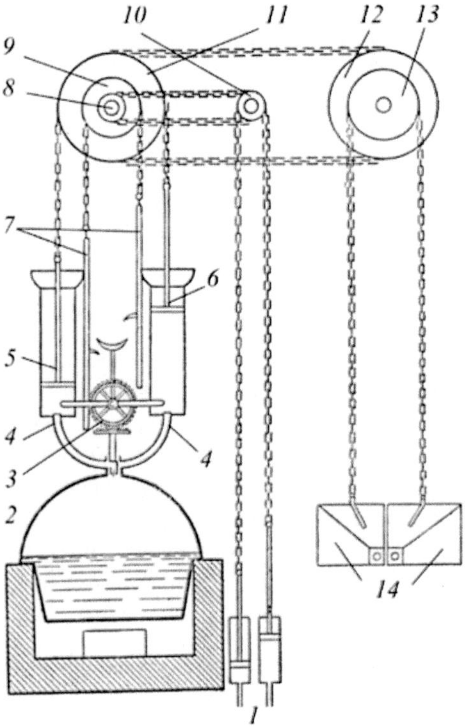

Fig. 1.4 Polzunov's steam engine scheme (1763). 1, 10, 12, 13—actuator to the blower bellows; 14, 2—steam boiler; 3—steam distribution valve; 4—pipes for injecting water into the cylinder cavity; 5, 6—pistons; 7—rods of steam-water distribution mechanism; 8—engine main shaft; 9, 11—drive pumps

Polzunov did not take a patent for his invention, so the world's first patent for a universal heat engine was issued in England by James Watt in January 5, 1769, six years later than the submission of the draft of Polzunov's machine was dated.

James Watt was a talented mechanic of Glasgow University; he maintained the relationships with many scientists, engaged in self-development in the field of heat. He studied the properties of the water steam, measured the water evaporation heat, determined the amount of steam, which turns out of water while boiling, and experimented with instruments of his own design.

In 1764, they turned to Watt with a proposal to improve the steam engine, proposed by the inventor Thomas Newcomen, in order to increase its effectiveness. Within 2 years James Watt created the industrial design of the new machine, which had a greater capacity almost doubled, he called it "Beelzebub". The increase of efficiency of this steam engine was promoted by the centrifugal governor application created by Watt. The owner of the ironworks John Rebec financed the work of Watt and assisted the embodiment of his ideas and the industrial production of steam engines.

After patenting his invention in London, Watt moved to Birmingham at the invitation of the famous industrialist Matthew Bolton, where the mechanical workshop has been fully provided for the work on the improvement of the steam engine.

In 1783, James Watt was elected to the Royal Society.

Watt was the first to develop the basic principles of a universal steam engine:

- cylinder, which is supplied by steam, must be constantly hot as steam, for which it is necessary to maintain permanently heated state, creating a shell around the steam;
- condensation process should be imposed for the cylinder boundary, steam condensation for the formation of a vacuum should be held at a temperature no higher than 30 °C;
- machine capacity can be increased if the piston does not affect the atmospheric pressure, but steam pressure;
- steam machine must not be connected to any power consumer; it should be a universal heat engine.

By the end of 1765, Watt had shown the work of his engine, in 1769 he built a more sophisticated experimental model, and in January 1769 he got a patent for his invention. The real triumph came when Watt established manufacturing steam engine for practical use in 1774 in the factory of the English industrialist M. Bolton (Fig. 1.5).

Cylinder without the top cover is placed in the steam shell, it connects with the condenser by a tube, where the balancing and steam-admission valves are disposed. When starting the machine, both valves are opened and the steam displaces air in the condenser, whereupon the valves are closed and the system is in equilibrium. With the opening of the bottom steam-admission valve from the space under the piston the

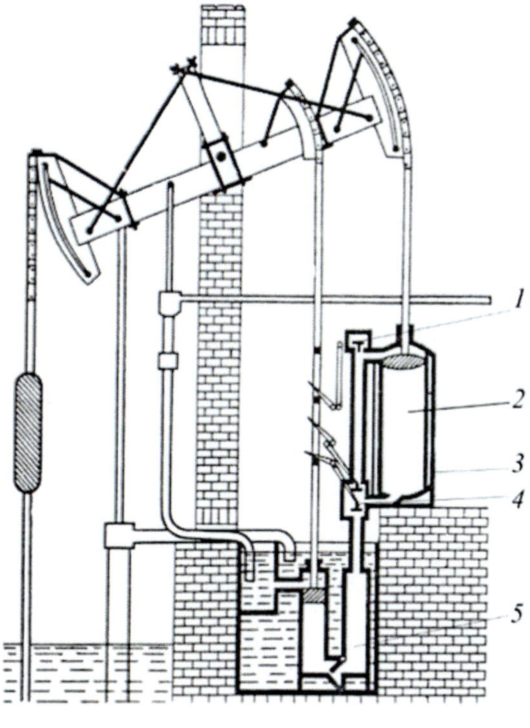

Fig. 1.5 Watt's steam engine: 1, 4—balancing and steam-admission valves, correspondently; 2—cylinder with the piston; 3—steam jacket; 5—condenser

steam moves to the condenser where it is condensed. Because of formed vacuum the piston under steam pressure, which comes from the boiler to the upper cavity of the cylinder, moves downwards, doing useful work that will be transmitted to the rocker via pump rod.

Next, the steam-admission valve is closed and the balancing valve opens. Fresh steam is supplied under the piston, the counter weight raises the piston, then the steam-admission valve opens and the cycle is repeated. According to economy, Watt's steam engines were two times better than the Newcomen's machine.

Efficiency of Watt's steam engine was increased to 4.1%. At the same time, it was not a completely universal motor.

In 1782, Watt got a patent for a machine in which both strokes were working. This machine, which was called a dual action machine, had already been continuously operating heat engine. Watt had to solve a more difficult task to convert a rectilinear reciprocating piston motion into rotary motion of the shaft. To this end, Watt creates a flat hinge mechanism—the so-called Watt's parallelogram.

By 1794—a quarter of a century after the invention of Watt—only in England there were 1500 steam engines, which replaced the work of 180,000 horses. To measure the power of steam engines D. Watt introduced the term "horsepower", which today still exists as a common unit of measurement.

People began to produce steam engines in France and Germany, Belgium and Sweden; in 1799, the first steam engine was installed in Russia after the Polzunov's engine.

If shipbuilders showed keen interest even to imperfect steam-atmospheric machines, then with the appearance of the steam engine the idea of a self-propelled vessel began to be rapidly realized.

In 1783, a French officer Claude Geaffroid'Abban built and publicly tested the steamship "Pyroscaphe" near Lyon after many unsuccessful attempts. This ship was driven by paddle wheels that rotated the steam engine; it could move against the current more than an hour.

Serious works on the creation of ships were carried out in America. In 1784, the inventor J. Ramsay presented his project of adjusting the Watt's steam engine for vessel traffic.

The project of Ramsay's a steam engine involved water jet propulsion device of an original construction. Its basis was a steam-atmospheric machine with a floating piston.

The inventor was disappointed with the results of tests: the power plant consumed a considerable amount of fuel and the ship's speed did not exceed three knots. After this project, Ramsay immigrated to England, where in 1798 he built a water jet boat "Columbia Maid", but his death prevented its testing [4].

At the end of the eighteenth century, the inventors attempted to apply paddles, endless chains with blades, paddle wheels in the construction of steam ships as the engines.

In 1787–1788, the American watchmaker John Fitch built three steam vessels, each of which was equipped with its original driving force. After a series of unsuccessful attempts to create steam ships with the original propellers (boat "Perseverance", boat "Experiment" etc.) D. Fitch tries to build a boat with a propeller, but the lack of funds did not allow him to finish the job [1].

Another attempt to apply Watt's steam machine for marine purposes was carried out by the British men Peter Miller and William Symington. They placed the paddle wheels in the inside of the hull; machinery shaft rotation of the wheels was transmitted via a chain drive. The tests, which were held in 1789, were successful, and the inventors have constructed a large vessel with a steam engine capacity of 12 hp. Unfortunately, frequent breakdowns of paddle wheels prevented its use.

An important step forward was that Symington, who built the towboat "Charlotte Dundas" in 1802 in England, used the Watt's steam engine with the capacity of 10 hp, which rotated the paddle wheel located at the stern by means of a crank.

The application of direct transmission from the piston rod to the shaft of the machine, as a part of this power plant, eliminated the bulky rocker.

The test demonstrated that the ship could tow two barges at a distance of 18 miles at a speed of 3 kt with a strong headwind. At the same time, concerns about the possibility of canal banks erosion by ship motion in waves led to the banning of its use.

The continuer of the ideas of D. Ramsay became Robert Fulton, who was born in the United States in a family of Irish emigrant.

After critically analyzing the drawings and calculations that remained after D. Ramsay, Robert Fulton conducted a research to determine the optimal characteristics of the vessel to reduce the resistance. He also investigated the features of interaction of the body with different propellers. Such actions of R. Fulton fully met modern ship design principles.

In 1796, Fulton moved to France, where he met the US envoy Robert Livingston, who became his partner and patron of the arts. This meeting led to the fact that already in 1802 the first Fulton's steam vessel realized the shakedown cruise on the river Seine in the presence of numerous spectators. For three hours, the vessel was going against the flow at a speed of 3 knots. Not having received the support of the French government, in 1806 Fulton returned to the US and started the construction of a new ship.

One year later, on August 17 1807, the ship "North River Steamboat of Clermont" was launched on the Hudson River. It was known as the "Clermont", a vessel with the displacement of about 80 tons, with five-meter paddle wheels that rotate at a frequency of 20 min^{-1}. A few days later it was reported in the newspapers about the opening of the regular steamer connection between New York and Albany, the distance between them was 150 miles. This distance was completed in 32 h, i.e. at a speed of about 5 knots. The length of the wooden vessel was 30.5 m, width—9.8 m. Board paddle wheels were driven by steam-cylinder engine producing 20 hp. The diameter of the engine cylinder was 610 mm, piston stroke length—1220 mm. The general view of the power plant of the steamer "Clermont" is shown in Fig. 1.6. It

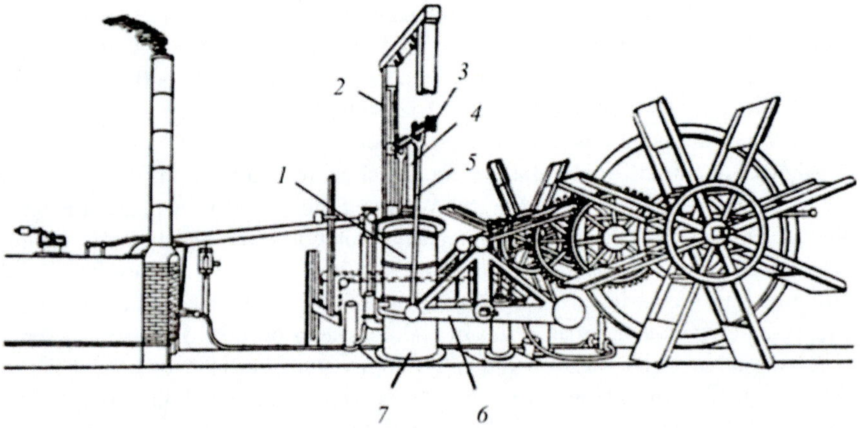

Fig. 1.6 The steam engine of the steamboat "Clermont": 1—piston; 2—guide; 3—crosshead; 4—rod; 5—piston rod; 6—rocker; 7—cylinder

was Fulton who not only used the steam engine, but also built a suitable boat for everyday use; he is considered to be the creator of the first steamboat.

During 1811, Fulton built four new steam boats intended for navigation on the Hudson; soon the steam boats appeared on the Mississippi and Ohio rivers.

The last project of the inventor was the first ever steam ship—steam-frigate "Demologos" with a displacement of 2000 tons, which was later renamed "Fulton-1". The typical for this class of ships sailing equipment was installed on this ship, along with the steam engine. The ship is made double-hulled; the boiler was located in one frame; the steam engine was in the other one. Accommodation of the paddle wheel between the frames protected it from combat and navigation damage. At the trials in 1815, "Demologos" reached the speed of 5 knots. All ships constructed by Fulton became the property of the first US cargo-passenger company.

In 1812, Henry Bell (England) built a steam ship "Comet", which was carrying passengers on the River Clyde.

The first Russian steamer, built in St. Petersburg at the Charles Byrd Mechanical Plant (now it is St. Petersburg Admiralty Association), was the boat "Yelizaveta", which began to carry out voyages between Petersburg and Kronshtadt (Fig. 1.7) in 1815. The speed of the ship was 5.8 knots, length—18.3 m, width—4.0 m, draft—0.6 m. The steam engine, with the capacity of 4 hp, moved the six-bladed side wheels, with the diameter of 2.4 m, width of 1.2 m, which rotated at a frequency of 40 min^{-1}.

Almost at the same time, the construction of ships began on Pozhevskiy plant on the river Kama, Russia. In 1817, the two ships were built with the engines of 36 and 6 hp. All the ships of that time sailed only on rivers or inland waters.

The first steamship to cross the Atlantic was the American steamer "Savannah", which in 1819 carried out a transatlantic voyage en route "New York—Liverpool—Petersburg" [5]. It was a three-masted steam-sailing ship with the length of 30.5 m, width of 7.9 m, and displacement of 350 t. The paddle wheels could be stripped and

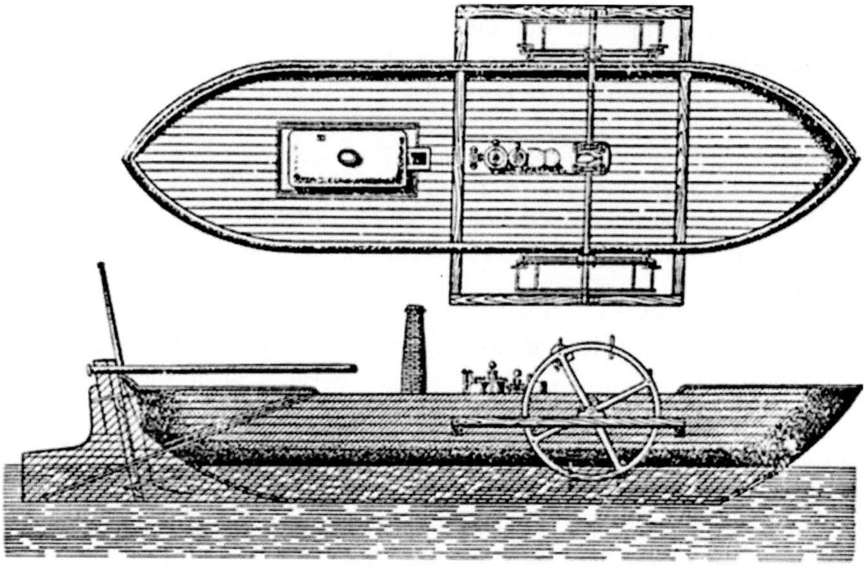

Fig. 1.7 The first Russian steam boat "Yelizaveta"

carried on deck. Fuel capacity of the steam engine provided work for 90 h. Most of the way the vessel was sailing. The inclined single cylinder steam engine had a cylinder with the diameter of 1035 mm, piston stroke of 1530 mm, the output shaft speed—16 min^{-1}. In its first voyage, "Savannah" left the US port on May 22, 1819 and in 29 days came to the Irish port of Kinsale, at that the last 80 h the vessel used just the steam engine. From Ireland the way was to Sweden, and then to St. Petersburg. Returning to America, "Savannah" exceptionally sailed.

In 1827, a wooden paddle steamer "Curacao" carried out the first crossing of the Atlantic from Europe to America without the help of sails. Its displacement was of 440 tons with a steam engine of 100 hp. When the English ship "Sirius" crossed the Atlantic in 1838 under the steam engine, it was given a grand welcome in New York. Hosswever, the celebrations were marred by objective circumstances. To be the complete leader in shipbuilding, the steam engine lacked a lot. Firstly, the speed of steam ships was smaller than the speed of sailing boats: when "Sirius" crossed the Atlantic Ocean in more than 18 days, the sailing ships covered that distance in 11–13 days. Secondly, the cruising range of sailing ships confined to food supplies and drinking water, while the ships needed a huge amount of fuel. In the furnaces of "Sirius", for example, about one ton of coal burnt every hour, for the transition it was 440 tons. At that, the tonnage of a ship did not exceed 700 tons. Third, the paddle wheels brought a lot of inconvenience by being completely or partially out of water while pitching, and then they dipped into the water again, which caused considerable tension in the elements of the ship's structure, and even the emergency tension.

The first steamships were exclusively wheel ships. The propulsion wheel is characterized by many weaknesses, so that it had a very low efficiency. A decent alternative to the paddle wheel was a propeller. Many inventors offered their designs. The most successful projects were the Czech Joseph Ressell's (patent in 1827), the Englishman O. Smith's and the Swedish D. Erickson's ones (patent in 1836). The practical application of the propellers began only in the 1840s.

The first screw steamer was built in England in 1838. It was the ship "Archimed" with the displacement of 237 tons, with a propeller, which was designed by A. Smith (Fig. 1.8).

The vessel had a low-speed machine, which was used to drive the paddle wheels. According to this, to increase the propeller shaft rotational speed that is required for effective use of the propeller, the designer was forced to input the multiplier into power plant, the loss of which led to a decrease in power transmitted to the screw [1].

The next step in building ships with propeller was the construction of steam-frigate "Prinstone" in the United States in 1843 by the Swedish shipbuilder D. Erickson. The ship was with the displacement of 950 tons. The propeller was driven by two speed steam engines with a total capacity of 400 hp with direct transfer to the screw.

Proof of recognition of steam ships in the middle of the XIX century was a decree issued in France in 1857, according to which all vehicles not equipped with steam engines were excluded from the lists of the fleet.

At the same time, the effectiveness of MPP with a steam engine was still very low; the efficiency of the best of them to the middle of the XIX century did not exceed 4–5%. Thus, out of each ton of fuel only 40—50 kg was expended for useful work, and the remaining amount was absorbed by loss.

This made the energy experts study and search for studies of processes that occur in boilers and steam engines.

The establishment of thermodynamics fundamentals as the science of engineering is inseparably connected with the name of Sadi Carnot, who showed that the presence of a hot spring of constant temperature T_1 and a cold spring of temperature T_2 provided the most effective cycle, which consists of two isothermal and two adiabatic processes; he also determined its efficiency.

The theoretical value of the cycle proposed by Carnot is that it provides an opportunity to assess the degree of perfection of all cycles used in thermal power plants. It is the ideal Carnot cycle, since the actual power plant worked, the heat losses were inevitable; thereby the proportion of the heat transformed into work was reduced.

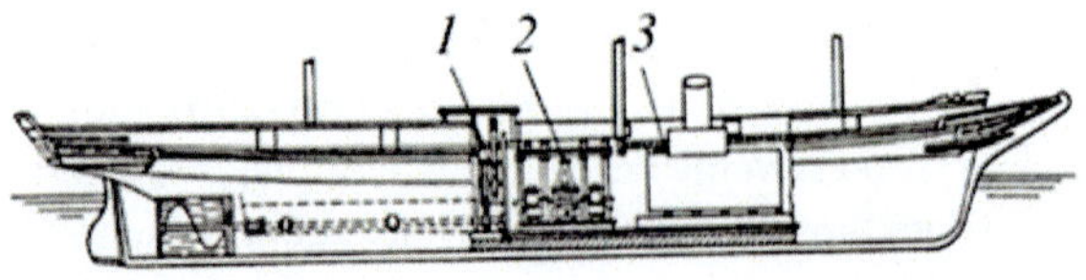

Fig. 1.8 Longitudinal section of the ship "Archimed" in the area of machine branch: 1—gearbox; 2—steam engine; 3—steam boiler

Real steam-energy plants operate on a different cycle proposed by the English engineer William Rankine and named after him. In such plants, the spent steam is completely cooled in a condenser, turning into the water, allowing using a compact feed pump instead of a bulky powerful compressor. Only 1–2% of setting power is spent to operate a compact feed pump, due to the incompressibility of water.

The thermal scheme of steam-energy plants was constantly improved, developed, stabilized until the end of the XIX century and essentially preserved up to the present time.

The development of power plants with steam engines was accompanied by improvement of their component parts. With increasing pressure steam up to 0.3–0.5 MPa (in the first steam engines the pressure did not exceed 0.13–0.15 MPa) twice, expansion steam engines were extended; they increased the degree of steam expansion, speed of pistons increased, and along with it the shaft speed of the machine increased too.

Significant complexity was brought by the exhaust steam mixing in a machine with seawater in a condenser, in which steam was cooled. To condense one kilogram of waste steam it needed to mix it with 20–25 kg of seawater. The mixture of condensate and seawater was pumped, much of it was removed over the side, and the remaining amount was fed to the boiler feed pump, which worked almost entirely on the outboard water.

In the late 60 s, people started to apply the surface condensers on the ships, in which the original cooling water washed the tubes and steam flowed inside the tubes. Later on, the constructive scheme of condensers changed and acquired a modern look: steam washed the tubes with pumped seawater, and then was cooled and condensed.

In the second half of the XIX century, the efficiency of steam-engine plants was about 8%, and the effectiveness of the propeller was estimated by the efficiency that did not exceed 30%. Thus, the effectiveness of conversion of fuel energy into vessel motion work was about 2.4%; that is almost 98% of the fuel energy lost.

In the intent to improve the performance of steam power plant, the designers paid attention to steam boilers, weight and dimensions of which largely determined the portion of displacement that was allocated for power plant. Weight of boilers reached 100–120 tons at that time, the number of installed boilers on one ship or boat reached several dozen units.

The most effective way to reduce weight and dimensions of the steam boiler is the intensification of the combustion process. To ensure that on ships, at the end of the XIX century they began to use forced draft instead of natural draft.

The use of artificial draft was the solution of many problems. Boilers with forced draft had twice smaller size and weight compared to the boilers of the same steam rate, but with natural draft. The ship pipes height reduction was essential for war ships.

Along with these, boilers began to be equipped with the tubular heat exchangers which were disposed in the gas-conducting part, where the air, which was supplied to the boiler, and feed water were heated, resulting in a significant increase in boiler efficiency.

The effectiveness of steam power plants contributed to the improvement of the construction of boilers. People used mainly fire-tube boilers on ships, and the name reflected the principle of their action. The gases passed into the flue tube from the furnace, gave the heat to water through the walls, surrounding the pipe, whereupon they left the boiler. Fire-tube boilers allowed increasing the steam pressure; they were more compact than the wall-sided ones. However, fire-tube boilers had disadvantages connected with the principle of their action, which was based on the use of a large volume of internal water with its rather weak circulation in the process of steam conversion.

To replace the fire-tube, boilers there appeared water tube boilers with substantially smaller volume of internal water and greater heating surface. In these boilers, water was heated in multiple water heating tubes of a small diameter, which were externally heated by the combustion products.

With the application of surface condenser, the steam and condensate which was formed out of it closed the circuit, which was called "steam-condensate cycle". This led to a complication of the scheme of the power plant with steam machines: there were auxiliary machinery with steam occasion.

At the end of the XIX century, the power of steam power plants reached high points; correspondently to this, the hour fuel consumption reached several dozen of tons. High capacity of coal hoppers significantly reduced the possibility of carriage of goods and passengers.

More often shipbuilders discussed the possibility of replacing the fuel, coal, with oil. Oil feed by the pump to the furnace radically simplified the stokers operation, reduced the number of them, and at the same time the combustion heat of fuel oil was higher by 40% compared with that of coal. Fuel oil is much easier to load on a ship, store and feed for marine users.

1.2 The Winners in the Struggle for Speed

Of course, the most important event of the naval parade, which took place in 1897 in honor of the jubilee of Queen Victoria, was the demonstration of the "Turbinia" vessel, which could easily bypass the fastest ships. The furor caused by this ship led to the fact that Charles Parson (the steam-turbine power plant "Turbinia" was created according to his ideas) got a lot of orders from the British Admiralty to develop similar projects.

A practical embodiment of the steam turbine idea was caused by the situation in the late nineteenth century, when the existing at that time steam engines, which successfully carried out their functions in all areas of technology, including in the shipbuilding industry, for almost 100 years, ceased to meet the demands that the industry put forward. Primarily, this was due to the lack of profitability and limited power. It was the steam turbine engine that was responsible for the effective spirit and objectives of the time (Table 1.1).

Table 1.1 Comparative analysis of marine steam engines and steam turbines

Parameter	Steam engine	Steam turbine
Useful work	Implemented by the potential energy of steam	Implemented by the kinetic energy of steam
Power	Expansion of steam below 0.02–0.05 MPa is not possible due to the excessive growth of the steam-engine low-pressure cylinder sizes. As a result, the specific steam consumption is significant and, consequently, the efficiency is low. Possible power of steam engines is limited	Steam turbine allows bringing up the steam expansion to 0.002–0.005 MPa, thereby providing a low specific steam consumption and high efficiency. Possible power of steam turbines is virtually almost unlimited
Operating principle	Cyclic changes in the speed of the piston, rod and slider with rectilinear translational motion gives rise to inertial forces that cause forced vibrations of the engine components, transmitted to foundation	Rotary principle. The only movable element is the steam turbine rotor. Impact loading from the turbine to the ship's foundation is completely absent
The center of gravity	High, especially in the most common vertical steam engines	Lower in comparison with a steam engine of equivalent power, the center of gravity position, which contributes to the stability of the vessel
Arrangement	The low output speed causes difficulty linking with electric	Perfectly assembled with electric
Steam contac with oil	Condensate fouling due to lubrication of parts that come into contact with steam	Complete absence of lubricated parts which come into contact with steam

The operating principle of the steam turbine is as follows. The steam produced in a steam boiler is supplied to the nozzle or group of nozzles, where it expands. As a consequence, the pressure decreases, and the ejection speed is increased, steam jets move to the curved blades. The resulting centrifugal force of the particles of steam puts pressure on the blades and converts into mechanical work, which rotates the turbine wheel, and with it the shaft on which it is fixed. There are two types of steam turbines: reactive and active.

In the turbines of the reaction type, the steam expansion occurs in both channels formed by guide vanes and in the channels formed by the rotor blades; in the stages of turbines of the active type, it occurs only in the interblade channels of the nozzle apparatus.

The collection of one row (circumferential) of the nozzle (active type turbines) or the guide (in the turbines of the reactive type) blades and a row of rotor blades is called a stage.

Single-stage active turbines are used in the active ship practice only as a subsidiary, i.e. for actuating the low-power auxiliary mechanisms. Single-stage reactive turbines are not used at all.

Main marine turbines, which rotate the propeller and auxiliary turbine to drive the auxiliary machinery, are always performed multistage. Multistage turbines may consist of the pressure stage (expansion of the steam from the initial parameters until the final pressure occurs in successive stages, each of which uses small heat gradients and pressure drops), or the speed stages (the kinetic energy of the steam flow obtained by expanding steam in the nozzles at the beginning of flow, is converted into mechanical work not on the one crown of rotor blades, but on several successive blades).

Schemes of different types of turbines, which are used in the ship practice, and the nature of the energy transformations, which occur there, are shown in Fig. 1.9. In the upper part of all images, the curves of the steam pressure p and speed C change while flowing through the turbine are shown. In the lower part of them, there are the sections expanded on a plane of the nozzle, guide and working channels.

In modern designs of the main gear-turbine drives, the association of speed and pressure stages frequently occurs within one turbine. The set of nozzles (guide) machines and working crowns at all levels constitutes the wheel space.

In many ways, the creation of a steam turbine is connected with the name of the talented Swedish engineer and inventor Gustaf de Laval.

In 1893, at the World Fair in Chicago, Laval showed his first active turbine with the power of 5 hp. The conversion of steam power to mechanical energy in the Laval's active turbine took place in two stages: the transformation of the potential energy in the steam nozzle into the kinetic energy and its transfer to blades, which are arranged in a circle placed on the drive shaft.

The nozzle had a conical shape that narrows and then widens towards the exit; it has become known in mechanics as the "Laval nozzle". Through this nozzle, at lower pressure of steam the turbine is accelerated to a speed exceeding the sound speed. Turbine speed rate was 30,000 min^{-1}, which corresponded to the circumferential speed of 460 m/s. By 1900, the Laval turbine power, which was shown at the World Exhibition in Paris, had already reached 350 hp.

Unfortunately, the Laval's turbine did not receive the further spread and the reason for this was as follows. Such turbines are used mainly for electric power generators, and the speed rate was much less than that of the turbines. Laval had to develop a mechanical transmission to them—the gearbox. The latter was several times larger in weight and dimensions than the turbine, and substantially reduced its efficiency (Fig. 1.10) [1].

To create an effective steam turbine, the Englishman Charles Parsons chose a different way.

The basic idea assumed in the development of Parsons was as follows: all the energy of steam that must work in the turbine should be divided into several parts and operated each in a single stage.

The epitome of this idea has become a multi-stage turbine, which consists of several stages, arranged in the course of steam. Each of these stages, like the Laval's

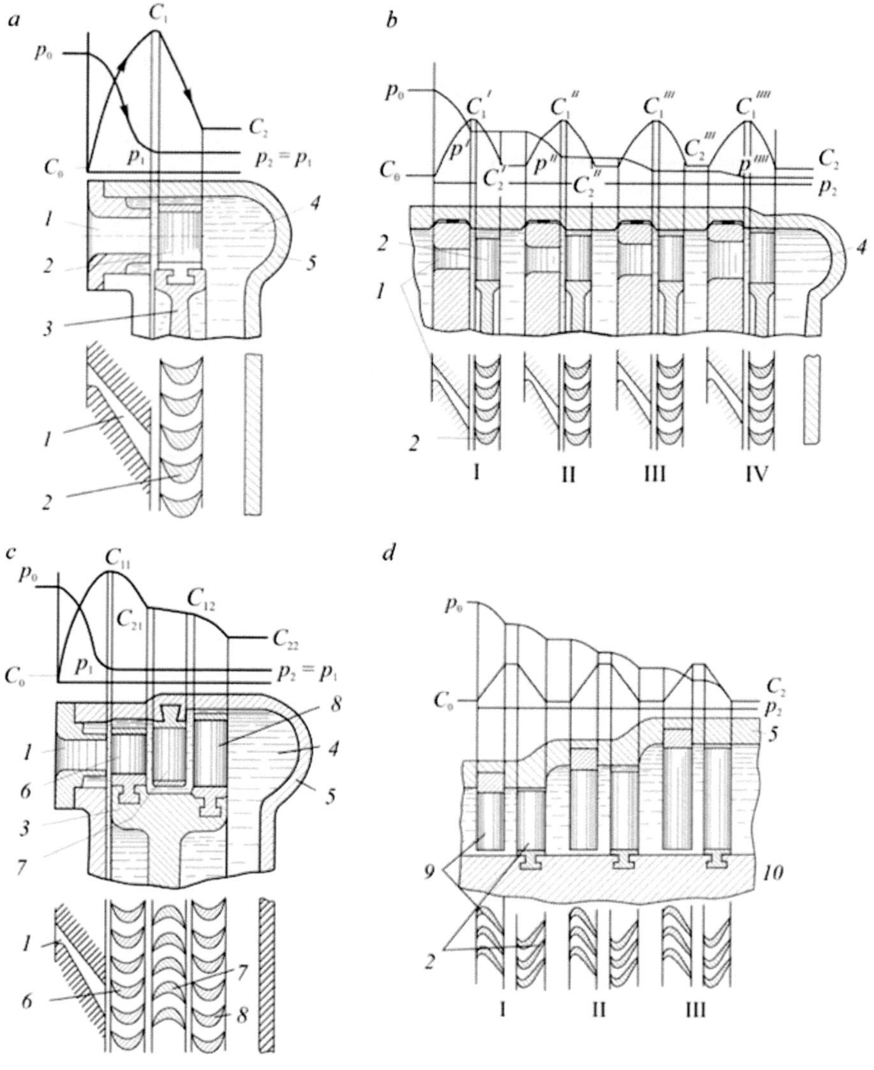

Fig. 1.9 Schemes of marine steam turbines of various types: **a**—active single-stage turbine; **b**—active turbine with four stages of pressure; **c**—active turbine with speed stages; **d**—reaction turbine. 1—nozzles, nozzle blades; 2—rotor blades; 3—drive, the impeller; 4—outlet blade; 5—turbine case; 6—the blades of the first working crown; 7—guiding crown; 8—the blades of the second working crown; 9—guiding blades; 10—the turbine rotor

turbine, consists of nozzle (nozzle unit) and a rotor disk with blades. All impellers are fixed on a common shaft. Due to this the energy of steam, which falls on one stage may be such that the speed of the steam from the nozzles, in terms of the maximum efficiency, and optimal peripheral speed was moderate. Thus, the multistage turbine allows varying the circumferential speed by changing the number of stages.

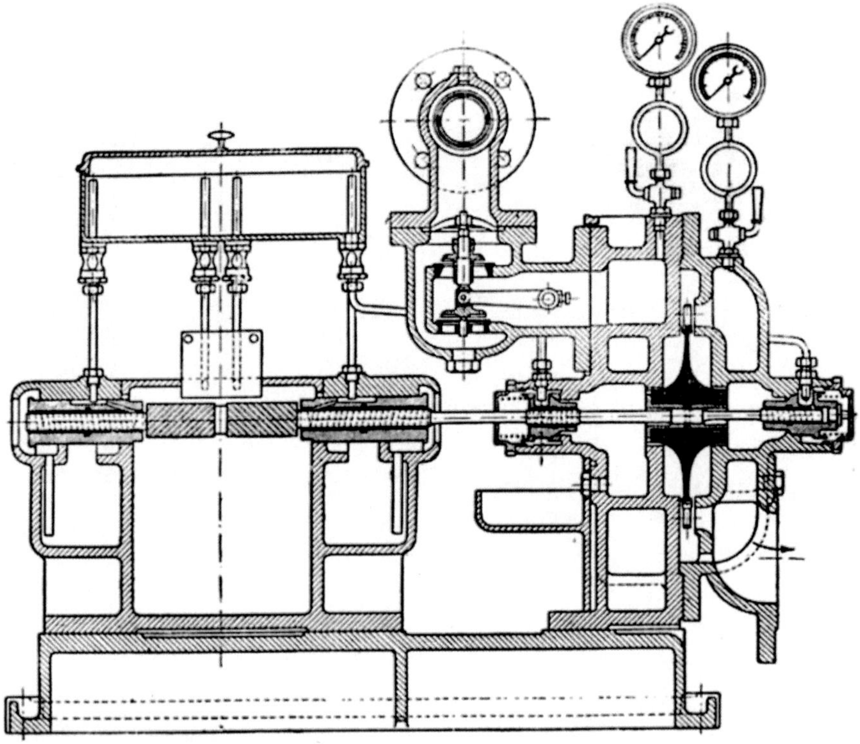

Fig. 1.10 Laval's turbine with the power of 30 hp

The first turbine, built in 1884 on the basis of Parsons patent at an engineering plant in Gateshead (England), had the power of 4 kW (by the way, Parsons first began to express power turbines in these units that were previously used only in electrical engineering) and the speed of 1000 min^{-1}. Relatively small rotations allowed directly connect the turbine and generator shafts, getting rid of the gear, leading to a substantial reduction in price of the turbine generator as a whole.

As of 1889, the number of built Parsons turbines at the plant in Gateshead reached 300 units, and the power unit increased to 75 kW.

The vessel "Turbinia" itself played an outstanding role in the history of steam turbines.

The vessel with the length of 30 m and displacement of 445 tons was built in 1894 by "Marine Steam Turbine Company" headed by Charles Parsons. He proposed an original solution—a three-shaft power plant with the drive of each shaft from the successive turbines of high, medium and low pressure. Three propellers were successively arranged on each shaft. Screw rotation speed was reduced to 2000 min^{-1}.

As a result, the turbine consisted of three groups of successively connected stages, each of which was a multistage turbine. Shafts of turbines and screws were connected directly without gears. The design of the propulsion system "Turbinia" was finally worked out in 1896. The vessel showed very high performance: its speed was reduced to 34.5 kt, with the turbine plant power of 2.400 hp.

British Admiralty, taking into account the benefits that the steam turbine gave to warships, decided to equip two new destroyers and a cruiser, launched in 1904, with the turbines. At the same time, three cruisers of the same type were manufactured, on which the steam engines were installed. Combined tests of ships confirmed the advantage of steam turbines.

In 1906, the battleship "Dreadnougt" was built in England. It had the displacement of 18,000 tons, with an incredible for that time speed—about 23 kt, which was provided by the steam turbine plant with the capacity of 24,700 hp.

The establishment of this vessel caused a sensation around the world, each country with a navy sought to purchase the Charles Parsons patent for marine steam turbines.

In the beginning of the nineteenth century, the steam turbine began to be used on merchant ships. The British passenger ship "King Eduard" became the first vessel with the steam turbine capacity of 3500 hp with displacement of 650 tons, which provided transportation on the River Clyde in 1901. Successful exploitation of this vessel helped to increase confidence in the steam turbines.

In 1906, the English turbine ship "Carmania" came out the Liverpool-New York transatlantic line. The tests were successful and in 1907, commissioned by the company "Cunard Line", the construction of two large passenger ships "Mavritania" and "Lusitania" of the same type began. These ships had the displacement of 38,000 tons. The speed of 25 kt was provided by steam turbines with the capacity of 68,000 hp. These vessels were designed to carry out transatlantic voyages. Moreover, "Mavritania" held the championship on these lines for speed for twenty years.

At this stage of development of marine steam turbines, they were connected directly to the propeller shafts. In order to achieve high efficiency, the propeller speed should be relatively low, for cargo ships it is 80–200 rev/min^{-1}. Such turbines had a considerable number of stages, and, accordingly, considerable weight and dimensions.

As it often happens in technology, the solution to reduce the overall dimensions of ship turbines was found almost by accident. During modernization of the power plant of the transport merchant ship "Vespasian", the slow-moving steam engine was replaced by a steam turbine, leaving the old propeller. As the propeller speed was much lower than the steam turbine speed, the designers were forced to use gearbox. Given the low speed of the steam turbine, the gearbox dimensions turned medium. The layout of the power plant turned out extremely successful, and since that time, such units dubbed "main gear-turbine drives" have been widely used in shipbuilding.

Marine steam power engineering industry was on the rise during the period preceding the First World War, when huge amounts of money were invested in the construction of new ships, most of which were equipped with steam turbine plants. "Parsons Ship Turbine Company" worked with more stress; during the war, the ships of the British Navy were equipped with steam turbines with a total capacity of 1 million hp. At the same time, the ship turbines were manufactured by other companies under the licenses of Parsons. Overall, the British fleet was supplied with Parsons steam turbines with total capacity of about 3 million hp.

The same turbines were fitted with navies of Great Britain, France, Italy, and Russia.

After the First World War, the steam power engineering industry continued to develop rapidly. Thanks to scientific and technological advances of the time, the level of technology and design of boilers increased so much that the steam pressure increased to 50 atm with a substantial increase in its temperature.

One of the disadvantages of steam turbines was a significant drop in effectivity as the load reduction, which significantly reduced the attractiveness of the steam turbine as a ship's engine. Parsons proposed a combined turbomachinery four-bowled plant, which employed the high-pressure turbine for inner shafts and low pressure turbines and located in front of them steam engines with the capacity of 100 hp for external shafts. Steam engines provide low speed and reverse. When the speed is higher than 13 kt, the steam engine is turned off and fed directly to the steam turbine.

Turbines of exhaust steam in a combined turbo-piston units are a special kind of marine steam turbines. It is not possible to carry out the expansion of steam to pressures below 0.15–0.20 atm in the reciprocating steam engines, and steam turbines are able to provide the expansion of steam at pressures up to 0.04- 0.05 bar. For more work and effectiveness increase of the steam power plant, the exhaust steam is additionally expanded in the turbine. Steam of the piston machine enters the turbine at the pressure of 0.55–0.60 atm, which allows the latter to develop up to 30% of the total plant capacity.

In 1907, the steamer "Otaky" with the displacement of 950 tons was built in England for New Zealand. It was equipped with a three-shaft combined plant with the capacity of 3350 hp. In this power plant, the steam engine and turbine were thermodynamically linked by the series connection of steam. The shafting, which was located on the sides, was driven by steam engines; the exhaust steam was fed to them in the turbine, which rotates the middle shaft. Compared with the same type of steamer fitted with a steam engine, the fuel economy on the "Otaky" was about 11%.

A little later, in the 1908–1912, these combined plants were installed on several vessels, including luxury transatlantic liners—steamers "Olympic" and "Titanic". Each of these vessels had three screws. Onboard screws were driven by reciprocating engines, and the middle one worked by means of the turbine exhaust steam.

At the end of 1920s and the beginning of 1930s, marine steam turbines were used with reactive stages and with one active control stage. Turbine units consisted of three turbines—of high, medium and low pressure—and a single-stage reduction gear. An example of the ship's steam turbine plant is the installation of a high power of the Italian passenger ship "Rex", which was built in 1932. The vessel was equipped with four gear-turbine drives with a total capacity of 100,000 hp that provided the speed of 27 kt [5]. Steam parameters were as follows: the initial pressure of 27 atm and the temperature of 380 °C. The turbine of high pressure had the active adjusting stage and 42 reactive stages; the turbine of medium pressure had 22 reactive stages, reversing turbine of high pressure was located in one case with the turbine of medium pressure, which consisted of an active wheel speed with two stages. The turbine of low pressure of front and rear moves moved the steam through dual flow. The lengths of the turbine casings of high, medium and low pressure were 3.7, 5.6 and 5.9 m, respectively, and the outer diameters of casings—1.6, 2.5 and 3.6 m.

Further increase of the initial parameters of steam and reducing the number of stages gave the possibility to reduce the size and weight of the turbines. In this regard, the active principle of the steam turbines received the increasing distribution. Active stages were first used in a high pressure turbine, and then in a low pressure turbine. Three-case units were rarely used due to the fact that all stages of their small amount can stay in the two cases.

The two-stage gears began to acquire significant distribution. This allowed performing more high-speed turbine in the two-stage transfer (3000–6000 min^{-1}), which meant it was lighter and more compact. The capacity of Individual steam-turbine units had reached 50,000 hp by 1940.

Electric transmission on ships with turbine units was used limited due to its inherent weaknesses. The main turbine units with electrical transmission were carried out as single-hulled.

Over the period from 1950 to 1960s, marine turbine units were manufactured mainly as double-hulled with a two-stage reduction gear. Turbines were executed with the active stages. Reversing stages were situated in the low pressure turbine casing. These aggregates were characterized by the following steam conditions: the pressure of 40–50 atm, the temperature 450–500 °C.

Further development of marine steam turbines is associated with increase of initial steam parameters and complexity of thermal schemes.

1.3 From Carnot Cycle to Diesel Cycle

The researches of D. Papin, T. Savery, D. Watt and others contributed to the fact that by the end of the XVII century the steam engine had become a universal motor, and it seemed that no replacement to steam could take place. To imagine the heat engine that would work differently than the steam engine at that time was very difficult. However, the most innovative engineers and inventors thought of an existing sequence of mechanical energy. Fuel combustion products were used to produce steam, which already carried out some work. They believed that it was much more effective to use the energy of the combustion products to get work directly. The problem was largely that the burning coal in the cylinder is the most common fuel while producing could not provide the required quality of combustion products.

There were many attempts to devise the necessary fuel. Thus, J. Gotdail in 1678–1682, and Christiaan Huygens in 1681 proposed a so-called naturally aspirated engine, where the piston is raised due to an explosion of gunpowder up and when lowering performed useful work. Problems arose due to the low frequency of the strokes in these engines [2].

The idea of replacing coal as a fuel to heat engines has been particularly acute in France, where, unlike in England, the coal was in complete deficiency.

In 1799, a patent for coal gas, which was extracted from sawdust or coal, was issued by the Frenchman Philippe Lebon. The patent on the heat engine, issued in 1801, supposed the compression of coal gas and air by separate pumps, mixing them in a special chamber and feeding the mixture into the working cylinder, where it ignited and expanded [2].

In search of a rational type of fuel, there were attempts to use hydrogen. In 1841, the English inventor James Johnstone received a patent for an engine that ran on a mixture of hydrogen and oxygen.

The combustion products were cooled, condensed, and the piston moved to atmospheric pressure. Nevertheless, hydrogen did not become the fuel for first engines, given the difficulties associated with its transportation and storage.

In 1823, Samuel Brown built an aspirated internal combustion engine on illuminating gas, where the vacuum was achieved by cooling (engine had a water jacket) gas residue in the cylinder after the release of combustion products. Ignition by open flame was carried out in the dead center.

In 1833, Valmont Wright built the engine with the cylinders cooling by a water jacket. According to the patent obtained in 1838, in the engine of William Barnet, the air and gas are compressed in some cylinders and the mixture is compressed before ignition in the working cylinder.

In 1842, the Englishman Drack patented and built an engine that worked on illuminating gas. Ignition of the mixture happened at half stroke. The cylinder had a box that connected to the middle of the stroke with a hot iron tube, through which the ignition of a working mix was carried out. Later Drack adapted his engine to run on kerosene. In 1847, at the exhibition in Philadelphia, the Drack's engine developed the capacity of 20 hp at the rotational speed of 60 min^{-1} [6].

In 1858, Christian Reitman converted his engine earlier built (in 1852) to work on illuminating gas, in which air and gas were compressed to 2–8 kg/cm^2, the mixture was ignited by an electric spark, and the combustion carried out the work in cylinder.

In the naturally aspirated engine with a free piston, which was built in 1857 by BarzantiMateucci, the flammable gas-air mixture was ignited by an electric spark at the bottom of the cylinder under the piston. Under the pressure of the combustion products, the piston moved upwards. It moved down under its own weight and the pressure of air. The piston rod brought into rotation the rod with the flywheel by this movement through a rack shaft.

In 1858, Degerin (France) suggested the gas engine with compression of the fuel mixture in the working cylinder.

All these examples did not lead to the creation of an engine that could create real competition to a quite brilliant for those times steam engine.

For the first time in the history of energy, Jean Etienne Lenoir developed a workable, suitable for practical application heat engine, introduced it into operation and achieved considerable success. In 1861, he received a patent for a two-stroke engine that runs on illuminating gas. At the World Exhibition in Paris in 1864, Lenoir's engine won the lead. At that time, 300 of his engines were made, which were manufactured in France and England. Lenoir even built a boat with its engine and after successful tests on River Seine sold it to a company providing water service between Paris and Charenton.

The first engines developed by J. Lenoir had the capacity of 3–12 hp at the rotational speed of 100 min^{-1}, and their effectiveness was about 4%. Despite these low figures, these engines were built for almost 20 years before the emergence and spread of Nikolaus Otto's engines [6].

The aspirated engine, built by N. Otto together with E. Langen, largely repeated the operation principle of the engine of BarzantiMateucci. Movement of the piston was converted into the output shaft rotation with toothed rack and pinion ball of freewheel. With capacity of 0.635 hp, the engine efficiency reached 15%.

Due to the high efficiency (almost 4 times higher than in the Lenoir's engine), Otto-Langen engines began to enjoy great demand, about 5 thousand units were released (with the capacity of 0.5–3 hp) with a total capacity of about 6 thousand hp. Some engines operated in the early twentieth century. The aspirated Otto-Langen engine largely became the precursors of modern diesel hammers.

In 1873, the American Brighton built and in 1876 demonstrated at the exhibition in Philadelphia the first gasoline engine that runs on air-gasoline fuel mixture, which was ignited by a permanent flame source [1].

Almost 20 years after the creation of a hydrogen engine, in 1873, H. Reitman built the first four-stroke engine, which worked on illuminating gas. He was the first in the world to use forced opening of the intake and exhaust valves in its engine. The engine had the capacity of 0.75 hp at the rotational speed of 200 min^{-1}. It provided the drive of drilling and milling machines in the workshop of the inventor for almost 8 years [1].

In 1892, the firm "Hornsby and Sons" built the engine that ran on oil, which was injected into the same cell as the compressed air into the engine; it evaporated and was ignited by the heated surfaces of the chamber, which was aggravated specifically before starting.

In 1884, the Russian navy sailor Kostovich Ignatius built a carbureted gasoline engine for the airship, which consisted of eight cylinders with the diameter of 120 mm during the piston move to 240 mm. According to experts, this engine developed the capacity of 50 hp.

The impetus for the further development of petrol engines was the desire to use them in the car. German engineers G. Daimler and W. Maybach played the decisive role in the creation of a gasoline engine. The main rate in the creation of an automobile engine was made on the two-stroke engine, the inventor of which is considered the Englishman Douglas Clark (1878). However, the hopes for a two-stroke engine did not come true, and almost all of the following designs of automobile engines were four-stroke.

The first car engine, the prototype of all modern car engines, was the engine of Daimler and Maybach, which developed the capacity of 1.5 hp and had the speed of 900 min^{-1}, with the cylinder volume of 250 cm^3. Daimler and Maybach's Engine began to be produced in significant amounts in different countries.

With the development of automobile engines, they became multi-cylinder, and rotational speed increased, leading to a reduction in the weight and dimensions.

It should be noted that there was a number of landmark achievements in the creation of efficient internal combustion engines.

Thus, in 1887–1890, James Hargraves produced several two-stroke engines with self-ignition of fuel—tar oil from high temperature of compression of air.

In 1885, Capitan made the engine that ran on kerosene, feeding it to the evaporator piston pump at the end of the exhaust stroke. The engine used automatic intake and exhaust valves drive. Capitan's engines were produced in significant quantities.

In 1890, the factory "Svidersky" in Leipzig built Capitan's engine with self-ignition of fuel oil from the high temperature of the compressed air. The fuel was atomized by compressed air.

Finally, the year of 1892 came and marked the beginning of the triumphant spread of a new type of engine.

In Germany, on February 28 1892, Rudolf Diesel received the first patent on the engine of his own design, which had all the hallmarks of a modern diesel engine. The patent assumed adiabatic compression of clean air in the cylinder to the pressure of 25 MPa with a high temperature achievement.

As the fuel, Rudolph Diesel applied carbon fine powder or a liquid fuel MIST.

Theoretical provisions governing the establishment of the engine were presented by R. Diesel in the book with the significant title "Theory and Construction of a Rational Heat Engine to Replace the Steam Engine and Combustion Engines Known Today", which was published a year after obtaining the patent.

In 1893, Rudolph Diesel received a new patent, additional to the main one, and built a heat engine on its basis. Its tests in 1895 were successful and showed its efficiency of 36%, i.e. 2 times higher compared with the existing best time to motors.

Diesel received a patent for his engine in all industrialized countries. In Russia, the firm "Ludwig Nobel" acquired this patent. The patents for the right to manufacture diesel engines were also purchased by "Sulzer" (Switzerland) and "Burmeister and Wain" (Denmark).

The merit of R. Diesel was not only that he proposed a new type of engine, but also that he was very persistent in implementing these ideas industrially.

In 1908, Diesel engines patents expired and plants around the world began manufacturing.

Symbolically, the engine that runs on heavy fuel oil with spontaneous ignition and the formation of a combustible mixture in the working cylinder got its founder's name "Diesel" (Fig. 1.11). The cross-sectional sample of the Diesel engine in 1896 shows the main elements of the application of its creator. This is a compressor, which is driven by an engine connecting rod through a balancer, and compressed air launcher through which kerosene is injected into the working chamber of the cylinder through a needle valve located on the geometric axis of the cylinder. The left side of the cylinder cover is a valve for the air intake and exhaust of the combustion products, which is actuated by a cam disposed on the arm to the right of the engine.

Rudolph Diesel passed away, but his engine was used increasingly, and also improved.

Russia's contribution to the improvement and the use of diesel engines is very impressive. Even in 1898, the St. Petersburg mechanical plant of the company

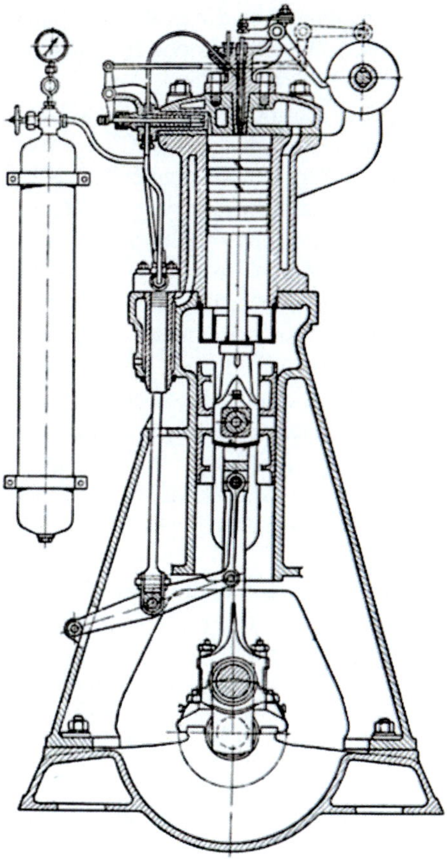

Fig. 1.11 Pilot Diesel engine (1896)

"Ludwig Nobel" acquired the patent and began manufacturing diesel engines. Rudolf Diesel first had serious doubts about the technical capabilities of the industry, but still started negotiating with the company "Ludwig Nobel", which belonged to the world famous Nobel family. The negotiations were successful.

The company's management was vitally interested in the creation and wide dissemination of the heat engine running on oil, since it would contribute to the expansion of consumption [1].

The Petersburg machine-building factory, owned by the company, immediately set a goal—to provide the engine work on crude oil. There were developed and implemented important structural improvements, which concerned the working process and the mechanical part of the diesel engine. And in 1899, the first engine passed the test; it was built in Russia, and it expended 221 g of crude oil on 1 hp/hour at the capacity of 20 hp, while foreign engines spent 240 g/(hp/hour) of more expensive fuel—kerosene. It put forward the Nobel plant into the leaders of engine manufacturing. Engines of this type were the main profile of the plant.

By the First World War, the total capacity of diesel engines produced at the Nobel factory and installed in power plants, industrial facilities, military and commercial ships had been 65,000 hp.

Special merit of the Nobel plant was the development of engines and equipment for ships. In the spring of 1903, the world's first ship "Vandal" built jointly by the Sormovo Plant (Russia) and the plant "Ludwig Nobel" was put into operation. It was the oil tanker of the lake type 75 m long, 10 m wide, with the displacement of 1.8 m and lifting capacity of 820 tons. The vessel speed of 7.4 kt was provided by three three-cylinder four-stroke diesel engines with the capacity of 120 hp at the rotational speed of 240 min^{-1}. To ensure forward move and implementation of the maneuvers, electrical transmission was used. Thus, the first steam ship in the world was at the same time the world's first diesel-electric.

In 1904, a similar ship "Sarmat" was commissioned in Russia; there, two engines with the capacity of 180 hp were installed. "Sarmat" used energy transfer, which was developed by the Nobel factory engineer Charles A. Del Proposto.

Already in 1908, the Nobel's factory built reversible three-cylinder four-stroke engines with the capacity of 120 hp at the engine speed of 400 min^{-1}, intended for submarines.

In 1907, the use of diesel on submarines in Germany began, and 355 submarines were already built and equipped with diesel engines at the end of 1918.

In Sweden in 1907, the schooner "Orion" with reversible diesel engine with the capacity of 60 hp was built. In these years, a significant contribution to the development of the diesel engine for the warships and transport vessels was made by the MAN plant (Maschinenbau Augsburg Nurnberg), which produced the six-cylinder engine with the cylinder diameter and stroke of 530 mm and the cylinder capacity of 200 hp. In future, the MAN plant increased cylinder capacity up to 300 hp and the number of cylinders to 10, i.e. engine power reached 3000 hp.

In 1910, the tanker "Vulcanus" was built in England. It had a diesel engine with the capacity of 400 hp. In 1911, the first Danish cargo ship "Selandia" with the displacement of 9800 tons and with the capacity of 2400 hp was put into operation.

In 1911, Nobel's factory built the first V-shaped eight-cylinder engine with the capacity of 200 hp.

In 1908, irreversible four-cylinder engine rated at 300 hp at the rotational speed of 240 $min^{-1}$was established on the river boat "Mysl", built by the Kolomna Plant (Russia).

To transport to the Caspian Sea in 1908, Russian shipbuilders built the steamer "Delo" (hereinafter "Valery Chkalov") with the displacement of 6000 tons, in which two four-stroke four-cylinder engine were installed, each with the capacity of 500 hp at the rotational speed of 150 min^{-1}.

Reverse of these engines was provided by transmission of the R. A. Koreyvo system. Its action principle was as follows: a clutch was turned on at astern running, which by means of a chain transmission transferred power to the intermediate shaft, then through a pair of gear-handed rotation of the propeller shaft, which was rotated in the direction opposite to the rotation of the motor shaft. Thus, the era of marine application of internal combustion engines began.

1.4 The Basic Steps for Creating Gas Turbine Engines

If we compare the steam and gas turbines as expansion machines, the difference between them is small. Their structure comprises a large number of common elements, especially in the flow section.

At the same time, other elements of gas turbine engines and steam turbine units are very different. Firstly—heaters. It is a boiler in the steam turbine plant, usually a large-size building, which significantly exceeds the turbine size. In a gas turbine engine combustion chamber, in which fuel is burnt in compressed air, it performs the function of the heater, and which is considerably smaller in size.

The environment performs the functions of a refrigerator in a gas turbine engine where the exhaust gases enter the turbine. Actually, the refrigerator as a structural unit in a gas turbine engine is generally absent. At the same time, the steam after the steam turbine is cooled by circulating water in a very bulky device—a condenser. Can a turbine plant operate without the condenser? Maybe (remember the steam engine locomotive), but it will have a negative impact on its profitability.

Thus, the gas turbine engine without the boiler and condenser, in which the phase transformations of the working fluid occur (water to steam, then steam into water), is much more lightweight and compact—this is its main advantage.

Nevertheless, the turbine engine has its disadvantages. If in the steam turbine plant the working fluid in the liquid state after condensation is compressed with little compression work in a steam cycle, then in the cycle of gas turbine engine the air is compressed, the compression work is large enough (normally about 40% of the work developed by the gas turbine). This leads to a reduction in the useful work cycle and makes high demands on the compression efficiency of processes in the engine compressor. At the same time, the thermal efficiency of the gas turbine engine, as well as any other heat engine, is defined by the relationship obtained by Carnot:

$$\eta_t = 1 - \frac{T_2}{T_1},$$

where T_1 and T_2—initial and final temperatures of the combustion products, respectively.

From this relation, it follows that the radical way of increasing efficiency is the ratio of the gas temperature rise before the turbine. Given that liquid and gaseous hydrocarbon fuel is used in combustion chambers of gas turbine engines, the maximum combustion temperature can be 2000–2200 °C. Considering limitations from the point of view of strength in the current conditions of GTE, the temperature T_1 is 1250–1350 °C.

Thus, to create a gas turbine engine that would be superior in efficiency to the steam turbine plant, it was necessary to ensure a high, much higher than that of the steam turbine unit, temperature at the beginning of the expansion process and to create a highly effective compressor for compressing air.

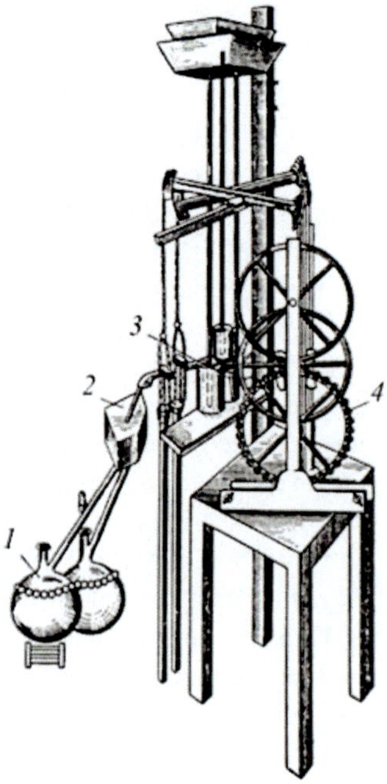

Fig. 1.12 Barber's gas turbine engine: 1—gas generator; 2—gas mixer; 3—combustion chamber; 4—gas turbine

For the first time in the history of machinery, the idea of creating a gas turbine was proposed by Englishman John Barber. In 1791, he received a patent for a heat engine, whose operating principle was almost completely meeting the principle of operation of modern gas turbine engine [2].

The Barber's engine (Fig. 1.12) was obtained by gasification of fuel from the coal, oil, wood. Barber's patent contained the main elements of the turbine engine, but it anticipated the technical possibilities of the end of the XVIII century, due to which the implementation of its basic idea was impossible.

In 1837, the Frenchman Bresson offered a gas turbine engine in which compressed air was a fan, but the design did not receive a practical application.

A significant contribution to the development and establishment of gas turbine engines was made by the German engineer Stolze, who patented the "fire turbine" in 1872, then built and tested it in 1900–1904. The peculiarity of this gas turbine plant is that it was very similar to the modern gas turbine engines (Fig. 1.13). In the Stolze's gas turbine, air was compressed in the 10-stage axial compressor, where the working fluid moved parallel to the rotation axis, and then heated in the regenerative heat exchanger and fed to the combustion chamber. The lower part of air of the combustion chamber, the so-called primary air, was mixed with fuel (fuel gas) and provided its combustion. Another part of it (secondary air) was mixed with the combustion products, whereby the temperature of the gas supplied to the turbine was reduced to acceptable values. It should be noted that this method is applied in the combustion chambers of modern engines for cooling the combustion products.

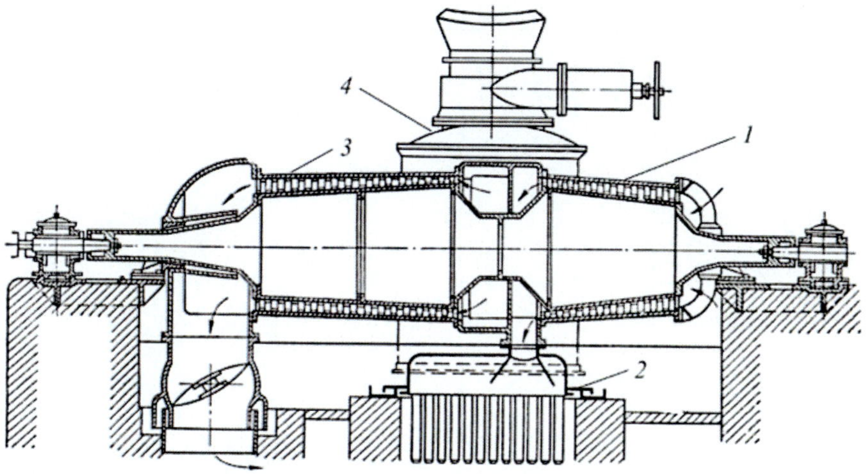

Fig. 1.13 Stolze's gas turbine engine: 1—axial compressor; 2—compressed air regenerative heater; 3—gas turbine; 4—gasifier

Fuel source was gasifier from coal, which developed flammable gas and fed it to the combustion chamber.

Unfortunately, the tests did not give the expected results.

The engine designed and built in 1897 by a mechanical engineer of the Russian fleet P. D. Kuzminsky can be regarded as a serious attempt to create a gas turbine engine. The engine of a small capacity was designed for use on a boat.

Its characteristic feature was the original design of the combustion chamber (Fig. 1.14).

In 1904, Rene Armingaud (France) created a gas turbine engine with combustion at constant pressure, which was working on gas. Due to the insufficient level of temperature and low efficiency of the basic elements, the engine could only run at idle.

Fig. 1.14 Kuzminsky's combustion chamber

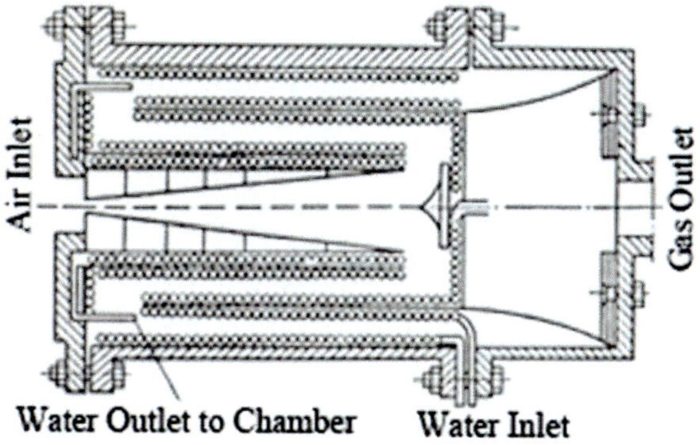

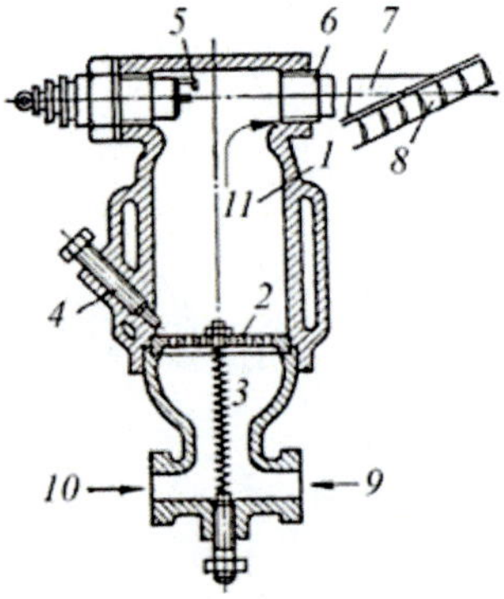

Fig. 1.15 Scheme of the Karavodin's turbine engine combustor. 1—combustion chamber body; 2—inlet valve; 3—spring; 4—adjusting screw; 5—incendiary candle; 6—gas pipeline; 7—guide vanes; 8—gas turbine; 9, 10—supply of fuel and air, correspondently; 11—combustion products to the turbine

Since on the way of the creation of effective gas turbine engines an item such as a compressor (or rather its low efficiency) was a serious obstacle, in the early twentieth century the inventors attempted to dispense with the compressor, creating incompressible engines. Incompressible turbine engines are those in which the combustion takes place at constant volume.

A striking example of gas-turbine engine with combustion at constant volume is the engine patented in 1906 and built in 1908 by the Russian engineer Vladimir Karavodin (Fig. 1.15).

The German engineer Holzwarth worked on the gas turbine engine with combustion at constant pressure. In 1906–1933, several gas turbine engines of the original design were developed and manufactured according to his project. For example, Holzwarth applied prior compression before supplying compressed air to the combustion chamber. To drive a compressor of pre-compression, he used a steam turbine.

The first Holzwarth's gas turbine engine with the estimated capacity of about 40 kW (1906–1908) had six combustion chambers equipped with shut-off valves. The pressure of combustion gases entering the turbine nozzle unit was 0.6 MPa [7].

Engines designed and built by Holzwarth in 1914–1927 had a horizontal axis of rotation (Fig. 1.16). Air was fed into the combustion chamber at the pressure of 0.2–0.3 MPa, combustion pressure at the end of the process was 1.2–1.4 MPa. Maximum achieved effective power was 200 kW.

The coefficient of efficiency of gas turbine engines was 13%, which was a major achievement at that time. The Holzwarth's gas-turbine unit with capacity of 2000 kW had the highest rates. It was launched in 1933 and worked on the blast furnace gas; the efficiency reached 20%, the degree of pressurization $\pi_K = 7$, and the turbine inlet gas temperature about 1000 K.

All further creation and improvement of gas turbine engines was based on gas turbines with combustion at constant pressure—the so-called direct-flow.

Problems with the creation of gas turbines were associated with insufficient scientific substantiation of the processes that occur in them. In this connection, it is difficult

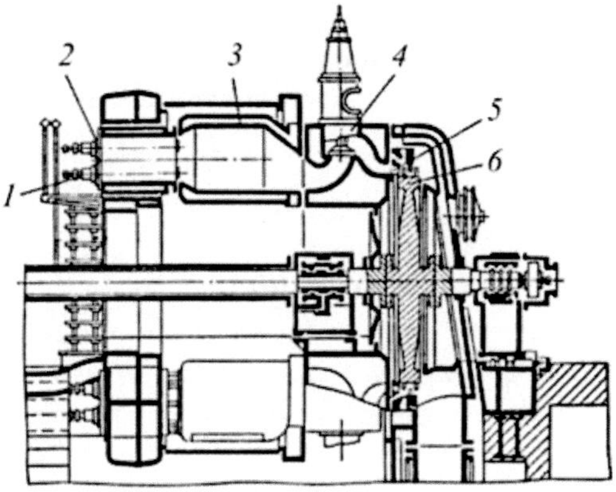

Fig. 1.16 Holzwarth's gas turbine unit on a horizontal shaft: 1, 2, 4—air, fuel and exhaust valves, respectively; 3—combustion chamber; 5—the turbine nozzle assembly; 6—turbine wheel

to overestimate the contribution to the turbine development of the Slovak scientist A. Stodola, who wrote a fundamental work "Steam and Gas Turbines" in 1924.

Two main problems stood on the way of creation of highly efficient gas turbine engines: the aerodynamic problem—development of a perfect blading of turbines and compressors, and the metallurgical one—creation of superalloys for the possibility of increasing gas temperature before the turbine.

The progress in turbomachinery aerodynamics achieved at the end of 1930s and creation of high-temperature alloys gave a powerful impetus to the development of gas turbine. The most intensive development took place for gas turbine engines for aviation. This is due to the fact that in this area the small weight and dimensions play a major role. In addition, a relatively small share (500–2000 h), which at that time was usually calculated for aircraft engines, significantly increased the limit of temperature of the gases leaving the combustion chamber. During this period, serious research towards the creation of GTE aircraft began in England, France, America and Germany. In Germany, the company "Heinkel" worked on this problem. The first research sample HeS-1 aircraft gas turbine engine, which worked on the reactive principle, started operating in 1937.

As a part of the turbojet engine, gas energy, which extended from the combustion chamber, was at compressor drive, the residue was converted to an output nozzle in the propulsion. Turbojet engine consists of a single-stage centrifugal compressor, a single-stage turbine combustor and evaporation. For this engine, the problem of materials for turbine components that operated at high temperatures was solved by using special materials of high price.

The first flight with the engine HeS-1 on the plane He-178 took place in August 1939, the same year the English designer F. Whittle created a turbojet engine, the works on the engine of the company "BMW" and "Unkers" were finished. After the Second World War turboprop engines appeared, in which most of the energy of the combustion products is used in the compressor and turbine drives a propeller, but much smaller—for propulsion. Intensive research and experience of operating aircraft gas turbine engines have greatly contributed to the spread of the use of gas turbine engines in other areas, and in some cases, especially aircraft engines with minimal modifications have applied to land transportation purposes. Thus, all the conditions for the practical application of gas turbine engines on ships have been created.

1.5 Nuclear Marine Power

The appearance and formation of nuclear energy. Of course, the determining factor in the event of nuclear energy can be considered the mysterious rays opening in November 1895 by the German physicist Wilhelm Conrad Roentgen, which are now called X-rays. The researchers suggested that X-ray radiation is accompanied by phosphorescence—the cold glow of some substances. Testing this hypothesis, the French scientist Henri Becquerel opened a new phenomenon on March 1, 1896—unknown types of rays, the source of which was a chemical element uranium. He called these rays of uranium.

Uranium was not the only element that was able to allocate the new rays. Compatriots of Becquerel, Pierre and Marie Curie, identified milligrams of previously unknown elements in dozens of tons of ore—polonium and radium, capable of a similar radiation, Then they conducted a detailed study of them. These researchers found the quality in thorium. The rays, discovered by Becquerel, became known as radioactive (from the Latin "radius"—ray), and the actual effect of their radiation—radioactivity.

The works of such famous physicists as O. Hahn and F. Strassmann, E. Fermi, K. Anderson, E. Rutherford, D. Chadwick consider the nature of radioactivity and the possibility of its practical application.

In December 1942, in Chicago (USA), under the leadership of the Italian physicist Enrico Fermi, a controlled nuclear chain reaction was carried out in the world's first atomic reactor. These works were performed under the highest security, since, unfortunately, their ultimate goal was to create a nuclear bomb.

Almost at the same time, the work on mastering the energy of the atom interrupted by the Second World War, headed by Academician I. V. Kurchatov, was resumed in the USSR. This work resulted in the creation of a nuclear reactor, in which a controlled nuclear reaction was also carried out on December 25 1946.

The first practical application of nuclear energy led to terrible consequences. In 1945, a US aircraft dropped the nuclear bomb developed by American scientists on two Japanese cities—Hiroshima and Nagasaki. Thousands of civilians were killed.

Further events related to the use of nuclear energy developed as follows. Already in 1954, the world's first nuclear power plant with capacity of 5000 kW started producing the energy in the Soviet Union. Since 1955, when the first nuclear submarine in the world "Nautilus" was built in the US, the competition for supremacy in the creation of underwater nuclear giants between the two countries—The United States and the Soviet Union—dragged for 35 years.

The nuclear power plant opened virtually unlimited possibilities for submarines, allowing radically solving the two problems decisive throughout the history of diving: underwater speed increase and ensuring the long-term submarine underwater location without surfacing to recharge batteries.

Over the period from 1955 to 1993, 464 boats were built worldwide, including 179 boats in the United States, and 243 boats in the USSR [8].

Parallel to the military use, the construction of civilian vessels with nuclear power plants took place. So, in 1959, the world's first nuclear-powered icebreaker "Lenin" (USSR) with the power plant of the capacity of 32,350 kW was launched.

In 1962, the dry cargo ship "Savannah" was built (US) with the nuclear power plant with the capacity of 14,700 kW, in 1968—the cargo ship "Otto Hahn" (Germany) with the nuclear power plant with the capacity of 7350 kW ETA.

In 1972, the Japanese shipbuilders launched the cargo ship "Mutsu" with the nuclear power plant with the capacity of 6600 kW ETA. Unfortunately, none of these vessels, except for the icebreaker "Lenin", could sustain a more or less prolonged period of operation. All of them, not even having operated a life cycle period were sent for recycling for various reasons.

Icebreakers with nuclear power plants showed high efficiency. The main reason for this was that it was in operation when icebreakers most clearly manifest benefits of such plants. After the successful operation of the icebreaker "Lenin", Soviet shipbuilders built a series of icebreakers, the head of which was the icebreaker "Arktika", and the last on construction terms was the icebreaker "50 Let Pobedy".

The project of a lighter carrier "SevernyyMorskoy Put" with the nuclear power plant (USSR) was supposed to be the head, and after the trial operation was supposed to bookmark a series of sister ships. In 1988, the "SevernyyMorskoy Put", built at the Kerch factory "Zaliv", was put into operation. However, it happened so that two years before the events, there was a man-made disaster in Chernobyl (Ukraine), which led to radioactive contamination of many countries. World community's reaction was cautious towards nuclear energy, in particular transport. There were problems with taps of the lighter carrier "SevernyyMorskoy Put" in many ports, despite the availability of all necessary certificates. At this time, the "SevernyyMorskoy Put" was mainly used in coastal trade between the ports of Russia. It is natural that there could be no question about the future construction of nuclear lighter carriers. Table 1.2 shows the characteristics of nuclear power plants of civil vessels.

Since 1974, the countries of the world (excluding the USSR) have not build a single commercial vessel with a nuclear facility (except Finland, which built icebreakers for the USSR). At the same time, ships were built with nuclear power plants for different purposes.

Since the mid-1970s, heavy nuclear missile cruisers have been built at the Baltic Shipyard in Leningrad (Russia). The head of them—"Kirov" (since 1992—"Admiral Ushakov")—transferred to the Naval Fleet of the USSR on December 30 1980, the second one did so on October 31 1984, the third—on December 30 1988, and the fourth—in 1996. Now the construction of cruisers in Russia is resumed.

As of 2010, US Navy has 11 nuclear aircraft carriers: "George Bush", "Ronald Reagan", "Harry Truman", "Jon Stannic", "George Washington", "Abraham

Table 1.2 Main characteristics of marine nuclear power plants

Parameter	Vessel name, country					
	The icebreaker "Lenin", USSR	Dry cargo vessels			The icebreaker "Arktika"	Lighter carrier "SevernyyMorskoy Put"
		"Savannah", USA	"Otto Hahn", Germany	"Mutsu", Japan	USSR	
Year of the vessel construction	1959	1962	1968	1974	1974	1988
Thermal power reactors, MW	3 × 90	74	38	36	2 × 171	135
Nuclear fuel	UO_2	UO_2	UO_2	UO_2	UO_2	UO_2
Fuel loading weight, t	3.0 × 1.7	8.06	2.98	3.03	–	–
Water pressure in the primary circuit, MPa	20.0	12.3	6.3	11.0	–	–
Water temperature at the inlet / outlet of the reactor core, °C	248/325	257/271	267/278	271/285	–	–
Steam parameters:						–
Temperature, °C	310	240	273	251	300	
Pressure, MPa	3.1	3.4	3.1	4.1	3.0	
Efficiency of the nuclear power plant	14.0	23.5	21.6	–	–	–
Power capacity to the propeller shaft, kW	1 × 14,400 2 × 7200	14,700	7350	6600	317,600×	129,400×
Mass steam generating plants in operation, pcs	3100	2500	2100	2920	3150	–

(continued)

Table 1.2 (continued)

Parameter	Vessel name, country					
	The icebreaker "Lenin", USSR	Dry cargo vessels			The icebreaker "Arktika"	Lighter carrier "SevernyyMorskoy Put"
		"Savannah", USA	"Otto Hahn", Germany	"Mutsu", Japan	USSR	
Year of the vessel construction	1959	1962	1968	1974	1974	1988
Mass nuclear power plants in operation, pcs	5850	3945	2550	3700	–	–
Specific gravity of steam supply, kg/kW	100	268	348	560	–	–

Note For all the mentioned nuclear power plants, the reactor type is pressurized water, pressurized reactor moderated and cooled—H_2O

Lincoln", "Theodore Roosevelt", "Carl Vinson", "Dwight Eisenhower", "Nimitz", "Enterprise". By the way, "Enterprise" with its full displacement of 94,000 tons and the capacity of nuclear power plant 4X47800 kW is the first nuclear-powered aircraft carrier in the world, which is still a part of the US Naval Forces. Nuclear fuel loaded in the reactor of the US aircraft carrier of the type "Nimitz" is enough for 13–15 years of operation; as for the aircraft carrier of the type "Ronald Reagan", it is assumed that these ships will operate without recharging the core for 50 years (!), i.e. the time of its life cycle [9].

The principle of operation and general arrangement of ship nuclear power plants. The main element of the ship's nuclear power plant is a reactor in which the energy, released as a result of nuclear fission of some heavy elements, is converted into heat.

Enriched uranium is used as a fuel for marine power reactors. The content of the isotope uranium 235 in natural uranium is only 0.7%, the other 99.3% is uranium isotope 238, which is heavily divided. Mostly, artificially enriched uranium is used in reactors; it contains uranium 235 in the amount of 3–5% or plutonium (rarely).

The process of uranium fission occurs under the action of neutrons, so that the nucleus of uranium 235 is split into two similar charges and approximately the same fragments. They repel each other and fly at high speeds. It is the collision of the fragments from the atoms of the surrounding medium, in which the nuclear fission and their kinetic energy is converted into heat.

For each neutron absorption in the fission of uranium nucleus, 2.5 neutrons are produced, which in turn cause the division of other nuclei of uranium 235, causing a chain reaction of fission. This reaction takes place in the part of the reactor, where the nuclear fuel is located—in the core.

For the purposes of energy, the reactors, in which speed is slowed down to the speed of their thermal motion (about 500–600 m/s) for more efficient interaction of neutrons with nuclei fuel neutron, are used. These neutrons are called thermal, and reactors are thermal reactors.

Hard (D_2O) and conventional (H2O) water, beryllium, and graphite can be used as the inhibitors. Water is widely used in marine reactors.

By the core type, nuclear reactors can be homogeneous, in which the nuclear fuel and moderator are a homogeneous mixture, and heterogeneous—with an arrangement in the nuclear fuel moderator in the form of rods or plates. These heterogeneous reactors today are the only type of reactors used in the composition of marine nuclear power plants.

Nuclear fuel in the reactor is concentrated in the fuel elements, which are sealed (to prevent ingress of radioactive fission products to the coolant), thin-walled zirconium tubs, aluminum alloys, stainless steel, filled with pellets of uranium oxide. The fuel elements are combined into fuel assemblies, or cartridges, placed in the technological core channels that organize the coolant flow. Spent fuel elements are replaced channel-by-channel.

The design of the fuel element and fuel cartridge of the reactor of the nuclear power plant of the vessel "Savannah" is shown in Fig. 1.17.

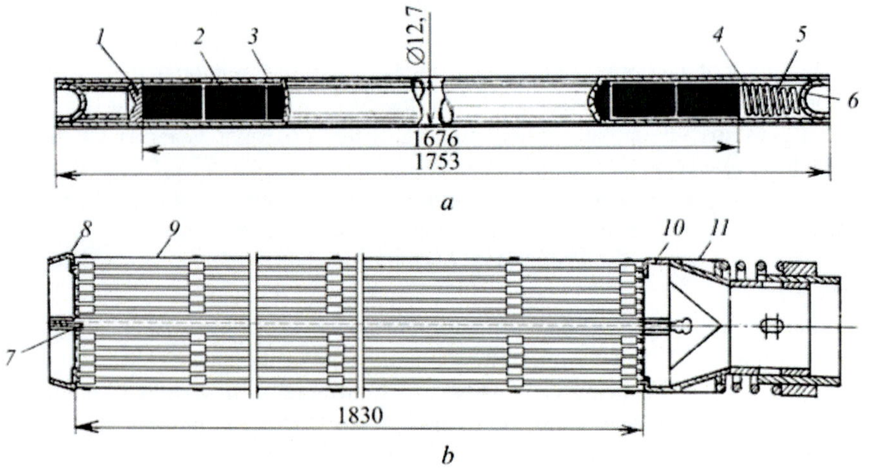

Fig. 1.17 Fuel element (**a**) and longitudinal section of the fuel cartridge (**b**) of the reactor of the vessel "Savannah". 1—plate; 2—nuclear fuel tablets; 3—shell; 4—gas void; 5—spring; 6—sealing plug; 7—central core; 8, 10—upper and lower terminals, correspondently; 9—fuel cartridge shell; 11—adapter

The reactor core is enclosed by a reflector, which reduces the loss of neutrons and the alignment of their flow. The same substances that are used as inhibitors are also used as the reflector. Retraction of heat from the core is produced by the refrigerant agent—coolant, the function of which can be performed by water, organic substances, liquid metals and their alloys (potassium, sodium, lead, bismuth) and gases (helium, nitrogen, carbon dioxide).

Regular, carefully purified water has been widely used in reactors of marine power plants as a coolant. Given that water is also used as a neutron moderator, such reactors are known as reactors of the water-cooled pressure type. Power control of the nuclear reactor, which is needed to ensure all spectrum of the marine power plant modes, is carried out by means of rods made of materials that readily absorb neutrons (cadmium, boron). Input and output of these rods from the core are regulated by the intensity of the chain reaction and release of energy.

If the rods are advanced deep into the core, the nuclear fission reaction slows down, and if they are moved further, it stops at all. For emergency stop of a nuclear reactor, there are rods of *emergency protection*, which should immediately stop the chain reaction in case of emergency.

In the operation of a nuclear reactor, the fuel burns down. When the fuel burns down to a certain point, replacement of the fuel elements is carried out. The time between these substitutions is called the *reactor campaign*.

When carrying out the nuclear reactions, about 80% of the energy is converted into heat, and the other 20% are allocated in the form of radiation, which can cause severe painful phenomena in the human body. That is why the reactor core is surrounded by a protective shield of stainless steel sheets, and there is an air–water-proof *biological protection* in the form of steel and concrete slabs of the thickness of 400–500 mm around the machinery and pipes where the radioactive water has flowed. It increases the weight of the nuclear power plant significantly. For example, the weight of the plant with the capacity of 10,000 kW of the US atomic submarine "Nautilus" is about 870 tons, about 740 tons of which (or 85% of the mass of the entire power plant) is the steam generating part with a biological protection [10].

Ship nuclear power plants can be carried out in a single- and double-circuit form. The single-circuit scheme of a nuclear power plant, in which the heat transfer fluid rejecting the heat in the core is a working fluid at the same time, provides that the water cooling the fuel rods of the reactor is converted to steam. Steam is directly expanded in the turbine and forwarded to the condenser, and then it turns into water, which is fed back into the reactor with the pump. The advantage of this scheme is its simplicity and the ability to use turbine steam of high parameters, which increases the plant efficiency. However, all single-circuit equipment of nuclear power plants becomes radioactive in operation. Therefore, its servicing be-comes much more complicated, especially when the fuel rods' shells can lose their impermeability is possible during operation.

The two-circuit nuclear power plants have gained the most widespread use on ships because they provide access to the equipment of the second circuit and are suitable for almost all known types of reac-tors, except for reactors with liquid metal coolants and liquid nuclear fuel reactors, which are forced to use three-circuit scheme.

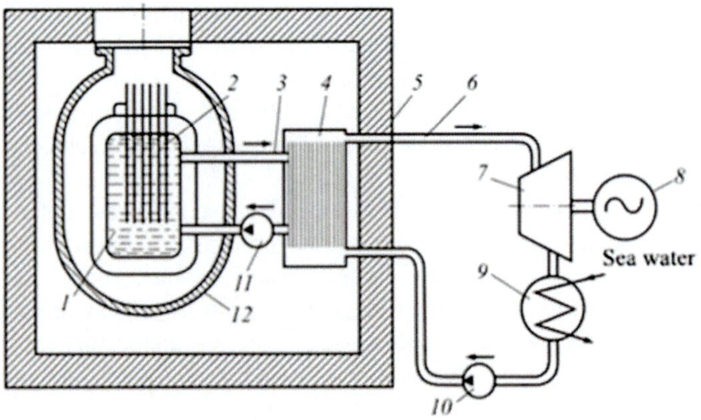

Fig. 1.18 Principal scheme of the double-circuit nuclear power plant: 1—reactor core; 2—fuel rod; 3—conduit of the primary circuit; 4—steam generator; 5, 12—screens for biological shield; 6—steam line of the secondary circuit; 7—steam turbine; 8—main generator; 9—condenser; 10—feed pump; 11—circulation pump of the primary circuit

Figure 1.18 shows the principal heat scheme of a typical ship nuclear power plant with the water-cooled reactor under pressure. Water is the coolant, which is called the first-circuit water; it is electrically pumped through the reactor and then it is fed into a steam generator tube, where it gives off the heat to the second-circuit water. The second-circuit water is heated, it boils, and is converted into steam entering the steam turbine, in which it performs useful work. Further, the steam is condensed, as in a conventional steam turbine plant, and the condensate is returned to the steam generator.

The water should have a sufficiently high temperature that exceeds the temperature of the steam to make the primary-circuit water heat the secondary-circuit water and to transmit enough heat to generate steam. At the same time, to prevent water boiling in the reactor, the temperature should be below the vaporization temperature at the pressure of the first circuit. Given these mutually exclusive conditions, the pressure of the first circuit is maintained at the level of 10–20 MPa [10].

The nuclear steam generating plant of the "Otto Hahn" vessel with a water-cooled reactor under pressure is shown in Fig. 1.19. It is a monoblock construction, in which the core is placed in the lower central part of the machine, and a once-through steam generator is located above the core. This kind of arrangement of the steam generator ensures constant operation of the reactor under the natural circulation of the coolant in the modes of up to 30% of the nominal one.

Of great importance in operation of the nuclear power plant is ensuring safety of the staff and the environment. The nuclear safety system of the mentioned water-cooled reactors provides for three physical barriers to ionizing radiation emissions that arise in their work. The first barrier is a completely hermetic shell of the fuel elements, in which nuclear fuel in the form of uranium pellets is placed. The second barrier is a first-circuit boundary, which also must be sealed in case of radiation

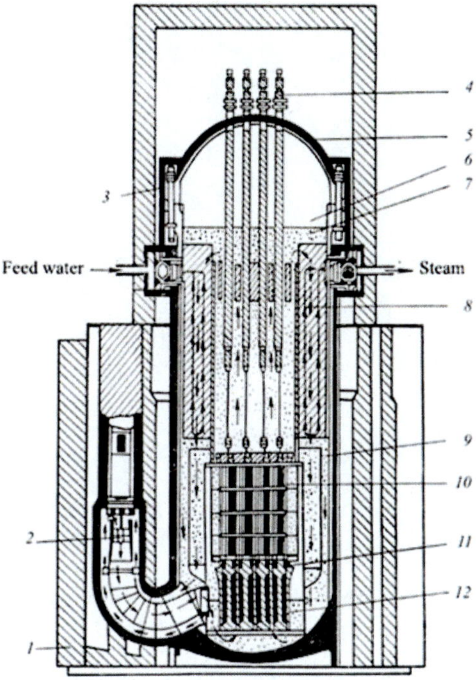

Fig. 1.19 Steam generating plant of the "Otto Hahn" vessel: 1—biological shield of the reactor; 2—axial flow pump; 3—body of the steam generating plant; 4—adjusting rods; 5—reactor upper head; 6—vapor blanket; 7—confinement; 8—once-through steam generator; 9—thermal protection; 10—reactor core; 11—support grid; 12—shanks of the adjusting rods

penetration from the fuel cell. And finally, the third barrier is a sealed shell, where the main equipment of the reactor compartment is located. The operational reliability of the ship nuclear power plant is provided by the unconditional observance of the necessary measures to ensure stability of all the three physical barriers to the spread of radiation.

The thermal scheme and the composition of the nuclear power plant are determined by the type of reactor and the working fluid, parameters of the steam produced in the reactor, and the plant modes that depend on the vessel type.

Figure 1.20 shows the principal thermal scheme of the nuclear steam power plant of the icebreaker "Arktika" [11].

Some of the Classification Societies (Russian Maritime Register of Shipping, English Lloyd, Norwegian Bureau Veritas) have developed the rules of construction of nuclear power plant vessels with the steam generating plant placed in a special sealed compartment of the ship or a lightweight container designed for a slight overpressure, which helps to prevent the radioactive gases and aerosols spread outside the compartment during the operation.

The special compartment is provided with a special ventilation system that maintains underpressure in it. The more dangerous the room is with respect to spread gases or aerosols, the higher the maintained underpressure is. These rules were used for arranging the nuclear power plant equipment on the "Arktika" icebreaker (Fig. 1.21)

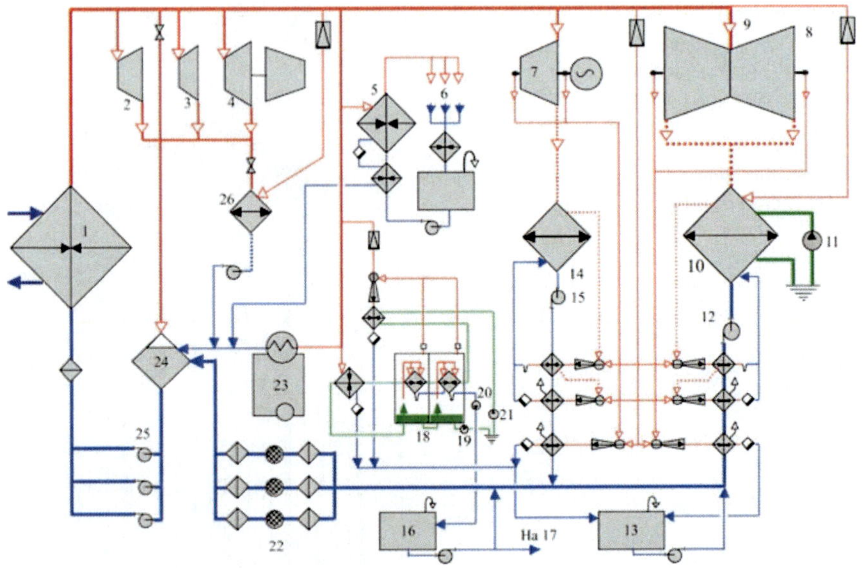

Fig. 1.20 Thermal scheme of the steam turbine plant of the icebreaker "Arktika". 1—steam generator; 2, 3, 4—turbo drive units of the feed pump, circulation pump and air compressor of the air body washing, respectively; 5—low-pressure steam generator; 6—general ship consumers; 7—turbine generator; 8—reduction cooling device; 9—main steam turbo electric unit; 10—main condenser; 11—circulation pump; 12—condensate pump; 13—collector of hot condensate; 14—auxiliary condenser; 15—auxiliary condensate pump; 16—distillate tank; 17—domestic needs; 18—water desalination plant; 19—brine pump; 20—distillation pump; 21—seawater pump; 22—block of ion-exchange filters; 23—auxiliary boiler plant; 24—deaerator; 25—feed pump; 26—parking condenser

[11]. The icebreaker's hull is divided by watertight bulkheads into eight compartments. The most important areas are set aside in separate watertight and airtight compartments, which not only ensure unsinkability of the ice-breaker in case of flooding of any two compartments, but also allow satisfying the radiation safety requirements completely.

The steam generating plant is located in the middle of the ship in a special airtight compartment provided with the light secondary biological shield. This special compartment is divided heightwise with the blocks of biological shield and sealed flooring into two rooms—the reactor (bottom) and hardware (top) ones.

The two reactor units are placed in the reactor room in the tank caissons of iron-water protection. The protection tank is separated from the airtight bulkheads by the cofferdams. The main engine room is located from the steam generating plant towards the bow, where the two main turbine generators with the capacity of 27,600 kW each are installed. Each turbine rotates three main series-connected alternators and two auxiliary ones with the capacity of 2000 kW each, as well as auxiliary machinery. The aft power plant is located in the aft of the steam generating power plant, which consists of three auxiliary turbine generators and one auxiliary diesel generator with

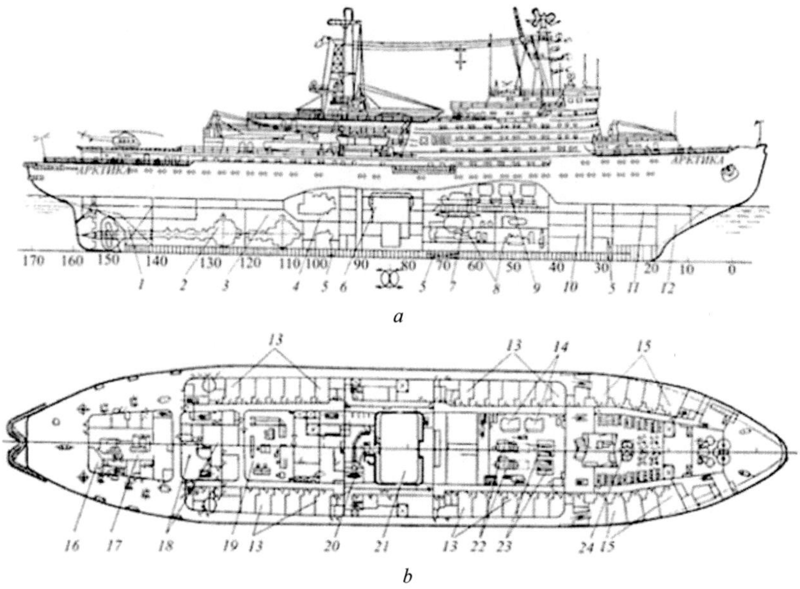

Fig. 1.21 Scheme of the nuclear power plant arrangement on the "Arktika" icebreaker: **a**—side view; **b**—upper deck; 1, 12—trim tank; 2, 3—roll and middle propulsion motors, respectively; 4—stern electric power plant; 5—ice box; 6, 21—reactor and hardware facilities of the nuclear steam generating plant; 7—main turbine; 8, 17—room of support mechanisms and towing winches, respectively; 9—main generators; 10—provision store; 11—cargo premises; 13, 15—single and double cabins, respectively; 14—desalination plant; 16, 19—aft and central control stations of the nuclear power plant; 18—medical unit; 20—sanitary inspection of central compartment; 22—low pressure steam generators; 23—auxiliary boilers; 24—crew canteen

the capacity of 1000 kW. Part of the equipment of nuclear power plant is located in the middle part of the superstructure: auxiliary boilers, evaporators, low-pressure steam generators, deaerator, and two emergency diesel generators with the capacity of 200 kW each.

Taking into account the increased demands on the safety of the nuclear power plants, the Regulations of Classification Societies are paying considerable attention to the placement of the reactor facility on the vessel and its protection from damage at colliding or grounding of the vessel. The regulations of all societies assume the existence of anti-shock protection and enhanced set in the area of the reactor compartment.

References

1. Moravskiy A. V., Fayn M. A. *Ogon v upryazhke, ilikakizobretayutteplovyedvigateli*[Fire in harness, or how heat engines were invented]. Moscow, Znanie Publ., 1990. 192 p.
2. Vitt P. *Gazovyeturbiny* [Gas turbines]. Moscow, Mashinostroenie Publ., 1965. 197 p.

3. Goloviznin A. M., Kuznetsov V. A., Pologikh B. G. *Sudovyeyadernyeenergeticheskieustanovki: ucheb. dlyavuzov* [Ship nuclear power plants: textbook for universities]. Moscow, Atomizdat Publ., 1976. 376 p.
4. Kartsev V. P., Khazanovskiy P. M. *Tysyacheletiyaenergetiki* [Thousands of years of power engineering]. Moscow, Znanie Publ., 1984. 224 p.
5. Akimov P. P. *Istoriyarazvitiyasudovykhenergeticheskikhustanovok* [History of marine power plants development]. Leningrad, Sudostroenie Publ., 1966. 187 p. Romanovskyi H. F., Ipatenko O. Ya., Patlaichuk V. M. *Teoriia ta rozrakhunokparovykh ta hazovykhturbin :navch. posib.* [Theory and calculation of steam and gas turbines: textbook]. Mykolaiv, UDMTU Publ., 2002. 292 p.
6. Belkind L. O., Konfederatov I. Ya., ShneybergYa. A. *Istoriyatekhniki* [History of technology]. Moscow, Gosenergoizdat Publ., 1956. 490 p.
7. Voznitskiy I. V. *Prakticheskierekomendatsiiposmazkesudovykhdizeley: ucheb. posobie* [Practical recommendations for the lubrication of marine diesels: textbook]. Saint Petersburg, Morkniga Publ., 2007. 129 p.
8. Horbov V. M. *Entsyklopediiasudnovoienerhetyky: pidruchnyk* [Encyclopedia of marine power engineering: textbook]. Mykolaiv, NUK Publ., 2010. 624 p.
9. Babich V. V. *Nashi avianostsynastapelyakhi v dalnikhpokhodakh* [Our aircraft carriers on the stocks and in distant voyages]. Nikolaev, Atoll Publ., 2003. 544 p.
10. Shirokov S. V. *Yadernyeenergeticheskiereaktory :ucheb. posobie* [Nuclear power reactors: textbook]. Kyiv, NGTU "KPI" Publ., 1997. 280 p.
11. Horbov V. M., Ratushniak I. O., Trushliakov Ye. I., Cherednichenko O. K. *Sudnovaenerhetyka ta Svitovyiokean :pidruchnyk* [Marine power engineering and the World Ocean: textbook]. Mykolaiv, NUK Publ., 2007. 596 p.

Chapter 2
General Information About Marine Power Plants

2.1 Purpose and Composition of Marine Power Plants

Marine power plant is a complex of functionally interconnected elements of power equipment, machines, components, and devices intended for the production, conversion, transmission and use of various types of energy necessary for the vessel functioning in accordance with its purpose [1].

Let us consider the main points of this definition. The marine power plant includes heat engines (machines), boilers (elements of power equipment); electric machines: generators, engines and converters; mechanisms: reducers, multipliers, pumps for various purposes (there are several hundreds of them), compressors, fans; cleaners (filters, homogenizers, fuel/oil separators); heat exchangers: coolers, heaters, evaporators, condensers; devices (air inlet, gas outlet, exhaust gas cleaning, shaft-rotating).

The main types of energy generated and consumed on the ship are *mechanical M, thermal T, electricE*. Mechanical energy is generated by heat engines and electric motors; electric power is generated by electric current generators; thermal energy is generated by steam-water boilers and heaters of various types.

Along with this, *hydraulic* (hydraulic systems of steering machines, deck mechanisms, hydraulic clutches, torque converters) and *pneumatic* types of energy (pneumatic drives of auxiliary mechanisms, compressed air systems) have their limited use on the ship.

Each marine power plant consists of *main* and *auxiliary* power plants.

As a rule, the main power plant is a part of the marine power plant which ensures movement of the ship. It is also called a propulsion plant, since it consists of the elements with the help of which energy is generated and transmitted to the ship propeller to move the ship. Transport vessels and ships spend up to 90% of the generated energy. At the same time, on some types of ships, for example, on the technical fleet ships (floating self-propelled cranes, dredges, earth-scoops, floating drilling rigs and drilling vessels), the total power of the technical means (cranes, drilling rigs, refueling pumps) often exceeds the power that is used to move the ship.

Z. Yang et al., *Marine Power Plant*,
https://doi.org/10.1007/978-981-33-4935-3_2

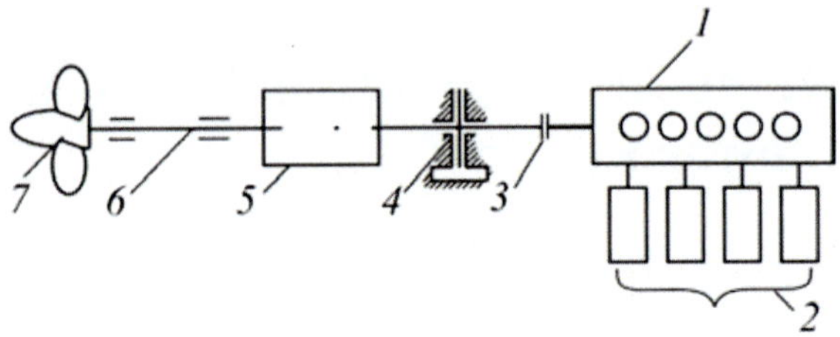

Fig. 2.1 Scheme of a propulsion diesel power plant

The main power plant (Fig. 2.1) consists of main engine *1* (one or several) with systems *2* that serve it; clutch *3* connecting engine with shafting *6*; main thrust bearing *4*, which accepts thrust from the pitch propeller (thrust is the driving force of the ship) through the shafting and transmits it to the ship through the foundation; transmission *5*, through which energy of the main engine is transmitted to the shafting. The transmission is designed to match the rotational speeds of the main engine and the pitch propeller; shafting *6* is used to transmit the torque to the pitch propeller and the thrust developed by the propeller to the ship hull; pitch propeller *7* provides conversion of the torque to the thrust.

Diesel and gas-turbine engines and steam turbines can be used as the main plants on ships. When steam turbines are used, the main steam boiler is also included in the main power plant.

The transmissions which are part of the propulsion system can be direct, mechanical, electrical and hydraulic

The propulsors are mainly used in the form of fixed (FPP) and controllable (CPP) pitch propellers, water jet and vane propellers.

The purpose of *auxiliary power plants* is to ensure the needs of the propulsion plant and general energy consumers of different types, as well as to create conditions for their effective functioning.

Such plants include a ship auxiliary electric power plant—marine power station, an auxiliary boiler plant, a water desalination plant, a comfortable air-conditioning system.

The marine power station is designed to provide the ship with electric power during its movement, at anchor and when performing mooring and loading and unloading operations by ship means. It consists of the following elements: auxiliary heat engine (it can be a diesel or a gas turbine engine or a steam turbine), electric generator, main distribution board, electrical devices (transformers, rectifiers, and converters), power cables leading to consumers and systems.

The auxiliary power plant may also include shaft generators that produce electric current at the expense of the power taken from the main engine. The shaft generators operate only during the main engine operation. The ship auxiliary power plant may consist of several (from two to ten) auxiliary electric power generators with the above-mentioned variants of the drive.

The required element of each auxiliary marine power plant is an emergency generator designed to ensure operation of vital ship facilities and systems when the ship is

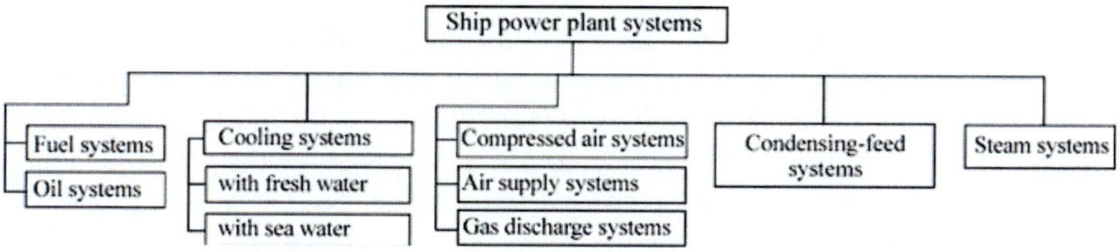

Fig. 2.2 Classification of the marine power plant systems

completely de-energized. The drive of such an electric generator on ships with any types of marine power plant is exclusively a diesel engine.

Power capacity of a marine power plant depends on the type and purpose of the ship. Passenger, fishing and research ships have the largest power capacity. There, it accounts for up to 60% of the main engine power capacity, while at universal dry cargo ships and boats it makes up 10–15% [1].

The purpose of *an auxiliary boiler plant* is to provide the ship and the power plant with steam and hot water. It includes auxiliary, recycling and combined boilers, heat exchangers, pumps, fans, boiler service equipment and systems, emergency warning and protection.

The water desalination plant provides production of fresh water necessary for technical and domestic purposes from seawater.

The comfortable conditioning system designed to maintain the necessary comfortable air parameters in the premises of the marine power plant and other premises where people can stay.

The elements of the marine power plant are combined into a single energy complex of the ship with the help of marine power plant systems.

The marine power plant system is a set of pipelines, mechanisms, apparatuses, devices, instruments and tanks designed to ensure performance of the certain functions by the marine power plant [1, 2].

The types of the marine power plant systems according to their purpose are shown in Fig. 2.2.

The purpose of *the fuel system* is to receive, store, deliver, pump, clean, heat and supply fuel to diesel and gas turbine engines and boilers.

The oil system provides processes of receiving, delivering, storing, cleaning and supplying oil for lubrication and cooling of machines and mechanisms.

The cooling systems are designed to supply water for cooling of the mechanisms, devices, and working media in heat exchangers of the marine power plant.

The compressed air system provides the marine power plant and the ship with compressed air of the parameters necessary for starting and reversing the main diesel engines, launching auxiliary engines, operation of automation and control systems, blowing out kingston valves, and household needs.

The air supply systems clean and supply engines and boilers with the air necessary for burning fuel and cooling the engines.

The gas-discharge systems discharge exhaust gases from engines and boilers.

The condensate-feed system is designed to extract condensate from the main and auxiliary condensers, to receive, transmit, store and supply feed water to steam-generating plants and units, as well as to the regulating and controlling bodies.

The steam systems supply overheated or saturated steam from steam-generating plants and units to the main steam turbines and auxiliary mechanisms, apparatus and equipment, as well as remove exhaust steam to heat exchangers.

A typical composition of the marine power plant system is shown in Fig. 2.3.

The type of the power plant and its composition are determined by the type and purpose of the ship on which it is installed. Figure 2.4 shows a classification of marine power plants according to the main common attributes.

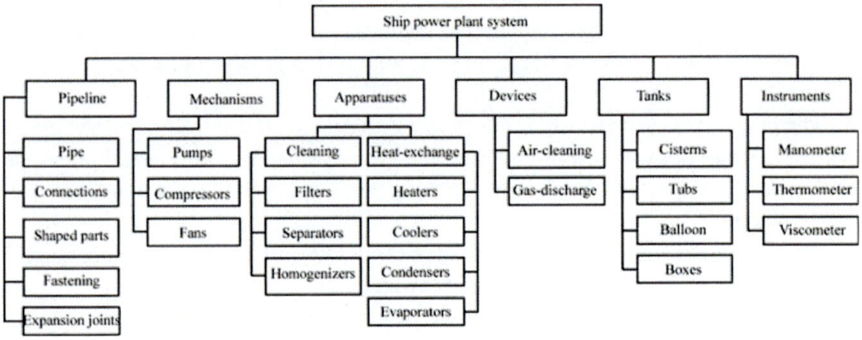

Fig. 2.3 Composition of the marine power plant systems

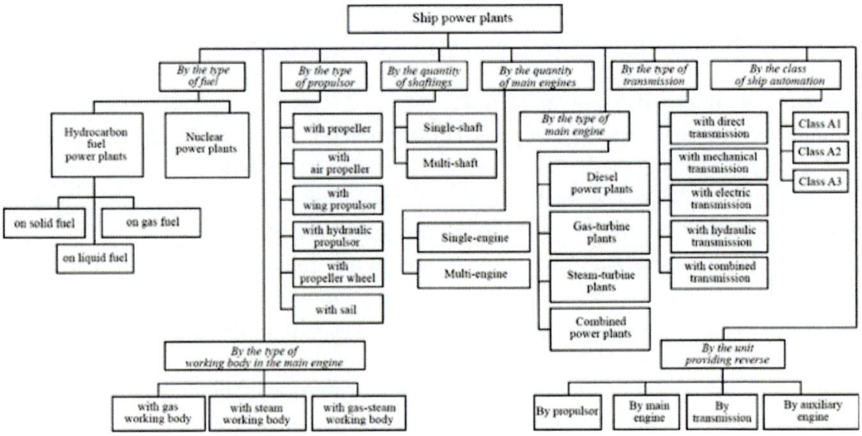

Fig. 2.4 Classification of marine power plants

2.2 Energy Conversion and Transmission in Marine Power Plants

2.2.1 Energy Flows in Marine Power Plants

The source of any kind of energy that is generated by a marine power plant is fuel of one or more species, which taken to the ship in the form of stocks in the amount required to carry out the voyage.

The processes that are connected with conversion of the fuel energy into the energy necessary for propulsion occur in the following elements of the marine power plant:

- *in the generator of the working fluid*, where chemical energy of the fuel is converted into thermal energy; it can also compress and heat the working fluid;
- *in the heat engine* that turns thermal energy into the mechanical one;
- *in the main gear,* required to convert rotational torque of the heat engine;
- *in the shafting,* which transmits the engine power to the propeller;
- *in the propeller,* which uses the transmitted power for the ship movement.

Implementation of the processes of conversion and transmission of the energy in the above-mentioned elements of the main marine power plant is associated with the energy losses that accompany them. Figure 2.5 shows the scheme of energy conversion and transmission in the main power plant.

Among the existing types of marine power plants, the working fluid generator in the form of a separate element—a boiler—is included only to a steam turbine plant. In the marine power plant with ICE, the processes which correspond to the working processes of both the working fluid generator and the heat engine occur simultaneously in the working cylinders. In the gas turbine plant, the functions of the working fluid generator are performed by the compressor, the combustion chamber and the turbine, the power of which is used to drive the compressor. The combination of the compressor, the combustion chamber and the turbine driving the compressor forms the gas generator. It can consist of one or two compressors, one or two turbines.

The measure of effectiveness of the marine power plant is the efficiency of the marine power plant and the coefficient of heat use in the marine power plant.

The effective efficiency of the marine power plant is

$$\eta_{\mathrm{e}} = \frac{M + E}{Q_T},$$

where M, E, T are the mechanical, electrical and thermal energy, respectively; Q_T is the amount of heat released during the fuel combustion in the elements of the marine power plant; the heat use coefficient is

$$\eta_{\mathrm{hu}} = \frac{M + E + T}{Q_{\mathrm{T}}}.$$

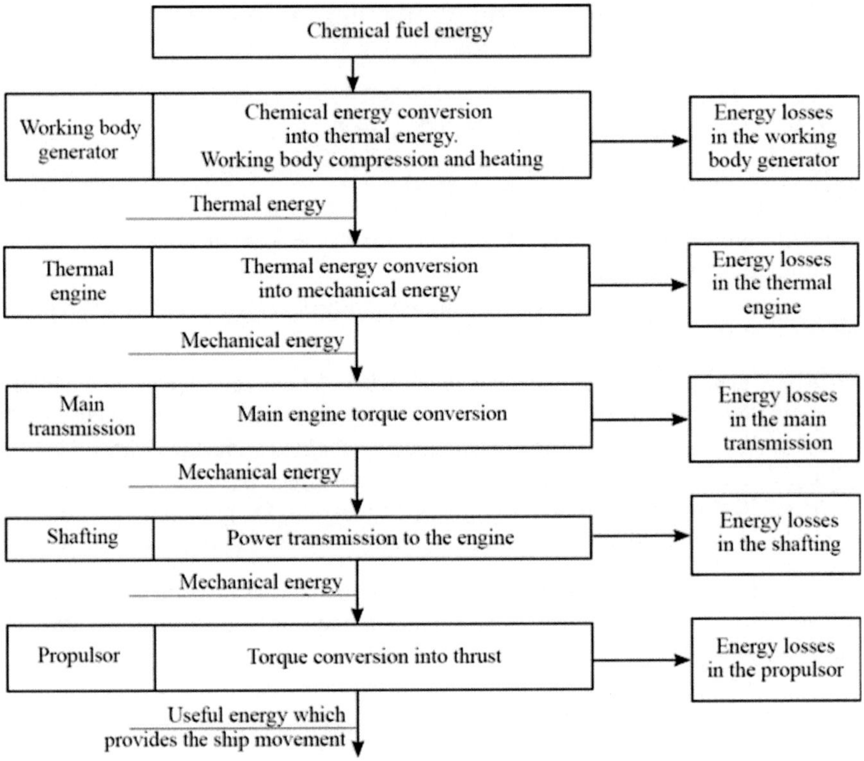

Fig. 2.5 Scheme of energy conversion and transmission in the main marine power plant

One of the main ways to improve the marine power plant efficiency, along with improvement of the energy production and transmission in the elements of the main power plant, is recovery. It is the use of the part of the heat losses that accompany all the processes of energy transmission and conversion to provide ship consumers with different types of energy. Modern marine ships with diesel power plants widely utilize recovery of the exhaust gases heat and cooling water. This is explained by the fact that up to 40% of the heat generated during fuel combustion is removed by exhaust gases and cooling water.

Heat recovery can be carried out in all types of marine power plants. It allows increasing the efficiency coefficient up to 45–47% and the heat use coefficient up to 80%. The recovered heat can provide ship heating (premises heating, hot water generation for general ship purposes), air conditioning, operation of water desalination and refrigeration plants, electric energy generation in turbo generators and obtaining additional power that can be supplied to the shafting. The most widespread plants on the modern ships are recovery boilers that operate on the exhaust gases heat and vacuum-type water desalination plants using the heat of the water cooling the diesel engine. Implementation of heat recovery is possible both in the main and auxiliary marine power plants—in the marine power station. Heat recovery of the main power plant can meet the needs of almost all the ship consumers without any additional fuel consumption.

The effect of heat recovery is greater when the efficiency coefficient of the main device in the marine power plant is smaller, and vice versa. The device uses the chemical fuel energy (of the main engine—the internal combustion engine, the gas turbine engine, the main boiler).

Heat recovery significantly affects the composition and layout of the marine power plant. The scheme of heat recovery of the marine diesel power plant with the shaft generator, recovery steam turbine and recovery power gas turbine is given in Fig. 2.6 [3].

Figure 2.7 presents a typical structural functional scheme of the transport marine power plant consisting of the propulsion complex, marine power station and boiler plant, in which the chemical fuel energy is converted into the corresponding types of energy. It suggests using the energy of the water that is cooling the main engine for

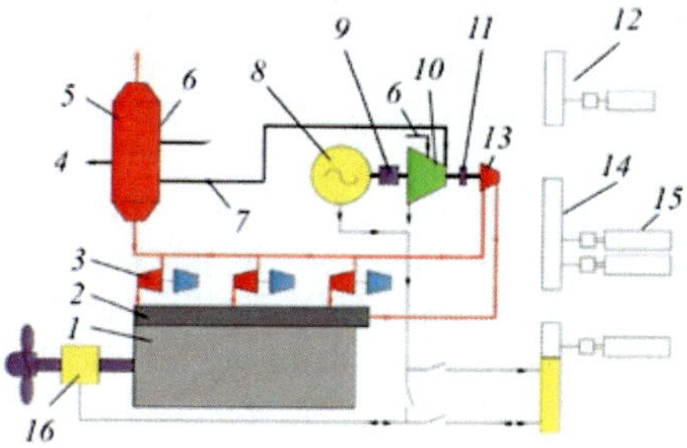

Fig. 2.6 Principal scheme of heat recovery in the diesel marine power plant: 1—main engine (ME); 2—main engine exhaust gases collector; 3—turbocharger; 4—steam for general needs; 5—recovery boiler (RB); 6—low-pressure steam; 7—overheated high-pressure steam; 8—generator of the recovery steam turbine; 9—reducer of the recovery steam turbine; 10—recovery steam turbine (RST); 11—reducer of the power recovery turbine with an overrunning clutch; 12, 15—emergency (EDG) and auxiliary (ADG) diesel generators, respectively; 13—recovery power gas turbine (RGT); 14—main switchboard; 16—shaftgenerator (SG)

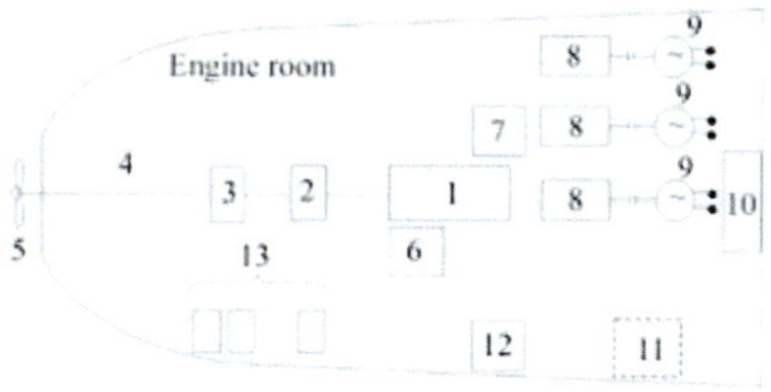

Fig. 2.7 Structural functional scheme of the transport marine power plant: 1—main engine; 2—clutch; 3—transmission; 4—shafting; 5—thruster; 6—desalination plant; 7—recovery boiler; 8—primary heat engine; 9—electric generator; 10—group switchboard; 11—emergency diesel generator; 12—auxiliary boiler; 13—main engine systems

operation of the vacuum desalination plant and the heat of the main engine's exhaust gases to produce steam in the recovery boiler.

Assessing the efficiency of the marine power plant with no regard for the operating conditions makes no sense, because the ship has several operating modes in which the power equipment used in the propulsion complexes works differently, and some elements cannot be used at all. For example, if three diesel generators are the part of the marine power station, then one, two or three units can operate in separate modes; the main engine does not work at anchor,and therefore the recovery boiler and the desalination vacuum plant do not work, but the auxiliary boiler and the marine power station will operate.

Operational modes of the transport marine power plant include the running mode, standby mode with cargo operations, standby mode without cargo operations, emergency mode. For a passenger ship, the standby modes of the marine power plant with or without passengers are typical.

When determining the efficiency of the marine power plant operation in each of the operational modes, one draws a table of the modes with designation of the elements of the given complexes operating in these modes (Table 2.1).

Let us consider the fuel energy conversion in the propulsion complex, marine power plant and boiler plant (Fig. 2.8, 2.9).

As a part of the propulsion systems of modern ships, the internal combustion engine is most widely used as the main engine due to the highest efficiency when compared to other heat engines. Therefore, further we consider the scheme of the process of energy conversion in the propulsion complex with the internal combustion engine as the main one.

When fuel is burned in the engine, the heat $Q_{\mathrm{h}}^{\mathrm{ME}}$ is generated, which ensures production of the useful power on the flange of the main engine—the effective power N_e^{ME}. This process is accompanied by the heat loss in the internal combustion engine

Table 2.1 Modes of operation of the marine power plant[a]

Equipment	Modes				Notes
	Running mode	Standby mode with cargo operations	Standby mode without cargo operations	Emergency mode	
ME (1)	+1	–	–	–	
ADG (3)	–	+2	+1	–	1DG—reserve
AB (1)	–	+1	+1	–	
WHB (1)	+1	–	–	–	
DP (1)	+1	–	–	–	
SG (1)	+1	–	–	–	
EDG (1)	–	–	–	+1	

[a]The "+" sign indicates inclusion of the MPP element in this mode. The figure with the "+" sign shows the number of the activated MPP elements. The "–" sign means exclusion of the element from operation

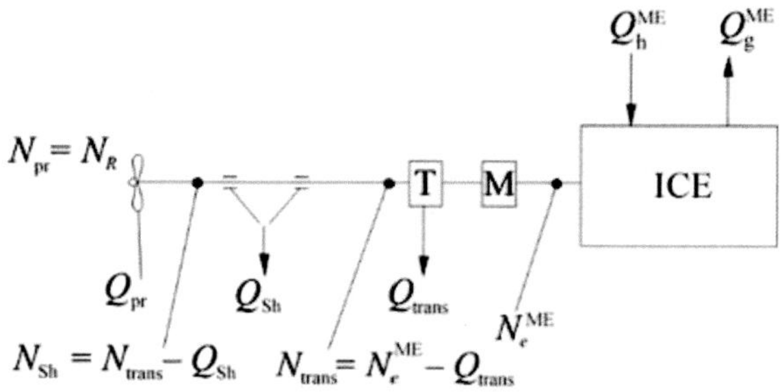

Fig. 2.8 Scheme of the processes of energy conversion in the propulsion complex

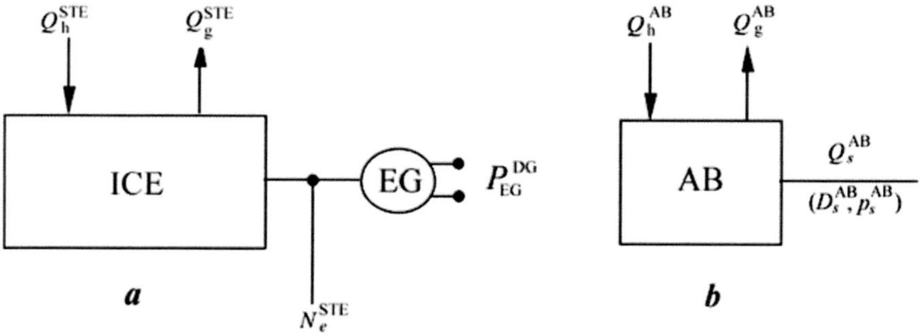

Fig. 2.9 Scheme of the processes of heat conversion in the marine power plant (**a**) and the auxiliary boiler plant (**b**)

Q_{ME}. These are the heat losses with waste gases, cooling water, oil, when cooling the charge air, residual losses. The value of Q_{ME} for modern engines is 45–50% of Q_h^{ME}.

From the main engine, power is transmitted through the clutch, ship transmission, and shafting, where there are also losses of mechanical energy converted into thermal energy: Q_{trans} and Q_{Sh}. Thus, the power supplied to the propulsor is

$$N_{Sh} = N_e^{ME} - Q_{trans} - Q_{Sh}.$$

The energy losses on the propulsor Q_P are determined by the propulsive coefficient.

The useful effect of the propulsion complex functioning, which takes place as the result of fuel combustion in the main engine, is the towing capacity N_R, which is converted into thrust:

$$N_R = N_e^{\mathrm{ME}} - Q_{\mathrm{trans}} - Q_{\mathrm{Sh}} - Q_{\mathrm{P}}.$$

The energy losses in clutches are taken into account only when using slip clutches; energy losses for all other types of clutches are neglected.

The efficiency coefficient of the propulsion complex is

$$\eta_e^{\mathrm{PC}} = \frac{N_R}{Q_{\mathrm{H}}^{\mathrm{ME}}}.$$

The schemes of heat conversion in the auxiliary power plants are shown in Fig. 2.9.

An internal combustion engine, a steam turbine, or a gas turbine engine can be used as the primary heat engine for the electric generator drive. Most often it is an internal combustion engine.

Conversion of the mechanical energy N_e^{ME} developed by the primary heat engine into the electric energy P_{EG} produced by the electric current generator is carried out with the energy losses Q_{EG}, which are taken into account with the help of the electric generator efficiency coefficient:

$$\eta_{\mathrm{EG}} = \frac{\mathrm{P_{EG}}}{N_e^{\mathrm{DG}}}.$$

The steam is produced in the quantity D_s^{AB} and with the pressure p_s^{AB} (MPa) in the auxiliary boiler (see Fig. 2.9b).

The heat $Q_{\mathrm{h}}^{\mathrm{AB}}$ that is released during the fuel combustion in the auxiliary boiler is distributed to the heat used to produce steam Q_s^{AB} in the auxiliary boiler and to the heat losses Q^{AB}:

$$Q_{\mathrm{h}}^{\mathrm{AB}} = Q_s^{\mathrm{AB}} + Q^{\mathrm{AB}}.$$

The heat use coefficient in the auxiliary boiler is

$$\eta^{\mathrm{AB}} = \frac{Q_s^{\mathrm{AB}}}{Q_{\mathrm{h}}^{\mathrm{AB}}}.$$

When considering the processes of conversion of heat of the exhaust gases of the main engine into thermal energy (the steam energy in the recovery boiler), the heat use coefficient is defined as follows:

$$\eta^{\mathrm{WHB}} = \frac{Q_s^{\mathrm{WHB}}}{Q_{\mathrm{g}}^{\mathrm{WHB}}}$$

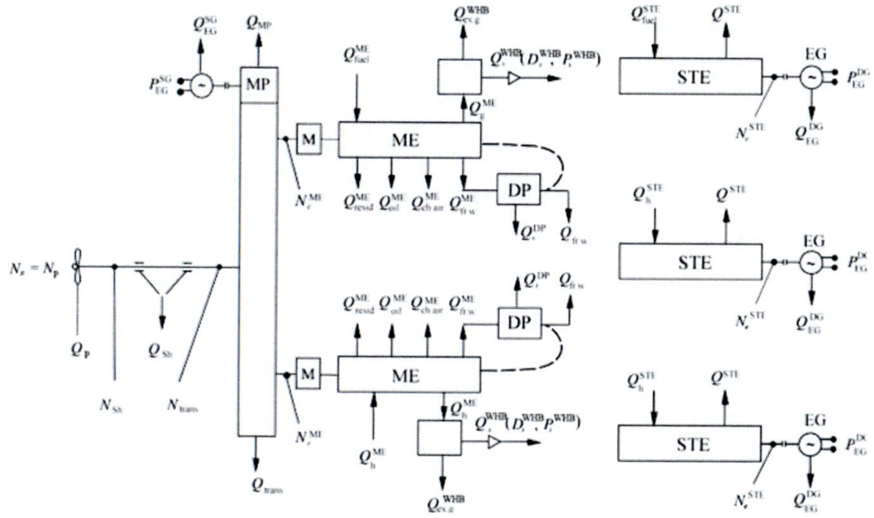

Fig. 2.10 Scheme of the energy flows distribution in the marine power plant on the running

where Q_s^{WHB} is the amount of heat required to obtain saturated steam; $Q_{\text{g}}^{\text{WHB}}$ is the amount of heat that can be transferred from the gases to feed water.

Let us consider energy conversion in the marine power plant of a specific composition. Figure 2.10 shows the scheme of energy conversion on the running mode for the marine power plant of a large freezing trawler [4]. The main power plant is single-shaft, diesel, two-engine with the power capacity of 2 × 2880 kW with power take-off to the shaft generator of 800 kW. It is supposed that on the running mode there operate two of three diesel generators, which are the part of the marine powerplant, each of 820 kW, two recovery boilers located on the main engine gas ducts, and two recovery desalination plants.

The scheme details the following components of the thermal losses of the main engine Q^{ME}: $Q_{\text{fr.w}}^{\text{ME}}$—the losses associated with fresh water; $Q_{\text{ch.air}}^{\text{ME}}$—the losses connected with the charge air cooling; $Q_{\text{oil}}^{\text{ME}}$—the heat losses with oil; Q_{g}^{ME}—the losses with the exhaust gases; $Q_{\text{resid}}^{\text{ME}}$—the residual losses; $Q_{\text{ex.g}}^{\text{WHB}}$—the losses with the exhaust gases from the recovery boiler; $Q_{\text{fr.w}}$—the losses with the fresh water that leaves the desalination plant; Q_s^{DP}—the heat spent on the seawater boiling and the steam formation in the desalination plant; Q_s^{WHB}—the amount of heat required to produce steam in the recovery boiler; D_s^{WHB}, P_s^{WHB}—the steam-generating power and the steam pressure in the recovery boiler; Q_{MP}—the losses in the multiplier.

The efficiency coefficient of the propulsion complex of the above-mentioned composition is defined as follows:

$$\eta_e^{\text{PC}} = \frac{N_R + P_{\text{EG}}^{\text{SG}}}{2Q_{\text{h}}^{\text{ME}}}.$$

The efficiency coefficient of the marine power plant on the running mode is

$$\left(\eta_e^{\text{SPP}}\right)_{\text{r.m}} = \frac{N_R + P_{\text{EG}}^{\text{SG}} + 2P_{\text{EG}}^{\text{DG}}}{2Q_{\text{h}}^{\text{ME}} + 2Q_{\text{h}}^{\text{DG}}}.$$

The heat use coefficient of the marine power plant on the running mode is

$$\left(\eta_{\text{SG}}^{\text{SPP}}\right)_{\text{r.m}} = \frac{N_R + P_{\text{EG}}^{\text{SG}} + 2Q_s^{\text{RB}} + 2Q_s^{DP} + 2P_{\text{EG}}^{\text{DG}}}{2Q_{\text{h}}^{\text{ME}} + 2Q_{\text{h}}^{\text{DG}}}$$

The visual form of representation of energy flows in the marine power plant is thermal diagrams. In accordance with the marine power plant composition, the thermal diagram of the energy flows distribution on the running mode is shown in Fig. 2.11 [4].

Similar thermal diagrams are also drawn for the other operational modes of the marine power plant. Analysis of such diagrams makes it possible to determine the efficiency and prospects of using secondary heat reserves in order to increase the efficiency of the marine power plant.

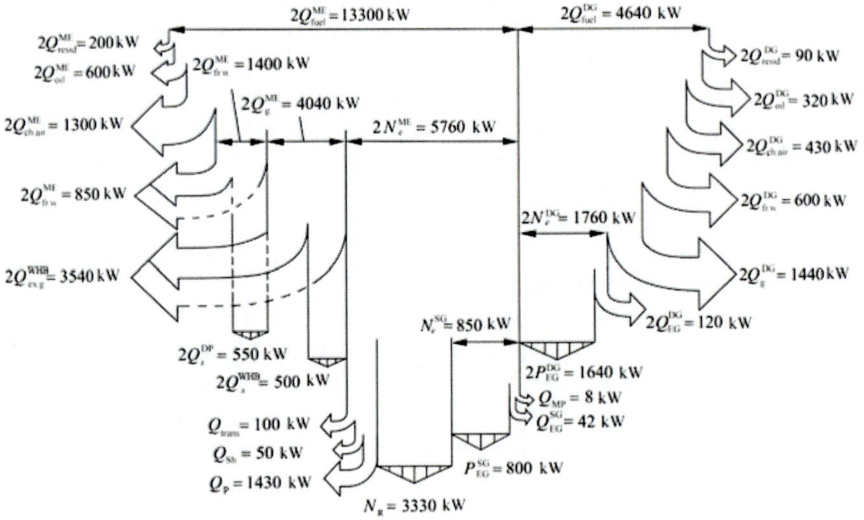

Fig. 2.11 Heat flows in the marine power plant on the running mode

2.2.2 Ship Shafting

The main elements of the propulsion complex includes the shafting, which brings power to the propeller, perceives the thrust created by the propeller, and transmits it through the main thrust bearing to the ship hull, ensuring its movement.

The main elements of the shafting (Fig. 2.12) are as follows:

- the shafts that transmit torque to the propeller and perceive the thrust that it develops (the rowing, intermediate, thrust and spacer shafts belong to such shafts), and their joints;
- the main thrust bearing that receives thrust and transmits it to the ship hull;
- the stern tube that ensures sealing of the shafting at its exit from the ship hull and at the same time is the prop of the shafting;
- the intermediate shafting bearings that are the shafting props between the stern tube and the main thrust bearing;
- the bulk seal that ensures sealing of the shafting at the point of its passing through the watertight bulkhead;
- the systems of lubrication, cooling of the stern tube and the steady bearings.

The turning of the shafting when the ship is at anchor is provided by a shaft-turning device. A braking device is used to stop the shafting line during its repair or ship towing, when the propeller rotation is not allowed. The braking device is installed mostly on a flanged connection, which in this case has increased diametrical dimensions. The braking device is activated when the ship is stopped, it is not used to stop the rotating shaft [1, 5].

In case of using an adjustable pitch propeller, rotation of the propeller blades is provided by the step-change mechanism, which most often has a hydraulic drive built into the shaft line. The shafting can be fitted with a tachometer sensor for measuring the rotational speed and a torsiometer—a device for determining the power.

The number of steady bearings is determined by the length of the compartments and the distance between the watertight bulkheads. The bearings are located in the places that are accessible for maintenance.

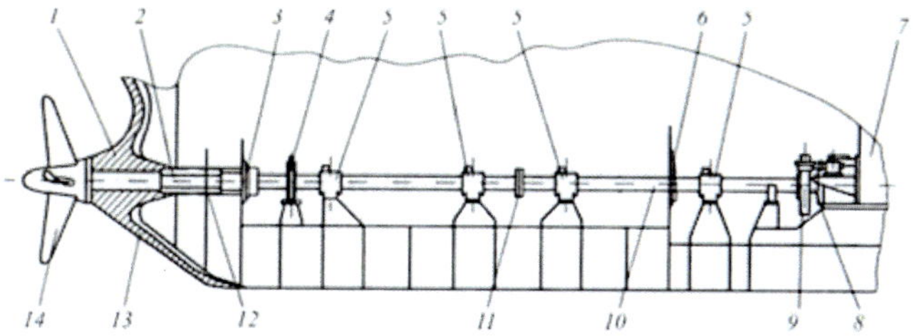

Fig. 2.12 Constructional scheme of the shafting: 1—stern post boss; 2—propeller drive shaft; 3—stern tube stuffing box; 4—brakes;5—steady bearings; 6—bulkhead stuffing box; 7—aft part of the main engine; 8—thrust block shaft; 9—shaft-turning device; 10—mid shaft; 11—shaft joint; 12—stern tube; 13—stern post; 14—propeller

The composition and dimensions of the ship shafting depend on the power capacity, rotational speed of the main engine and its location on the ship. The engine room is located in the aft part on most transport ships; in this case, only one mid shaft is included in the shafting along with the propeller shaft.

As of 2016, the maximum power transmitted by the shafting is 80 MW (the container ship “Emma Maersk”); the diameter of the propeller shaft made of low-carbon or alloy steel is 1400 mm. The specific weight of the ship shafting with the propeller (in relation to the total weight of the power plant) is 5–18%, the relative cost is 7–9% [6].

Figure 2.13 shows the general layout of the shafting of the single-propeller bulk carrier with deadweight $DW = 52450$ tons, length $L = 201.6$ m, draft $T = 12.3$ m. The main engine is 8L60MC with the effective power $N_e = 15360$ kW and the rotational speed $n = 123$ min^{-1} [6].

Stern tube devices employ bearings made of nonmetallic materials with seawater lubrication or metal stern bearings lubricated with oil.

An important element of the stern tube devices which include the latter is the seals, which have to ensure reliable sealing, exclude oil leakage from the aft and head parts of the stern tube. Most of the seals provide sealing with rubber cuffs of different designs.

A typical construction of a stern tube sealing with a hydraulic lock is shown in Fig. 2.14.

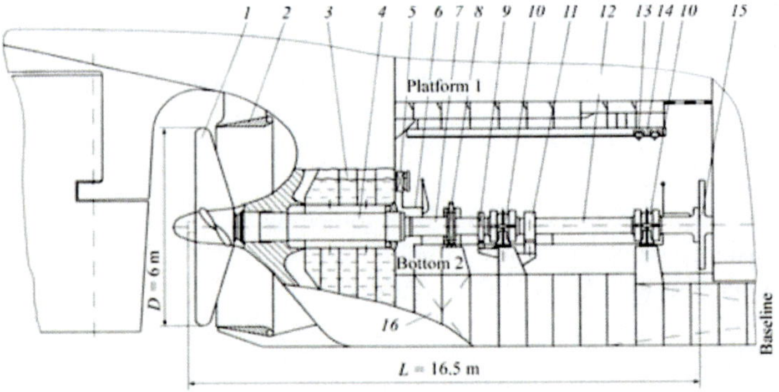

Fig. 2.13 Shafting with oil lubrication of the stern tube bearing: 1—propeller; 2—propeller shrouding; 3—lowest level of water in the afterpeak; 4—stern tube device; 5—tank for cooling of ship head sealing; 6—oil pipeline; 8—worm reduction brakes; 9—tachometer converter; 10—steady bearing; 11—contact-brush device; 7, 12—mid shaft; 13—lifting trolley; 14—monorail; 15—main engine drive shaft; 16—tank for draining oil from the stern tube

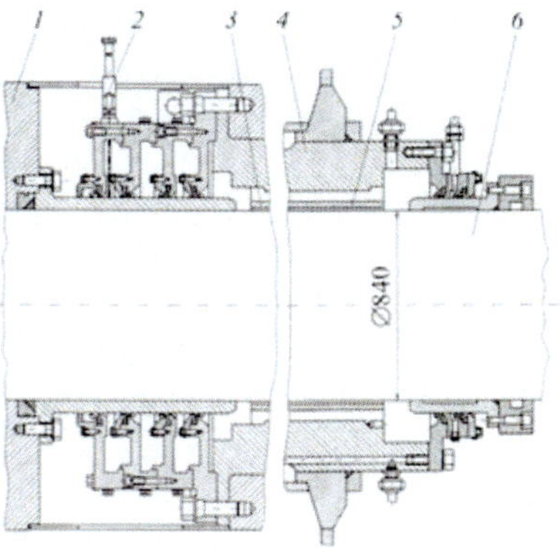

Fig. 2.14 Stern tube sealing "Neptune": 1, 6—propeller and propeller shaft, respectively; 2—propeller shaft subsidence measuring device; 3, 5—aft and head stern tube bearings, respectively; 4—stern tube

2.2.3 *Marine Power Plant Transmission*

The marine power plant transmissions are divided into main and auxiliary ones. The former are used to transfer energy from the main engines to propellers, while the latter transfer energy from auxiliary engines (diesel engines, turbines, electric engines) to machines and auxiliary mechanisms (electric generators, compressors, pumps).

The following transmissions have found their application in marine power plants (Fig. 2.15):

- direct (immediate);
- mechanical (reduction);
- hydraulic (using the hydraulic clutches and torque converters, the hydraulic pump driven by the engine, and the hydraulic motor that works for the propeller);
- electric (diesel or turboelectric plants with the main diesel or turbine generators and the propulsion electric engines);
- combined (mixed).

Direct transmissions are mainly used with low speed diesels at n_{eng} = 60–250 min^{-1}. Along with this, they have found their application on river ships, fishing trawlers and tugboats, where medium speed engines with n_{eng} = 275–400 min^{-1} are installed.

Diesel plants with LSE (Low Speed Engine) and direct transmission have gained widespread application in the transport fleet—ships with such plants make up about 60% of the world's ships with the displacement of more than 2 thousand tons. It is explained by a number of their positive qualities: high efficiency coefficient of transmission, constructional simplicity and high reliability of transmission, high ecological compatibility of the main diesel engines, long service life, low maintenance costs [1, 7].

At the same time, diesel installations with direct transmissions have some disadvantages, such as high weight and dimensions, complexity of the auxiliary machinery driving, and reduced survivability.

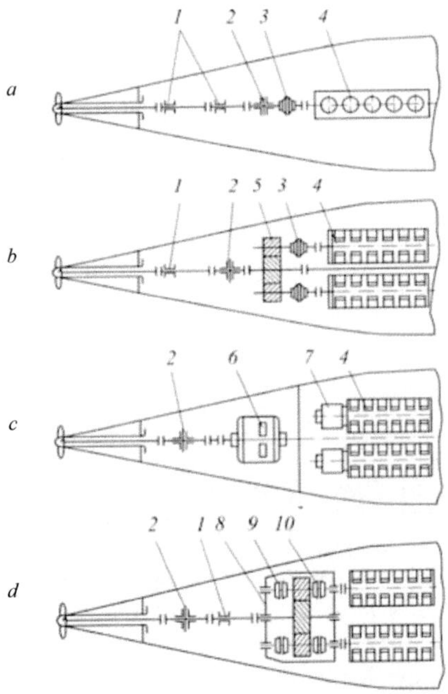

Fig. 2.15 Scheme of diesel engines with different power transmissions: *a*—straight; *b*—mechanical; *c*—electrical; *d*—hydraulic; 1, 2—steady and thrust bearings, respectively; 3—clutch; 4—main engine; 5—reducer; 6—main electric engine; 7—electric generator; 8—reverse hydraulic transmission; 9, 10—hydraulic torque converters of the astern and forward strokes, respectively

Mechanical transmissions are the train gears with a constant gear ratio (the gear ratio $i = n_1/n_2$, where n_1 and n_2 are the rotational speeds of the input and output shafts, respectively).

Modern train gears have reached a high degree of perfection. They are characterized by moderate weight and dimensions, high efficiency coefficient (97–98%) and a substantial service life.

The train gears that reduce the rotational speed are called reducers, and the gears that increase it are called multipliers.

With the help of a reducer, a propeller, a two-propeller drive from one engine, and a drive of various auxiliary mechanisms (electric generators, pumps) can be driven from one or several main engines.

The train gear of the reducer consists of a main drive pinion that is driven by the engine, and the driven gear that rotates the shaft or the driving mechanism. The diameter of driven gears is 4.5 m, and the module is 10.

The types and designs of ship reducers are diverse. They are performed in the single-stage form (for $i \leq 15$, because the wheel is large in case of large gear ratios) or the multistage form (mostly two-stage in modern marine power plants) with the gear ratios up to 30–50.

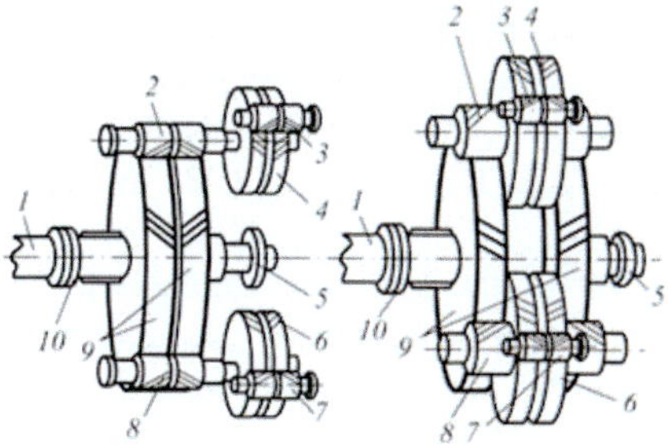

Fig. 2.16 Two-stage reducers: 1—shafting; 2, 8 and 3, 7—driving gears of the second and first stages, respectively; 4, 6—driven wheels of the first stage; 5—crest of the main thrust bearing; 9—driven gear of the reducer; 10—flange

According to their kinematic schemes, double-reduction (two-stage) gears are divided into two types: *articulated gears* and *nested gears.*

In the two-stage articulated reducers (Fig. 2.16a), the first and the second gear stages are arranged one by one. The wheels of the first stage in the nested reducers are located in the axial interval between the gear rings of the main wheel (see Fig. 2.16b). In the world practice, both types are used, but the articulated gears are more common.

The desire to reduce deformation of gears *2* of the first stage has led to creation of the gears with bifurcation of power (dual tandem gear trains) (Fig. 2.17). In such a transmission, each of the first-stage driving gears transmits torque to not one but two of driven wheels *4*. There are four driving gears *3* of the second stage that rotate gear *4*. The bifurcation of power makes it possible to reduce the size of the gears and wheels, the entire transmission becomes more compact as a whole. The transmission with bifurcation of power is widespread in modern ship practice.

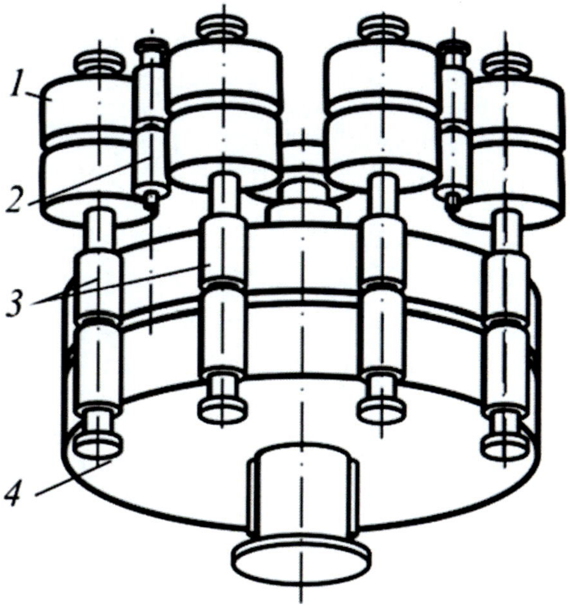

Fig. 2.17 Dual tandem gear train

The gears mentioned above are called *parallel shaft gears*. Turbo-gear units of considerable power implement two-stage reducers, which include *planetary gearing*. The weight and dimensions of such reducers are 30% smaller, the cost is 20% lower and the efficiency coefficient is 0.2–0.5% higher than that of other types of gears [1, 6].

The planetary gear schemes are shown in Fig. 2.18. The overall positive feature of such transmissions is the axiality of the engine and the shafting. The planetary gear consists of three main elements: central sun pinion *1* mounted on the drive shaft, the epicycles—inner-gear wheels *4* and carrier *2* with an axis on which pinion gears *3* (satellites) rotate. Transmission usually follows one of the following two patterns: with the fixed epicycle (see Fig. 2.18a) or with the fixed carrier (see Fig. 2.18b). Differential stages can also be used as a part of the reducer; this is a modification of the planetary degree where all the links are in rotation and transmit power.

Ship parallel shaft gears (reducers) mainly employ *cylindrical helical gears*. Along with the cylindrical gears, the marine power plant reducers with the dynamic principle of maintaining—on hydrofoils and on the air cushion—include *bevel gears*. Their purpose is to drive superchargers and propellers of air-cushion ships or propulsors of hydrofoil ships.

Gear wheels consist of a shaft, gear rims and a middle part that connects the shaft to the rims. Most often, the gear rims are manufactured separately and are mounted on the middle part of the wheel. High-quality carbon or alloy steel is used for the rim blanks.

Pinions are made in the form of long shafts with central drilling, which is designed to reduce weight and control quality. The pinion material has higher strength and hardness than the material of the gear rims of the wheel. The most common material for making gears is high-alloy steel.

Protection of gear transmission from shock loads in the reducers is provided by *torsion* (flexible) shafts that pass through the hollow gear and transmit torque from the turbine rotor to this pinion.

Oil is supplied to the engagement points to reduce heating and wear at the pressure of 0.15–0.20 MPa in such a way that it is drawn into the engagement with the rotating teeth.

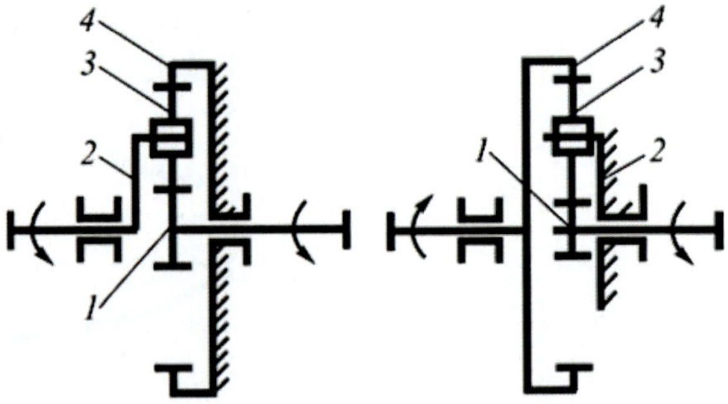

Fig. 2.18 Planetary gears

The gear wheels are located in a close gearbox, which is made horizontally detached, and its upper and lower parts are interconnected by flanges on the bolts.

The gearbox is welded from sheet steel; the gear bearings and the wheel shaft are placed in it.

Parallel shaft gears of single-machine units are performed in the single-stage form; their gear ratio does not exceed 7–10. A typical construction of a single-stage reducer with the displacement of the driving and driven shafts in the vertical plane is shown in Fig. 2.19.

If it is necessary to ensure axial arrangement of the engine and the shafting, even in case when the gear ratio enables using a single-stage reducer, two-stage reducers are implemented (Fig. 2.20).

The main steam-turbine aggregate of the "Krym" tankers includes a reducer with a three-stage planetary-parallel gear. The reducer construction provides for the separation of the torque of each of the two turbines (HPT—high pressure turbine and

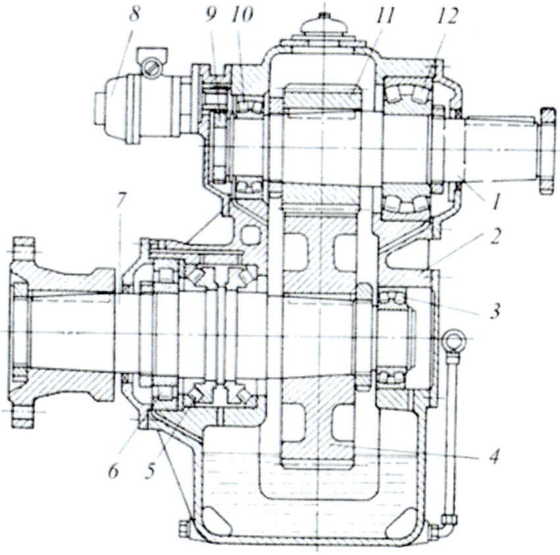

Fig. 2.19 Single-stage reducer with the shaft arrangement in the vertical plane: 1, 7—driving and driven shafts, respectively; 2—the body; 3, 5, 6, 10, 12—thrust bearings; 4, 11—wheel and pinion, respectively; 8—oil pump; 9—oil pump drive

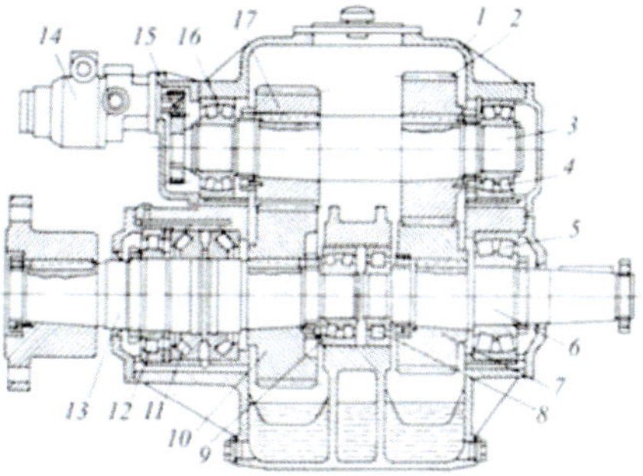

Fig. 2.20 Two-stage reducer with axial input and output: 1—body; 2, 10—wheels; 3, 6, 13—mid, driving and driven shafts, respectively; 4, 5, 8, 9, 11, 12, 16—thrust bearings; 7, 17—pinions; 14—oil pump; 15—pump drive

LPT—low pressure turbine) of the unit into three flows and the summation on the large gear wheel of six torque flows.

The general view of the planetary-parallel reducer is shown in Fig. 2.21 [6].

Designs of the reducers in the modern marine power plants are diverse. Their features are related to the ship type, as well as the type and composition of the marine power plant.

Figure 2.22 [8] provides a kinematic scheme of the reducer of the power plant on the "Kapitan Smirnov" ship with the horizontal cargo handling (roll-on-roll-off). The M25 power plant of these ships is a gas-steam turbine. The main engine is a reverse GTE; its waste gases are used for operation of the recovery boiler, the steam from which goes to the recovery steam turbine. The reducer sums the power of GTE (14000 kW) and ST (3200 kW) and transfers it to the shafting.

The reducer is non-reversible, two-stage, of the parallel type. Input *2* is from the GTE, with further bifurcation of power into two driven gears, while input *1* is from the recovery ST. The cam-friction clutch provides connection of the recovery ST to

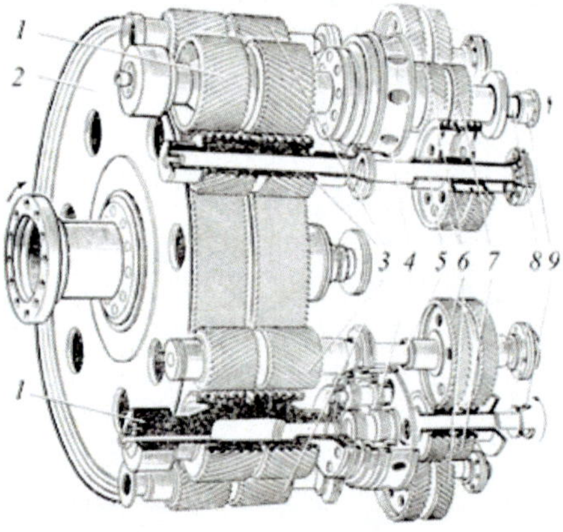

Fig. 2.21 General view of the planetary-parallel reducer of the main- turbo gear unit of the "Crimea" tanker: 1, 3—gears of the third parallel stage; 2—main driven gear wheel; 4—differential (first planetary gear stage); 5, 8—torsional shafts; 6—wheel of the second stage; 7—gear of the second stage; 9—gear rim

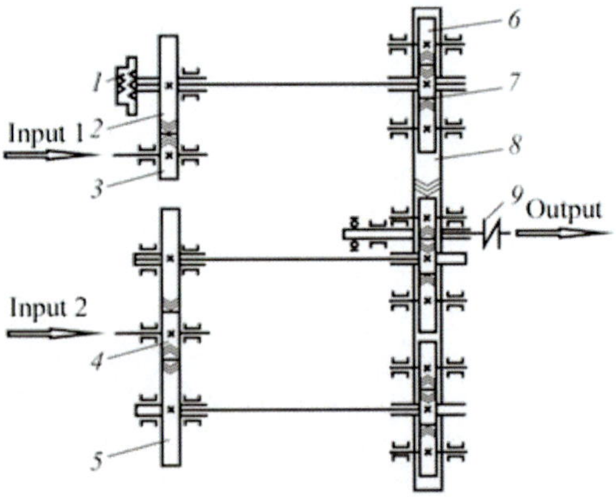

Fig. 2.22 Non-reversible summing reducer: 1—cam-friction clutch; 2, 5 and 3, 4, 7—driven and driving gears, respectively; 6—mid pinion; 8—driven wheel; 9—elastic clutch

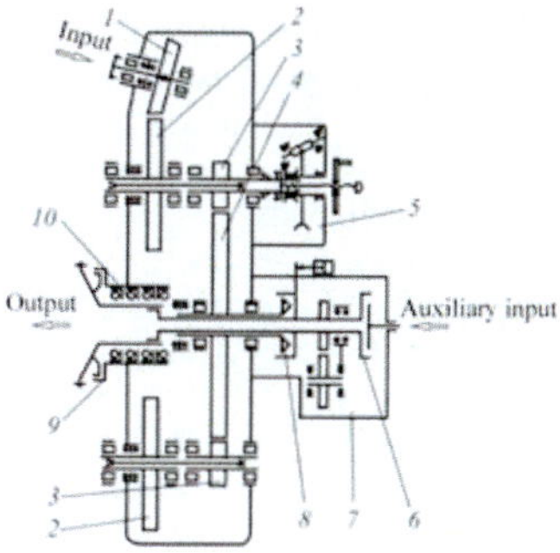

Fig. 2.23 Reducer of the hydrofoil marine power plant: 1, 3—driving pinion; 2, 4—driven wheels; 5, 9—turning and braking devices, respectively; 6, 8—gear and remote clutches, respectively; 7—oil unit; 10—thrust bearing

the power circuit of the reducer. The weight of the reducer is 55 t; the dimensions are 4.6 × 3.6 × 4.0 m.

The reducer of the gas turbine power plant on the hydrofoil ship "Cyclone" (Fig. 2.23) [8] is angular with bifurcation of power in the first stage and its summation in the second one. The first stage is a high-speed hyperboloid gear train. The second stage is a cylindrical helical gear. The angle between the axes of the input and output shafts of the reducer is 10° in the vertical plane, which makes it possible to arrange the power plant with a one-way power supply and output from the reducer. The reducer provides the possibility of additional connection of one more engine to the common shafting.

In case when a single main engine is used, the reducers of the air-cushion vessels and ships are basically a complex that consists of three gearboxes: a distributing reducer that provides distribution of the main engine power to the air propeller and the fan that pumps the air under the bottom, a propeller reducer and a fan reducer.

The distributing reducer (Fig. 2.24) [6] is cylindrical-conical with two power flows in the bevel gears (conical gear wheels). (Figure 2.25)

Features of the reversible reducer construction can be illustrated by a reducer designed to sum the power of two GTEs with the power of up to 18000 kW each (Fig. 2.27) [6]. Such a reducer was used on the large antisubmarine ship "KomsomoletsUkrainy", where two gas turbine units were installed with the world's first two all-mode GTEs with the power capacity of 13000 kW each [8].

The reverse reducer is two-stage, of the parallel type. The reverse is carried out by means of the complex consisting of a hydrodynamic reverse clutch, forward-drive friction and cam clutches. The friction clutches are designed to equalize rotational speeds of the cam clutches.

Hydraulic transmission is a set of hydraulic mechanisms, through which the energy of rotation of the driving shaft (engine shaft) is transmitted to the driven shaft—shafting with the propulsor. Depending on the principle of operation, there are distinguished hydrostatic and hydrodynamic transmissions.

In *hydrostatic transmission*, the energy of rotation of the engine shaft in the hydraulic pump of the volumetric type is converted into the hydrostatic head of the working fluid (mineral oil or special synthetic liquid) supplied through the pipeline to a hydraulic engine, where it is converted into the propulsion energy. The efficiency

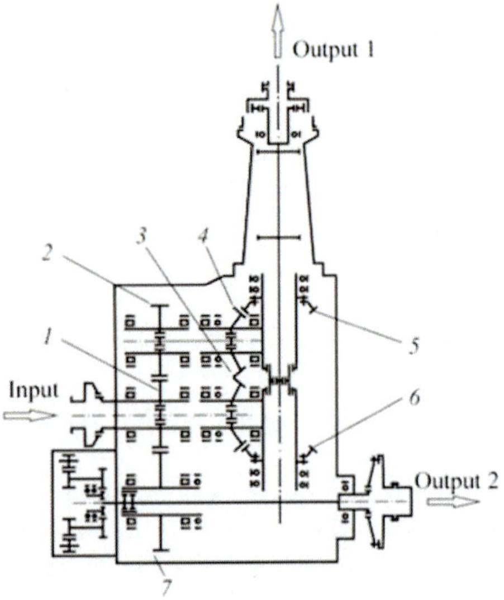

Fig. 2.24 Distributing reducer of the gearbox of the air-cushion marine power plant: 1 and 2, 7—driving and driven pinions, respectively; 3—conical pinion; 4—driving conical pinion; 5, 6—conical wheels

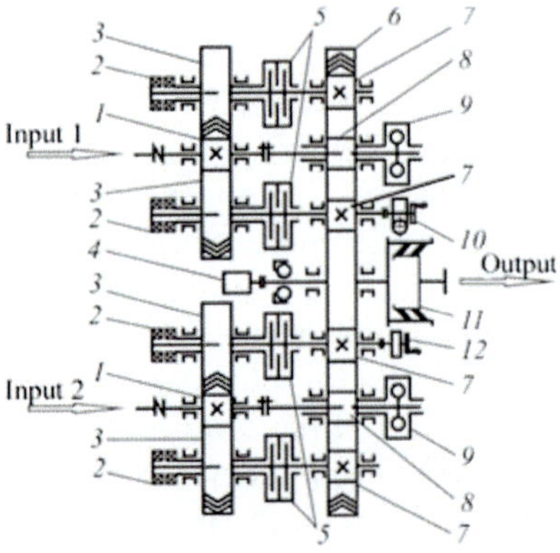

Fig. 2.25 Scheme of a reversible summing reducer: 1, 8—driving and reverse pinions, respectively; 2, 5, 11—cam, friction and soundproof clutches, respectively; 3, 7—overrunning pinions; 4—oil pump drive; 6—driven gear wheel; 9—reverse clutch; 10, 12—shaft-turning and braking devices, respectively

coefficient of hydrostatic transmission is 0.83–0.88. Hydrostatic transmissions are used to drive thrusters, wing propulsors in the main power plants with the power capacity of 1000–2000 kW on ships characterized by the frequent maneuvering (ferries, small fishing vessels) [1].

In *hydrodynamic transmission*, energy is transmitted from the driving shaft to the driven one by means of the dynamic (high-speed) head of the circulating working fluid (usually oil). The hydrodynamic transmission comprises a centrifugal pump and a hydraulic turbine, which are arranged so that their wheels form a toroidal cavity filled with working fluid. Depending on the construction, hydrodynamic transmissions are divided into *hydrodynamic clutches* (hydraulic clutches) (Fig. 2.26) and *hydrodynamic transformers* (hydrotransformers) (Fig. 2.27). The main difference between the hydrotransformer and the hydraulic clutch is that the former has a motionless

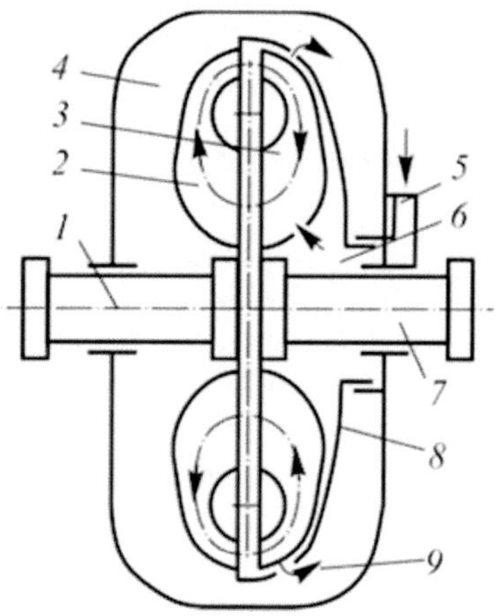

Fig. 2.26 Scheme of the ship hydrodynamic clutch: 1—driving shaft; 2—pump wheel; 3—turbine wheel; 4—auxiliary cavity; 5—oil input channel; 6—filling chamber

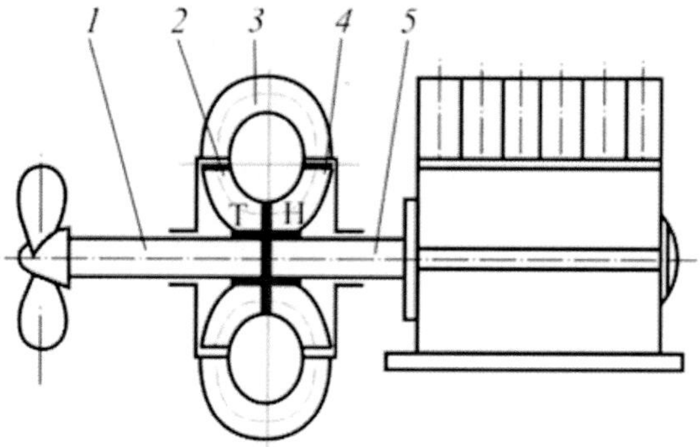

Fig. 2.27 Scheme of the ship hydrodynamic transmission: 1—propeller shaft with the hydrotransformer; 2—hydrotransformer; 3—directing device; 4—impeller of the hydrotransformer; 5—driving shaft

directing device rigidly connected to the gearbox between the pump and turbine wheels.

The efficiency coefficient of the hydrotransformer is 0.88–0.92, and that of the hydraulic clutch is 0.96–0.98.

Electric transmissions. In electrical transmissions, a double conversion of energy occurs: the mechanical energy of the ME is converted into the electric energy in the electric generators, the electric energy is transferred to the propeller engine through the cable lines, which convert it into the mechanical energy of rotation of the propeller shaft and propeller. There are electric transmissions of direct and alternating current, as well as the combined ones: with alternating-current generators and direct-current electric engines [1]. They allow summarizing the power of several high-speed main diesel or turbo generators for driving one or several low-speed propulsion electric engines. This increases survivability of the marine power plant and expands possibilities for application of low- and medium-speed engines. Since there is no mechanical link between the main power electric generators and the propeller engine, the primary (thermal) engine is protected from external influence of the propeller, and the arrangement of the power plant elements in the engine room of the ship can be considered rational.

The possibility of using the main generators for powering other electric consumers of the ship is another advantage of electric transmission. The direct-current electric transmission has high maneuverability, fast reverse (if compared to the other types of transmissions), and high starting moments. Full stopping of the propeller at this transmission is carried out within 5–16 s, and the ship run-out at its nominal initial speed does not exceed 6–7 ship lengths. Propulsion plants with the direct-current transmissions are easily controlled.

Electric transmissions, especially direct-current ones, have significant advantages, but they are also characterized the following disadvantages: relatively low efficiency coefficient (0.84–0.88 for direct current, 0.88–0.93 for alternating current); large weight and size characteristics; high cost of electric engines and generators; direct-current machines are less reliable than alternating-current machines.

Taking into account the advantages and disadvantages mentioned above, the direct-current electric transmissions are used on ships that need frequent changes in speed and modes of operation with speed changes (icebreakers, ice navigation ships, tugboats). Meanwhile, the alternating-current transmissions are used on ships where the propellers work longer on the constant modes without changing the speed, as well as on other ships in combination with adjustable propellers [1].

2.3 Fuels and Oils for Marine Power Plants

2.3.1 Liquid Petroleum Fuels

Technical and operational characteristics of fuels. Fractional composition. Liquid fuel consists of carbon C, hydrogen H, oxygen O, nitrogen N, sulfur S, ash A and moisture W. The fuel contains 85–89% of C, 10–14% of H, 0.05–4.50% of S, 0.05–0.80% of (O + N), 0.05–0.30% of A, and 0–3% of W. Carbon, hydrogen, oxygen, nitrogen and sulfur belong to the combustible mass of the fuel. The non-combustible part of the fuel, which consists of ash and moisture, is the fuel ballast. The combustible mass with the ballast is called the working mass of fuel.

Density (d) is defined as the mass of the fuel volume unit, measured in kg/m^3.

High-quality straight-run distillate fuels such as Gas Oil have the density of 830–880 kg/m^3. The density of heavy residual fuels is 870–970 kg/m^3. Cracking residues can have the density of 970–1030 kg/m^3. If the fuel density is above 975 kg/m^3, it is partially or completely composed of cracking products [9].

Viscosity is the resistance that the layers of fuel create to their mutual displacement under the influence of an external force. Viscosity determines the energy consumption for fuel transportation through pipelines, the conditions for transportation and pumping, the efficiency of the oil sprayers of heat engines and boilers. The speed of precipitation of the mechanical impurities largely depends on the viscosity during storage, heating and transportation of liquid fuels, as well as its ability to settle out of water.

The most applicable characteristic is kinematic viscosity, which is measured in Stokes (St) or centistokes (cSt). The viscosity of distilled water at 20.2 °C is 1 cSt (1 St = 10^{-4} m^2/s, 1 cSt = 10^{-6} m^2/s).

In addition to kinematic viscosity, the following viscosity measurement units have found particular use in different countries: Saybolt Universal, SaiboltFurol Viscosity, Engler Viscosity, Redwood Viscosity. The difference between them resides in the methods for determining the viscosity. Analytic or graphical dependences are used for practical recalculation of viscosity units [9, 10].

Viscosity of fuel is a classification parameter; on its basis, petroleum fuels are divided into *distillate fuels* (2.5–14.0 mm^2/s) and *heavy fuels* (30–700 mm^2/s). The viscosity index determines the conditions for receiving fuel on the ship, storing it, preparing it and supplying it to the engine.

The value of viscosity essentially depends on the temperature. Viscosity is normalized at 40 °C in the fuels for high-speed diesel engines (distillate fuels) and at 50 °C for low-speed diesel engines. As the temperature of the fuel increases, the viscosity and the density decrease, so the fuel is heated to improve its spray. The choice of the heating temperature depends on the requirements of engine manufacturers, which set the level of the required viscosity of the fuel to the engine. Usually this indicator is in the range of 10–15 cSt. Fuels that have a different specific viscosity require different heating temperatures.

Flash point is the lowest temperature at which fumes of the liquid fuel mixed with the air flare up when approaching open flame. The International Standard (ISO 8217) and the Rules of Classification Societies require that the flash point of the fuel received on board is not below 60 °C [11].

Pour point. The temperature that corresponds to the loss of the fuel mobility is called *the pour point.* For light fuels, it is-5— + 10 °C, and for cracking fuels it reaches +30 °C and even up to +45 °C.

Sulfur content. Modern marine fuels are characterized by the increased content of sulfur and its compounds in marine fuels. The sulfur content in light fuels is 0.2–2.0%, in heavy fuels—up to 4.5%.

The greatest harm to marine diesel engines is caused by the sulfur corrosion of the cylinder-piston group parts.

In recent years, there has been a worldwide tendency to create special areas—the control zones for sulfur content in the exhaust gases of marine power plants (SO_x Emission Control Area—SECA), where the sulfur content of the fuel should not exceed 1.5% by mass. Otherwise, an exhaust gas purification system with a guaranteed level of purification from emissions of sulfur oxides to a level that does not exceed 6.0 g SO_x/kWh should be used on the ship.

Vanadium and sodium content. The increase of vanadium content in heavy residual fuels, which can reach 300–350 ppm (parts per million), is the reason for a sharp decrease in the service life of the ICE exhaust valves, which is associated with burning of the landing surfaces.

The content of organic sodium compounds rarely exceeds 30 ppm, but when seawater gets into the fuel, the amount of the sodium compounds increases by 100 ppm for every percent of increase in its content.

Vanadium and sodium oxides cause high-temperature corrosion of the metal exhaust valve discs, so that the density of the valve fit on the seat in these areas decreases, and there is a breakthrough of gases and even burnout of the valve disc.

The quality of fuel largely depends on the number and size of mechanical impurities that affect fuel equipment.

The International Standards limit the content of such impurities by the following values: for diesel fuels—0.01–0.02%, for heavy fuels—0.1%.

The *aluminium silicon* content limit was introduced into the international certificates for marine fuels in 1996 due to the spread of accidental wear of fuel equipment and the cylinder-piston group of diesel engines, which was caused by a high content of aluminium and silicon in the fuels. The ISO certificate allows the content of Al and Si in the fuel not to exceed 80 ppm.

The content of coke and asphaltenes. The content of coke—the carbon residue—is the percentage remainder of the sample weight that undergoes heating without access to the air. In distillate fuels, the carbon residue does not exceed 0.1–0.3%. For Marine Diesel Oil (MDO), it increases to 2.5%, which is the evidence of the presence of a certain amount of residual fractions in the fuel. For heavy straight-run fuels and crack-fuels, the carbon residue can reach the value of 18–25%. When burning such fuels, there is a significant carbon deposition in the area of the piston rings and exhaust ports, and the smoky discharge is observed.

Asphaltenes in fuel are insoluble and contained as a colloidal solution. Their density reaches 1160 kg/m^3, the melting point is 200 °C. The content of asphaltenes in heavy fuels is 3–12%. The presence of asphaltenes leads to sludging and precipitation in tanks, and to the instability and incompatibility of fuels.

The impurities of various salts and oxides that make up the fuel composition form the mineral residue—ash—after the fuel combustion. The amount of ash-forming substances in fuels does not exceed 0.15% by the weight.

Seawater, which inevitably gets into the fuel during transportation, storage and bunkering, increases the ash content of the fuel and its aggressiveness to the parts of fuel equipment. The ISO International Standard limits the water content in the fuels delivered to the ships: 0.3% for distillate fuels and 1% for heavy fuels.

One of the most important characteristics of the fuel is the amount of heat released when the combustion of 1 kg of fuel is complete, which is called *combustion heat*. The combustion heat, which determined under the condition that the water steam generated by the fuel combustion is not condensed, is called *low heat value*. It is measured in kJ/kg and is designated as Q_i^r. This is a very value that is used in heat engineering calculations. The high heat value is greater than the low one by the amount of heat of steamgeneration which returns upon condensation of moisture.

Stability and compatibility of fuels. Stability is the ability of the fuel to retain its properties for a long time, to prevent precipitation, sludging and layering during storage, pumping and heating. The main causes of precipitation are the resins, asphaltenes, carbonates, carboids and water contained in the fuel. Destabilization and increase of the precipitation rate is facilitated by the fuel heating and mixing with other types of fuels.

Compatibility of fuels is determined by maintaining the stability (there is no subsequent precipitation) of the mixture after mixing of the components. The incompatibility is the property due to which the fuels that were stable before mixing lose stability after mixing.

Self-ignition and combustion. Self-ignition is a characteristic of the fuel that determines the duration of the process consisting of the stage of preparation of fuel for combustion after its injection and the stage of development of the subsequent combustion in the cylinder. Formation of the combustible mixture in the diesel ICE is accompanied by a number of physical and chemical processes, the development of which takes time estimated from the moment of the fuel injection to the moment of its self-ignition. This time is called the self-ignition delay period.

Cetane number is the indicator of self-ignition of distillate fuels, which is estimated by the duration of the self-ignition delay period. It determines the cetane content (in per cent by volume) in the reference fuel, which is equivalent by flammability to the fuel under consideration. The distillate fuels of the paraffin series have the cetane numbers in the range of 45 to 60, the fuels with a predominant aromatic content—of about 35, and the fuels with a certain amount of naphthenic compounds have the cetane numbers between these limiting values.

Microbiological infection of fuels. During oil refinement, hydrocarbon fuels are heated to high temperatures, which makes them sterile. But in the process of transportation, storage and pumping, they can be infected with microorganisms and bacteria from the atmosphere.

The bacteria do not multiply while there is no water in the fuel, but their intensive reproduction occurs if any watering is provided.

Microbiological infection of fuel can be avoided by preventing watering of the fuels and setting the appropriate temperatures in the fuel tanks.

Types of oil fuels

The oil fuels that are used in ship engines and boilers are divided into two groups:

distillate oils, which consist of light fractions obtained by distillation in straight-run or cracking plants and are characterized by the viscosity values (at 40 °C) in the range of 1.4–14.0 mm^2/s;

heavy oils, which are the mixtures of heavy residual fractions with gas oils; the viscosity of such fuels (at 50 °C) is in the range of 30–700 mm^2/s.

Until November 2005, the characteristics of marine fuels had been determined by specifications of the International Standards Organization (ISO)—ISO 8217:1996 (Specification of marine fuels) and the British Standards Institution (BSI)—BSMA 100:1989 (Petroleum fuels for marine oil engines and boilers).

In 2005, the new international standard ISO 8217:2005 was introduced, which defines the characteristics of marine fuels. In this standard, some changes and clarifications are made in comparison with the standard ISO 8217:1996. It makes provisions for four types of distillate fuels (Table 2.2) [9].

Table 2.2 International Fuels Standard ISO 8217:2005 (E). Requirements for marine distillate fuels

Characteristic	Limit	Distillate fuels Category ISO-F			
		DMX	DMA	DMB	DMC
Density at 15 °C, kg/m^3	max	–	890.0	900.0	920.0
Vicosity at 40 °C, mm^2/s	min	1.40	1.50	–	–
	max	5.50	6.00	11.00	14.00
Flashpoint, °C	min	43	60	60	60
Pour point, °C: —winter quality —summer quality	max max	– –	−6 0	0 6	0 6
Cloud point, °C	max	−16	–	–	–
Sulfur, %	max	1.00	1.50	2.00	2.00
Cetane index	min	45	40	35	–
Carbon residue, %:	max	0.30	0.30	0.30	2.50
Ash, %	max	0.01	0.01	0.01	0.05
Total sediment, existent, %	max	–	–	0.10	0.10
Water, %	max	–	–	0.3	0.3
Vanadium, mg/kg	max	–	–	–	100
Aluminium plus silicon, mg/kg	max	–	–	–	25
Used lubricating oil: zinc, mg/kg phosphorus, mg/kg calcium, mg/kg	max max max	– – –	– – –	– – –	15 15 30

The DMX fuel is a high-quality distillate, which due to its low pour point and flash point is intended solely for use in lifeboat engines and emergency diesel generators.

The DMA fuel is a high-quality distillate called *Marine Gas Oil (MGO)*. The DMB fuel is the basic grade of distillate fuel which is used in marine diesel engines not equipped with a heating system. It contains traces of residual fractions, so it has a darker color compared to the previous grades. In the marine practice, it is called *Marine Diesel Oil (MDO)*. The DMC fuel is also part of the MDO group, but it contains a substantial amount of residual products and is therefore recommended for use under the conditions of fitting the marine power plant with the necessary fuel preparation means.

Heavy fuels are conventionally divided into two groups: *intermediate fuels*, which are a mixture of residual products with the distillate fractions and have a viscosity up to 180 mm^2/s, and *heavy residual fuels* (boiler fuels and fuel oils). The fuels of the second group consist of the residual products of straight-run processes of oil refining and cracking and have a viscosity higher than 180 mm^2/s.

The International Standard ISO 8217:2005 makes provisions for 10 grades of heavy residual fuels from RMA to RMK (Residual Marine) in the viscosity range from 30 cSt (RMA 30) to 700 cSt (RMK 700). The characteristics of such fuels are given in Table 2.3.

The fuels of the A30 and B30 groups are used at low ambient temperatures; they have a lower density and the minimum viscosity, which provide good self-ignition.

The fuels of the D80—H700 groups are used in the presence of fuel heating systems up to the necessary and maximum permissible engine temperatures.

The fuels of the K380 and K700 groups are used exclusively in the plants equipped with special separators designed to process high-density fuels.

2.3.2 Alternative Fuels

The use of traditional oil fuels for the marine power equipment leads to a number of issues, the main of which are the environmental and economic ones.

The emissions from the sea ships account for up to 14% of all the emissions from combustion of fossil fuels and up to 16% of the sulfur emissions from the products of refinement [4]. Along with this, carbon oxides, high-molecular aromatic hydrocarbons and solid carbonaceous particles enter the atmosphere.

Along with the environmental problem in the development of ship navigation, there are economic difficulties associated with the reduction of the world oil resources. The beginning of the twentieth century is characterized as the end of the era of cheap oil. The reduction of oil reserves is reflected in the persistent trend towards the rise in price of oil fuels. The economic problems associated with the use of traditional motor fuels on ships are relevant for all the modern countries.

One of the ways to solve the above-mentioned problems is the use of alternative fuels for operation of marine power plants.

There are distinguished the following groups of alternative fuels [12]:

emulsions of oil fuels with oxygen-containing compounds, such as alcohols, ethers, water (their performance properties almost coincide with the properties of traditional oil fuels);

synthetic liquid fuels from organic raw materials (fossil or renewable), which are similar in performance to oil fuels;

oil-derived fuels (alcohols, liquefied natural gas (LNG), liquefied petroleum gas (LPG)), which in their properties are different from traditional liquid fuels. The use of the fuels of this group requires modification of engines and fuel storage systems.

Let us consider the main types of alternative fuels, which can be applied in heat engines and boilers on the ships for the foreseeable and more distant future.

Liquefied petroleum gas (LPG) is by 90–95% a mixture of propane and butane with the impurities of heavier hydrocarbons. By its basic characteristics, it is close to liquefied natural gas. The liquefied petroleum gas is a high-quality product of

Table 2.3 International Standard ISO 8217:2005 (E). Requirements for marine residual fuels

Characteristic	Limit	High-viscosity fuels (Category ISO-F)									
		RMA 30	RMB 30	RMD 80	RME 180	RMF 180	RMG 380	RMN 380	RMK 380	RMH 700	RMK 700
Density at 15 °C, kg/m^3	max	960.0	975.0	980.0	991.0				1010.0	991.0	1010.0
Kinematic viscosity at 50 °C, mm^2/c	max	30.0		80.0	180.0		380.0			700.0	
Flashpoint, °C	min	60									
Pour point, °C,: winter quality summer quality	max max	0 6	24 24	30							
Carbon residue, %	max	10		14	15	20	18	22			
Ash, %	max	0.10				0.15					
Water, %	max	0.50									
Sulfur, %	max	3.50		4.00	4.50						
Vanadium, mg/kg	max	150		350	200	500	300	600			
Total sediment potential, %	max	0.10									
Aluminium plus silicon, mg/kg	max	80									

refinement of oil and an oil-containing associated gas. It has an advantage over other types of gas motor fuels: the butane-propane mixture at a normal temperature and pressure of 1.6 MPa becomes liquid.

When transferring from the liquid to the gaseous state, the propane volume increases by 270 times, and the butane volume increases by 230 times. At the same time, LPG has the following disadvantages: the process of evaporation slightly reduces its starting qualities, special heaters are necessary to operate the engine at low air temperatures, and the associated gas production costs are by 6–8 times higher than those for natural gas.

In general, there are sole examples of the use of LPG in the role of fuel.

Liquefied natural gas (LNG) is a cryogenic liquid that is a mixture of the hydrocarbons of the series C_1–C_{10} and nitrogen with a predominant (85–99%) share of methane. LNG is produced from natural gas by cooling it to the temperatures of—162—130 °C. The boiling point at the atmospheric pressure is—162—160 °C. When LNG is transferred to the gaseous state (with gasification), its properties completely correspond to the properties of natural gas. The density of LNG, depending on the pressure and the component composition, reaches 370–430 kg/m^3. Fire and explosion safety concentration of the gasified LNG in the air under normal conditions is (5–15)% of the volume (by methane). The minimum flash point of the mixture of the air with the gasified LNG, depending on its composition, is 450–600 °C. Gasification of 1 m^3 of LNG gives almost 600 nm^3 of natural gas.

The predominant share of methane in LNG determines the proximity of its main characteristics to the indicators of pure methane. The low heat value of the gasified LNG at the atmospheric pressure averagely makes up about 50 MJ/kg, which is much higher than that of liquid oil fuels.

Natural gas under normal conditions is a light, colorless, flammable gas without smell. The relative density (by air) is 0.56. The cold fumes of LNG are heavier than the air; it evaporates at leaks, spills and depressurization.

The ships employing natural gas as the fuel are divided into two groups:

- ships carrying LNG or performing its primary processing: specialized gas carriers, production ships, storage and unloading ships for servicing FPSO (floating production storage and offloading vessels), which are floating plants where natural gas is liquefied and reloaded to other ships or delivered to the port. Gas carriers and FPSOs can utilize natural boil-off gas and additionally—forced boil-off gas;
- ships where the use of LNG is determined by economic and stringent environmental requirements in the area of navigation, as well as by the possibility of bunkering with this fuel, i.e. ships where LNG is used exclusively as the fuel.

On the ships of the second group, liquefied natural gas is placed in special cryogenic tanks. The tanks are located vertically or horizontally. They consist of two chambers—internal and external; the distance between them is 120–300 mm. Deep vacuum is maintained between the chambers for isolation of the liquefied gas, which is stored at—162 °C. All the gas pipelines and valves are housed in carefully ventilated casings equipped with emergency sensors that signal possible leaks.

Before its use, liquefied gas must be regasified and supplied to the engines at the temperature close to 20 °C under the pressure of about 0.5 MPa. This process usually takes place in the evaporator, the heat source for which is hot water from the main engine cooling system, steam or electric energy.

The use of LNG as the fuel suggests higher capital costs for the ship, while the emissions of nitrogen oxides and carbon dioxide in the exhaust gases of the MPP are reduced significantly.

Thus, the annual emissions of nitrogen oxides from "Viking Energy", the ship for servicing gas production platforms with its MPP running on LNG, are reduced by 200 tons compared to the ship that uses oil products [12].

Hydrogen. The advantages of hydrogen include a high heat value and the absence of sulfur oxides and solid particles in the combustion products. The main disadvantage is its high cost. According to the forecast of the European Commission which is developing a program for introduction of alternative fuels, by 2020, hydrogen will have accounted for about 5% of the total consumption of the motor fuel.

Hydrogen can be obtained by hydrolysis of water. Natural gas, oil fuels, biodiesel fuels, and methanol can serve as the source of hydrogen; they are converted to a hydrogen-rich gas directly on board of the ship.

These fuels inside or outside the fuel element (FE) are converted to hydrogen and carbon monoxide [12]. The serious problem that limits the use of hydrogen in transport is its storage. Hydrogen can be stored in the liquefied, compressed form or using an intermediate carrier as part of special materials—metal hydrides, which can retain a considerable amount of hydrogen due to their its structure.

Hydrogen is often used as part of the FE on small ships in the marine transport. The fuel elements are the devices that generate electricity directly onboard through the process of reverse electrolysis: water and electric current are formed due to the reaction of hydrogen and oxygen. Hydrogen is supplied to the FE anode, and air oxygen is supplied to the cathode. The functions of the electrolyte between them are most often performed by a membrane made of polymer coated with a thin layer of a noble metal, which ensures the passage of protons. The gases are brought under the pressure of 0.15–0.27 MPa.

The consortium FellowSHIP (Fuel Cells for Low Emissions Ships) has been created in Europe for the introduction of FEs in the marine transport. It includes the classification society Det Norske Veritas (DNV), the companies Eidesvik Offshore, MTU CFC Solutions, and Vik-Sandvik Wartsila Automation Norway.

In the USA, Fuel Cell Energy develops the power plants running on fuel elements with the power of 500 kW for marine use that operate on kerosene and diesel fuel. The power plants are created in a modular design, which allows them to be used both on ships and vessels. The purpose of this project is to increase the efficiency of onboard power supply of sea ships [12].

Alcohol fuels. Methanol and ethanol are used as alternative motor fuels.

Methanol CH_3OH is the alcohol which can be obtained from coal, oil, and biomass (wood residues). Currently, methanol is used as a motor fuel in limited quantities.

The energy value of methanol is twice lower than that of petrol, which increases the specific fuel consumption. Compared to oil fuels, when burning methanol, less nitrogen and carbon oxides are released, while sulfur oxides and solid particles are not released at all.

Methanol is toxic; its use causes corrosion of the parts of the fuel system. It can be used on ships as a source of hydrogen production, which is necessary for the operation of fuel elements.

Ethanol C_2H_5OH is more effective as an additive to oil fuels than methanol. Stability of the ethanol-fuel mixtures is higher than that of the methanol-fuel mixtures. Ethanol is less toxic and corrosive than methanol; the level of the emissions is almost the same.

The environmentally positive moment is that ethanol is a renewable energy source; even waste can be used for its production.

In the USA, ethanol as a "clean" fuel is used in 21 states, and the ethanol-petrol mixture make up 10% of the fuel market.

The use of ethanol in the marine power equipment is currently very limited.

Fuels of coal origin. Coal as the main fuel for the ship boilers in the nineteenth and at the beginning of the twentieth centuries is practically not used today in marine power engineering. There are single examples of the use of crushed (15–35 mm) coal in the boiler furnaces of the modern ship STP [6].

Among the new coal technologies, the technology of water-coal fuel is of considerable interest. It belongs to the new generation of fuels and is an artificial type of the composite fuel consisting of coal and water with the addition of plasticizers. The technology is based on mechanical-chemical activation, in the process of which the structure of coal as a natural rock mass is destroyed. Water is also subject to the transformations resulting in formation of an active dispersed medium.

In the process of production of the coal-water fuel, mineral impurities are removed from the coal (up to the ash content of 0.5–1.5%), it is ground to the particle size of 10–500 μm, mixed with the water and subject to the cavitation treatment. The mass fraction of water is 30–50% When using the water-coal fuel, emissions of nitrogen are reduced by 80–90% if compared to coal (due to a lower temperature in the combustion zone—1000–1250 °C), of sulfur oxides—by 70–85%, of solid particles of the micron fractions—by 80–95% [10].

Water-fuel emulsions. A significant increase in the efficiency of marine power equipment can be achieved by means of combustion of liquid fuels (fuel oils and diesel fuels) in the form of emulsions. Performance of the equipment using the water-fuel emulsion (WFE) improves, since the water evenly distributed in the form of small droplets throughout the mass of fuel contributes to intensification of the atomized fuel combustion [10].

Water is a constant companion of oil during its extraction, processing, storage, and supply of the oil products to engines and boilers. With the uniform distribution of the moderate amount of water in the fuel (up to 15–30%), the combustion process is activated, the concentration of harmful substances in the combustion products decreases, and the carbon formation decreases. The positive influence of WFE on the combustion process in the engines and boilers furnaces is confirmed by the examples of its application in the stationary and marine power engineering [13, 14].

Preparation of the water-fuel emulsion takes place directly on the ship. Emulsions with the water content up to 20% are usedfor the engines, up to 30%—for the steam boilers [14]. Application of WFE in MPP is multifaceted. Along with the use for the engines and boilers, it is recommended to combust the oil waste formed during storage, transportation, preparation and application of petroleum fuels, oils and lubricants as part of the water-oil emulsion. The combustion should take place in the boilers.

The use of WFE for operation of power equipment on ships allows [14]:

- reducing the carbon formation, improving the reliability of the cylinder-piston group, gas outlet, fuel equipment, increasing the time between the fuel filters cleaning;
- reducing the temperature of the exhaust gases and, consequently, reducing the heat stress of the parts of the cylinder-piston group;
- reducing toxicity of the exhaust gases (on average, nitrogen oxide emissions decrease by 30–50%, carbon monoxide—by 50–80%, sulfur—by 20–60%);
- utilizing waste water contaminated with oil products as part of WFE.

Biodiesel fuel (BD) is derived from the plant oils. It is a mixture of the complex ethers of higher carboxylic acids which are formed by catalytic transesterification (*aithör*—ether + Lat. *facere*—to make) of the plant oils of various origin with methanol or ethanol. During transesterification at a normal pressure and the temperature of 60 °C, there interact ten parts of oil and two parts of methanol (or ethanol) with an addition of potassium hydroxide or sodium hydroxide. The characteristics of the obtained product are as close as possible to the characteristics of the diesel fuel obtained from oil.

The raw materials for the production of biodiesel fuel are fatty or essential oils of various plants including rape, canola (a kind of rape), palm oil, jatropha (oil culture), coconut oil, soybean, castor oil. Biodiesel fuels can be also produced from the waste vegetable edible oils, animal fats, fish oil [15].

A promising source of the raw materials for the production of biodiesel is seaweeds. There are technologies for growing seaweeds in bioreactors located near heat and power plants. Their exhaust heat can cover up to 77% of the heat demand necessary for growing seaweeds [6].

An important property of biodiesel fuel is that it does not damage plants and animals when it enters water. It undergoes almost complete biodegradation in the soil or in the water; microorganisms process 99% of the biodiesel over the period of 28 days. When burning biodiesel, sulfur oxides are almost completely absent, and there is release of the amount of carbon dioxide equal to that consumed by the plant—the primary raw material for the production of oil—over the whole period of its life.

The biodiesel fuel B100 or its mixtures in different proportions with diesel fuel (B5—mixture of 5% BD with diesel fuel, B20—mixture of 20% BD with diesel fuel) is used in small and medium-sized ships of the coastal or inland navigation (ferries, passenger ships, cruise ships, fishing vessels, research vessels, coastguard vessels, yachts) [12].

Among the fuels considered, the real alternative to traditional fuels that are used in MPP are biodiesel, liquefied natural gas and water-fuel emulsions.

2.3.3 Oils for Marine Engines and Mechanisms. Purpose and Characteristics of Oils

Oils that are used in the internal combustion engines, gas turbine engines and steam turbine aggregates are designed to perform the following functions:

- reduction of friction and load in the moving units;
- cooling of the friction units by removing heat;
- removal of the wear products formed in the friction units;
- sealing of the cylinder-piston group in the area of the piston rings (only for ICE).

Since internal combustion engines are more stress-bearing than turbine engines, the oils used for performing these functions should have the following characteristics and properties [17]:

- the required viscosity, stable viscosity-temperature characteristics in the friction zones, the presence of high anti-wear properties to ensure long-term operation of the cylinder bushings, piston rings, and bearings;
- high thermal stability and oxidation resistance to prevent formation of high-molecular compounds in the oil and their deposition in the engine;
- the presence of a sufficient reserve of alkalinity in the oil to neutralize acids that cause the sulphurous corrosion;
- anticorrosion properties to prevent rusting and chemical corrosion of the bearing alloys and the shaft necks.

The main characteristics of oils are density, viscosity, neutralizing ability, flash point and pour point.

Density of oils lies in the range of 860–930 kg/m^3. The oil density changes during its operation: a reduction indicates its rarefication with a lighter fuel, and an increase indicates the oil contamination with soot and oxidation products.

Viscosity is one of the most important characteristics of oil. With an increase in viscosity, the bearing capacity of the oil film rises, and it withstands higher specific pressures. At the same time, an increase of viscosity is accompanied by an increase in frictional losses, deterioration of the oil spreading, and complication of its movement in narrow channels and cracks.

Viscosity of oils largely depends on the temperature: it decreases with the temperature increase and increases with the temperature decrease. The viscosity-temperature properties of oils are characterized with the viscosity data usually given for two temperature values—40 and 100 °C. For these purposes, the viscosity index (VI) is used. This index is a dimensionless value that reflects the viscosity-temperature dependence of the oils. A high viscosity index (100 or above) demonstrates a slight decrease in viscosity with the temperature increase, while a low index points to a substantial decrease in viscosity.

The oil viscosity can be estimated by the SAE (Society of Automobile Engineers) categories, whose numerical values range from 0 to 400.

The alkalinity property is the ability of the oil to withstand corrosion when exposed to the sulfuric acid formed in the cylinders and the oil oxidation products (organic acids). These acids are neutralized by introduction of additives into the oil, which provide it with the alkaline properties estimated by the total base number (TBN). The unit of measurement of the total base number is mg KOH/g of oil.

The pour point temperature is the lowest temperature at which the oil loses its mobility. Oils that are recommended for the circulating systems of the marine diesel engines have a pour point in the range of 9—15 °C.

The flash point temperature is the lowest temperature at which the heated oil evaporates and the vapors form a mixture with the air that flares up on approaching flame. It is determined in open or closed crucibles. The flash point temperature of the oil is 200–230 °C.

The oil evaporation depends on its fractional composition and flash point. The lower the flash point is, the lighter the fractional composition is. At a lower temperature, the light fractions boil off (their boiling point is 200–500 °C), which leads to an increase in oil consumption.

Composition of oils. When being used in modern high-powered diesel engines, especially those running on heavy fuel, the oils quickly lose their properties, and it results in intense corrosion, wear, and decrease in service life.

To prevent such consequences, *additive complexes* are nowadays brought in the *basic body of oils*. They are intended to significantly improve the quality characteristics of the oils.

The basic body of the oil for ship engines can be mineral (consisting of oil refinement products) or synthetic (synthesized from chemical compounds that provide the desired properties). Mineral oils have found the predominant use in the ship engines; their advantage is affordability and a low cost.

Synthetic oils are irreplaceable for operation of the engines in the northern latitudes and in the Arctic.

Additives are synthetic chemical compounds that are brought in the basic body for improving the motor properties of the oils and ensuring their efficient operation in the ship engines.

The additives are brought in almost all the oils in amount of up to 25% by weight. They can improve physical properties of the oils or have a chemical impact. The *additives package* is carefully selected in the production of oils for each grade and development of the conditions for its use.

Classification of oils. Oils that are used in the marine diesel engines are divided into *motor* (circulating) and *cylinder* oils. The former are intended for use in the engine lubrication system; they circulate in a closed circuit and are subject to periodic replacement. The latter provide lubrication of the cylinder mirror.

Cylinder oils. In addition to the above-mentioned qualitative indicators, the cylinder oils for the LSE should also have the following specific properties:

- stickiness (oiliness), which should be high, so that the oil is not blown off the surface at the cylinder blowing;
- the ability to spread out so that the oil disperses along the cylinder and covers its entire surface;
- viscosity properties of the oil should ensure preservation of the oil film in the area of the piston rings (at high pressures and temperatures);
- high reserve of alkalinity for neutralization of the acid formed during combustion of high-sulfur fuels;
- antiwear properties, which should prevent formation of burrs in the cylinder-piston group.

When choosing the oil designed to lubricate the cylinders of the LSE, it is necessary to take into account the recommendations of the manufacturer. Special attention should be paid to the ratio of the oil alkalinity and the sulfur content of the fuel.

In case of continuous operation on a low-sulfur fuel with S < 1.5%, one should adhere to the recommendations of the engine manufacturers when selecting the oil alkalinity (BP Oil 155 can be used) (Table 2.4). The choice of the oil alkalinity is also influenced by the magnitude of the load, since prolonged operation at reduced loads causes a decrease in the temperatures of the cylinders and an increase in the amount of the formed sulfuric acid; thus, a higher reserve of the oil alkalinity is required to neutralize it.

This became the basis for the development of a new generation of cylinder oils. Along with high neutralizing properties, they have enhanced detergent and antiwear properties, as well as the ability to retain a strong film of a sufficient thickness on the cylinder mirror. BP CL 656 and Castrol CYLTECH belong to these oils (Table 2.5). The cylinder oil consumption in the LSE is 1.0–1.2 g/(kW·h) [16].

Oils for the circulating lubrication systems. Circulating (system) oils of the ship crosshead diesel engines should have the following properties:

- ability to provide fluid lubrication in the bearings (oil wedge);
- ensuring effective lubrication of crosshead bearings under the conditions of high specific pressures;
- high thermal stability.

The level of alkalinity of the circulation oils of crosshead engines is low; it is 3–6 mg KOH/g of oil.

In addition to the functions of lubrication, *circulating oils* of the medium-speed throne engines should ensure effective neutralization of the sulfuric acid on the walls of the cylinders, taking into account the fact that lubrication of the MSE cylinders is carried out by the same oil that is thrown onto the walls of the cylinders (spray lubrication).

Therefore, the MSE motor oils should have a higher alkalinity. So, it is recommended to choose an oil with TBN 20 for an MSE running on fuels with the sulfur content of less than 1.0–1.5% and an oil with TBN 30 or 40 for the sulfur content of above 1.5–2.0%. The viscosity is recommended to be within SAE 30 or 40.

The recommended oil brands are listed in Tables 2.4 and 2.5.

Table 2.4 Characteristics of British Petroleum (BP) oils for marine diesel engines

Oil grade	Viscosity		Alkalinity, mg KOH/g	Recommendationsforuse
	by SAE	Kinematic at 100 °C, cSt		
Low-speed engines Cylinder oils				
EnergolCLO 50 M	50	19.5	70	When using heavy fuels with sulfur content of up to 4%
CL 1005	50	19.5	100	When using heavy fuels with sulfur content of above 4%
CL 656	50	24.0	85	Forlong-strokeengines
CL 155	50	18.5	15	For engines that run on the fuel with S $\leq$ 1.5%, and for the run-in after replacement of piston rings and bushings
Low-speed engines Circulating oils				
EnergolOE-HT30	30	11.5	6	For all types of crosshead engines
Medium- and high-speed engines on the fuels of the MDO type Circulatingoils				
EnergolDL-MP30	30	11.5	9	Multipurpose oil, can also be used for LSE and ship mechanisms
DL-MP40	40	14.0	9	Forheavy-loadedengines
DS3-153/15	30/40	11.5/14.0	15	For heavy-loaded engines, complies with API CD
Medium- and high-speed engines on heavy fuels with the sulfur content of 1.5–3.5% Circulating oils with high detergent and antideposition properties				
Energol IC-HFX203/204	30/40	11.5/14.0	20	This series of oils is designed to select the optimal ratio of TBN oil/S% of fuel
IC-HFX303/304	30/40	11.5/14.0	30	
IC-HFX403/404	30/40	11.5/14.0	40	
IC-HFX503/504	30/40	11.5/14.0	50	

The consumption of the circulating oil for a modern ICE is 0.11–0.14 g/(kW·h) for crosshead engines and 1.4–2.5 g/(kW·h) for trunk engines [16].

The holding capacity of lubrication systems is the highest for the LSE employing oil cooling of the pistons and makes up 2–3 l/kW. With water cooling of the pistons, the holding capacity of lubrication systems is 1.2–1.8 l/kW. The oil circulation rate is defined as the ratio of the oil circulating pump supply to the system capacity. It is

Table 2.5 Characteristics of "British Petroleum-Castrol" oils for marine diesel engines

Oilgrade	Viscosity		Alkalinity, mg KOH/g	Areas of application, characteristics
	by SAE	Kinematic at 100 °C, cSt		
Castrol CDX-30	30	11.1	5	Circulating oil for crosshead engines
Castrol CYLTECH 80	50	19.0	80	Cylinder oil for high-powered engines
Castrol S/DZ70	50	19.0	70	Standard cylinder highly alkaline oil
Castrol MLC 30/40	30 40	11.7 14.0	12 12	Circulating oil for trunk engines
Castrol MXD 153/154	30 40	11.5 14.0	15 15	Circulatingoil forMSE
Castrol TLX 203/204	30 40	11.5 14.0	20 20	Circulating oil for MSE on heavy fuel
Castrol TLX 504	40	14.0	50	Alkalinereserveoil

4–8 in the LSE. This means that all the oil passes through the engine 4–8 times in one hour of work.

The holding capacity of the lubricating systems of the powerful MSE is in the range of 0.8–1.5 l/kW, the circulation rate is 15–20.

Turbine oils and oils for auxiliary machinery and devices. The purpose of *turbine oils* is lubrication of the turbine bearings and removal of the heat released under the influence of high temperatures of combustion products or steam. Accordingly, oil must have a high thermo-oxidative capacity; it must withstand the influence of high temperatures.

In the turbo compressors of the superchargers with the built-in lubrication systems, given the limited amount of oil that they contain, it is recommended that it is replaced after 1000–1500 h.

Two types of turbine oils are used in the marine gas turbine plants of the UGT type [6]:

- low-viscosity oils—oils for marine gas turbines, oil MN-7.5U—standardized petroleum base oils with complex of additives;
- medium-viscosity oils—oils Tp-46 (with additives) and Tp-30, which are lubricating mineral oils from sulfur waxy crude oil with deep selective treatment.

The oil for marine gas turbines intended for lubricating gas turbine engines of the light, aircraft type is made of transformer oil with the addition of 0.8–1.2% of an antiwear additive and 0.5–8.0% of an antioxidant additive.

The oils of the T-46 and Tp-46 types are used for lubrication of heavy GTE, GTP reducers, the main and recovery steam turbines, gas turbine superchargers.

The following grades of foreign turbine oils are recommended: the THB68 oil produced by British Petroleum—Energol equivalent to T-46 and Tp-46 oils; the TC-S68 oil (68 means the viscosity in cSt at 40 °C) and the synthetic Shell Turbo Oil T68. The oil for marine gas turbines is replaced with Aero Shell Turbines Oil 2, BP Aero Turbines Oil 1010, Castrolaero 1010.

Compressor oils. The main functions of oil in the compressors are to reduce the friction losses and wear of friction surfaces, to remove heat from friction pairs, to increase the sealing impact of the piston rings and glands. When choosing oils for the air compressors, their type, piston speed, pressure and discharge temperature must be taken into account.

The following oils are recommended for air piston compressors: BP-RC68, 100 and 150 are mineral oils for the compressors at the discharge temperature of up to 220 °C; BP ENERSIN RX100 is a synthetic oil for piston compressors that operate at high compression temperatures [6].

The choice of oils for the compressors of refrigeration plants has particular features. Unlike air piston compressors, refrigerant compressors are characterized by contact of the oils with the refrigerant, mixing with it and entering the low-temperature zone. The oil carried out from the compressor cylinders by the refrigerant forms bonding gum residues in the compressor, while wax crystals are evolved from the oil on the evaporator surface, which significantly worsens the processes. This requires additional requirements for the oils designed for operation in refrigeration plants.

The BP ENERGOL LPT and LPT-F oils belong to the class of mineral oils. The LPT 46 and LPT 68 oils are recommended for the piston and rotary compressors that operate on Freon 12. The LPT-F32 and LPT-F46 oils are recommended for use in the Freon 22 systems; they are characterized by a low temperature of wax deposition.

Oils for hydraulic systems. Hydraulic oils are used as the working fluid for the energy transfer in the hydraulic systems of steering vehicles, hatch covers, electrohydraulic cranes, ramp drive mechanisms, deck mechanisms, propeller pitch changing mechanisms. Taking into account the operating conditions, hydraulic oils must have a high fluidity to provide minimum resistance during the flow through the pipeline and the ability to reduce friction in the moving parts of hydraulic mechanisms and their wear. These properties largely depend on their viscosity.

The maximum permissible viscosity depends on the type of the pump, rarefication at the pump inlet and the permissible pressure loss in the hydraulic system. It makes up 850 cSt for blade and gear pumps, 500 cSt for screw pumps and 200 cSt for piston pumps.

Thermal oils. In recent years, special mineral oils have been used along with the water steam as heat carriers on ships. The liquid coolants have gained particularly widespread use on tankers, where there is always the need for a considerable amount of heat for heating the cargo. The liquid that is used as a heat carrier must have the following qualitative characteristics [6]: high stability, high heating capacity, high thermal conductivity, low viscosity, low level of vaporization, compatibility with the structural materials.

A typical representative of thermal oils is the BP Transcal N thermal fluid designed for the liquid-phase closed thermal systems with the working temperature range up to 320 °C and the minimum permissible temperature of 10 °C. At the system start, it is necessary to heat the pipes that are in the temperature zones below—10 °C with the help of steam or electric satellites to prevent a stop of the movement of this thermal fluid due to wax accumulations on the walls of these sections of the pipes.

The service life of the thermal fluid can be measured in years, but its overheating and degradation can lead to the need of replacing it in a few months.

2.4 Mechanisms and Equipment of Marine Power Plants

Main mechanisms and equipment of the marine power plants that provide operation of the systems of various purposes include pumps, fans, compressors, heat exchangers, filters and separators.

2.4.1 Pumps, Fans and Compressors

The pump is a hydraulic machine designed to move fluid, in which the mechanical energy of the drive is converted into the potential energy of the pumped fluid. According to the principle of operation, the pumps are divided into *volumetric, lobe* and *jet pumps*. They are used in general and special ship systems, marine power plant systems, auxiliary installations and complexes.

In volumetric pumps, the increase of the potential energy of the fluid (pressure increase) is caused by pushing it to the discharge chamber. Depending on the construction of the displacement body, the volume pumps can be *piston* (Fig. 2.28) with reciprocating motion of the displacing pistons and *rotational* with rotational

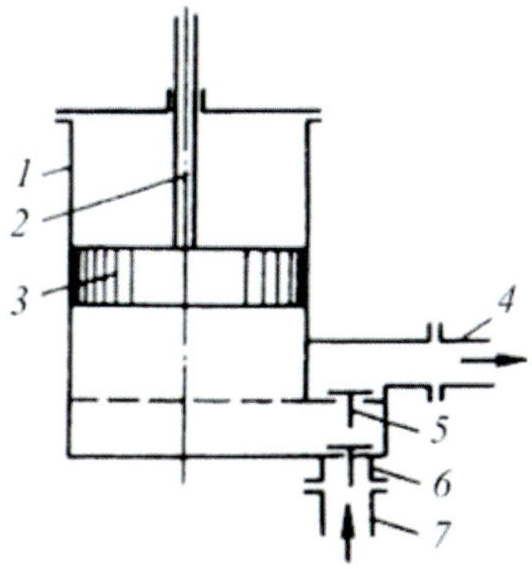

Fig. 2.28 Scheme of the piston pump: 1—cylinder; 2—pump rod which connects piston; 1—body; 2, 3—driving and driven pinions; 3 with the drive engine; 4, 7—pressure and suction pipelines; 4, 5, 6—working pump chambers; 5, 6—discharge and suction valves A—suction cavity; B—discharge cavity

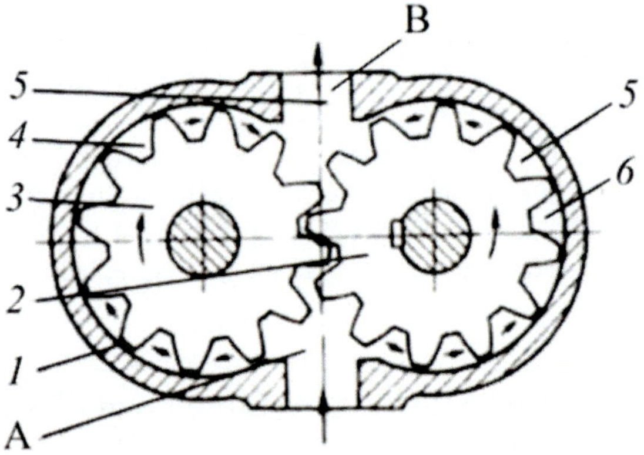

Fig. 2.29 Scheme of the gear pump: 1—cylinder; 2—pump rod which connects piston; 1—body; 2, 3—driving and driven pinions; 3 with the drive engine; 4, 7—pressure and suction pipelines; 4, 5, 6—working pump chambers; 5, 6—discharge and suction valves A—suction cavity; B—discharge cavity

motion of the displacing rotors. Gear and propeller pumps are the most commonly used rotary pumps in MPPs; they are implemented to move viscous fluids.

Piston pumps can be used for feeding, drying, oil-refueling and stripping, but their use in modern MPPs is limited.

Gear pumps (Fig. 2.29) are used as oil and fuel pumps. They have the feed rate of 0.5–500.0 m^3/h at the discharge pressure of up to 10 MPa. The working bodies of such pumps (Fig. 2.31) are made in the form of pinions which transmit the torque.

Screw pumps (Fig. 2.30) are mostly made with three screws. The medium screw is the driving screw; the others are the driven ones. The pressure created by the screw pumps can reach 25 MPa.

The scheme of another rotational pump is shown in Fig. 2.31. In body *1*, two rotors *2* and *4* driven from the engine rotate in different directions. The shafts of the rotors are connected by the gear wheels. The rotors take the fluid from receiving cavity *3* and transfer it to discharge cavity *5*.

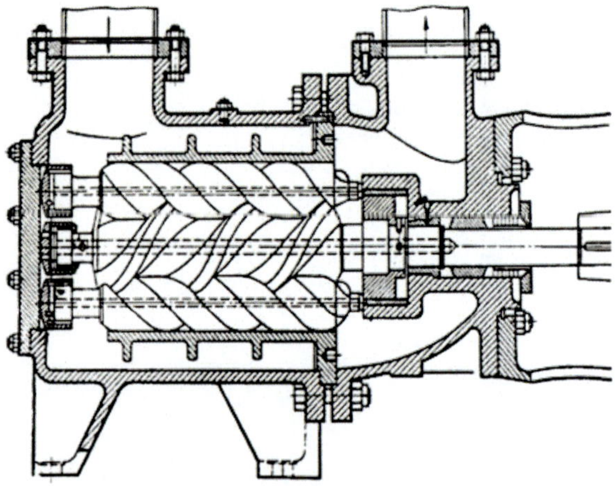

Fig. 2.30 Screw pump

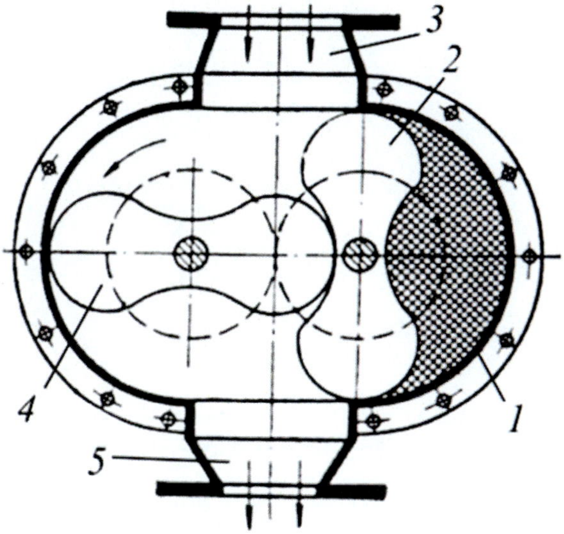

Fig. 2.31 Rotational pump

The volume of the sucked fluid or gas is shown by hatching. The pumps of this type were widely used in two-stroke diesel engines as the blowing ones.

The operating principle of *lobe pumps* is based on the increase of the kinetic energy of the fluid being transported and further conversion to the potential energy in the diffuser.

The pumps of this type include centrifugal and axial pumps (Fig. 2.32).

Single-stage centrifugal pumps are used for pumping fluids with relatively low pressures of about 0.3–0.4 MPa. If it is necessary to achieve high pressures, multistage pumps are used. In the ship plants, the single-stage pumps are used in the diesel engine cooling systems, while the multi-stage pumps are used in the fire-fighting and condensate-feed systems.

Fig. 2.32 Schemes of lobe pumps: **a**, **b**, **c**—centrifugal pumps: **a**—single-; **b**, **c**—two-stage; **d**—axial pump

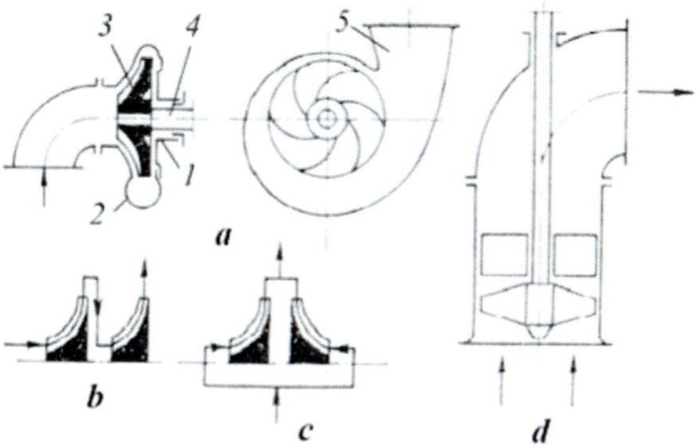

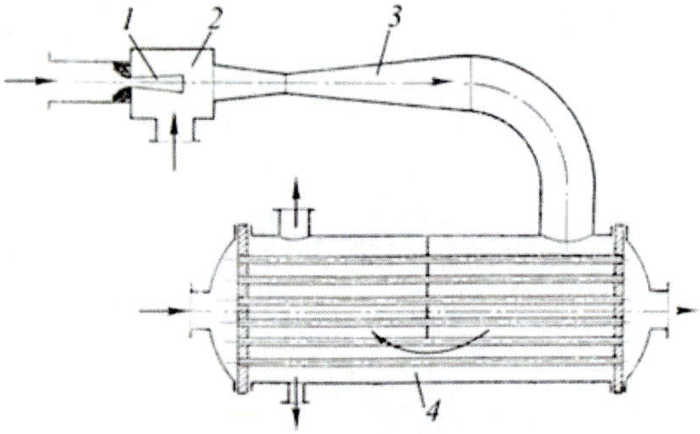

Fig. 2.33 Scheme of the steam jet ejector arrangement: 1—nozzle; 2—mixing chamber; 3—diffuser; 4—refrigerator

Axial pumps are used for pumping large quantities of fluid at low pressures of 0.05–0.15 MPa (for pumping the main condensers of the steam turbine plant, in the water jet propulsion systems).

The principle of operation of jet pumps deals with the transfer of the kinetic energy of the working medium (liquid, steam or air) to the transported fluid. Figure 2.33 shows the scheme of the arrangement of a steam jet ejector, which is a suction jet pump with water steam as the working fluid. The air steam jet ejectors are used to remove air, or rather the steam-air mixture from the condensers that operate with rarefaction, and also to suck off the steam-air mixture from the outer seals of steam turbines.

The steam with the pressure of 1.3–2.0 MPa and the temperature of 250–300° is used as the working steam in the steam jet ejectors. Steam jet ejectors can be performed with two or three stages.

Fans are the mechanisms designed to discharge air at low pressures (not higher than 15 kPa or 1500 mm water column), which corresponds to the degree of pressure increase (i.e. the ratio of the discharge pressure to the suction pressure) not more than 1.15.

By the principle of operation, there are distinguished centrifugal and axial fans. General arrangement of a centrifugal fan is presented in Fig. 2.34. The air is continuously fed into impeller *1* through port *2* in its center, passes between the channels formed by the wheel blades from the axis to the periphery, and is then directed

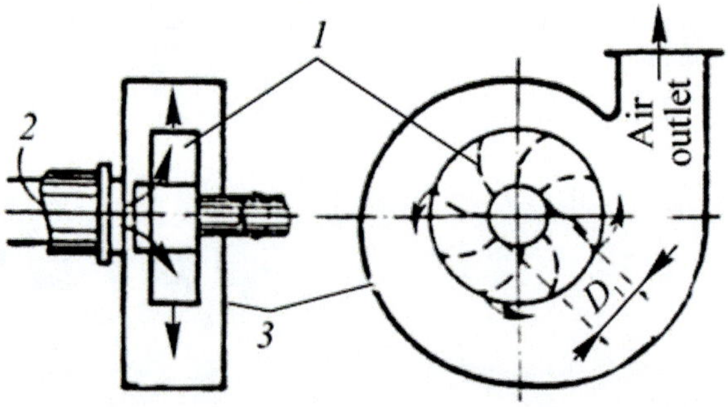

Fig. 2.34 Scheme of the centrifugal fan

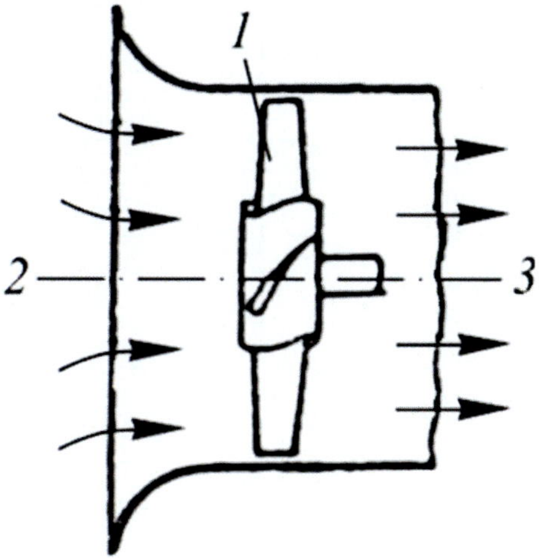

Fig. 2.35 Scheme of the axial fan

through casing *3* to the discharge pipeline. The increase in pressure in the centrifugal fan is caused by the impact of the centrifugal forces that arise when the air moves in the channels between the wheel blades.

On ships, centrifugal fans have found their application as the blowing fans of boiler plants, as smoke exhausters, and for ventilation of living and staff areas as well.

The structure scheme of the axial fan is shown in Fig. 2.35. The air that is sucked through connecting pipe *2* flows to impeller *1* along its axis, and is then sent to discharge pipe *3*. The fans of this type create slight pressure, they are used on ships for ventilation of holds, living and staff areas. Along with that, they have received considerable distribution as air blowers for hovercrafts.

Compressors are designed to compress and move the air or the other gaseous medium at the degree of pressure increase more than 1.15. By the principle of operation, compressors are divided into the volumetric and lobe ones. Piston compressors can provide the air pressure up to 5–20 MPa; they are used in the compressed air systems of MPPs. To ensure a high air consumption at relatively low degrees of pressure increase, centrifugal and axial (less often) compressors are used.

The scheme of the stage of the centrifugal compressor is shown in Fig. 2.36. The main elements of the stage are snail *1*, blade diffuser *2*, impeller *3*, and inlet connecting pipe *4*.

Through the inlet connecting pipe, the working fluid enters the rotating impeller, in which it moves under the action of the centrifugal forces from the center to the periphery and is compressed. In the blade diffuser, an additional increase in pressure occurs due to conversion of a part of the kinetic energy into the potential energy determined by the speed at which the working fluid leaves the wheel. The snail discharges the flow from the compressor with the minimum possible losses of energy. The impeller blades can be flat or curved.

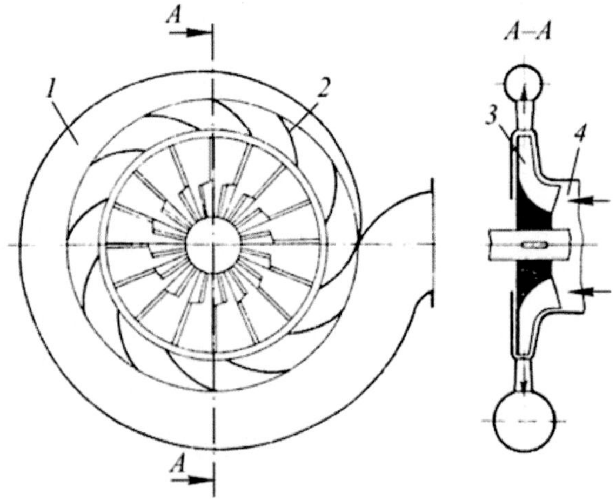

Fig. 2.36 Constructional scheme of the centrifugal compressor stage

The fixed guiding device can be installed in front of the impeller with the aim of a shockless air inlet.

The pressure increase of 4.0–4.5 is achieved in one stage of the centrifugal compressor, so most of these compressors are single-stage.

Axial compressors are mainly used as the components of the gas turbine engine.

The main characteristics of the pumps, fans and compressors are feed and charge.

Feed is the amount of liquid or gas that moves per unit time. If the feed is measured in volume units, it is called the volumetric feed (m^3/s); if it is measured in mass units, it is the mass feed (kg/s).

Charge is the amount of energy that is transferred to the mass unit of the transported medium, J/kg. In this regard, the charge's linear dimension is meter and it physically represents the height at which the medium moving due the energy transferred to it can be lifted.

The engine pressure ratio is the ratio of the final pressure behind the compressor in the outlet section of the discharge connecting pipe P_c to the pressure in the inlet section of the intake connecting pipe P_{inlet}.

2.4.2 Heat Exchangers

The working media of the MPP, such as fuel, oil, water, or air, are subject to thermal impact in various elements of the plants. For example, the fuel oil, which has the pour point of 10 °C, is impossible not only to burn, but also to pump without preheating it. Obligatory preheating is the condition for high-quality spraying of high-viscosity fuels.

The water which cools the engine, the oil and the charge air need cooling, while the feed water of steam boilers needs heating. In addition, the feed water must be subject to technical degassing (deaeration) in a special device called deaerator for reduction of the amount of dissolved oxygen which leads to corrosion of the elements of steam boilers.

Heating, cooling and deaeration take place in the corresponding structural devices. These devices, which are parts of the main and auxiliary energy complexes, are *heat exchangers* designed to transfer heat from one coolant to another.

Such devices can be *recuperative (surface)* and *regenerative (mixing)*. In the former, coolants are separated by solid walls forming the heat exchange surfaces for heating or cooling depending on the purpose of the device, and the latter do not have the separating solid surface. The surface heat exchangers (heaters, coolers) have found wide application on ships, while the mixing heat exchangers (deaerators, steam-gas condensers) have found limited application.

Most often heat exchangers are made in the shell-and-tube form. The main elements of such devices are the cylindrical body (casing), inside of which there are pipes forming the heat exchange surface. The pipes are mostly smooth, sometimes ribbed (in the charge air coolers). The pipes can be straight, fixed in two tube boards, or serpentine, loop-like, spiral, with both ends fixed in one tube board.

Increase in the efficiency of the heat transfer processes is possible through the use of the multi-pass shell-and-tube devices with cross baffles installed in their casing.

Devices with coolants of the "steam-condensate" or "steam-oil" type employ brass pipes or stainless steel pipes. Coolants of the "liquid fuel-steam" or "liquid fuel-condensate" type require using pipes made of steel of wide application. The outer diameter of such pipes is 10–20 mm with the wall thickness of 1–2 mm. The ends of the pipes are hermetically fixed in the tube boards. The bodies of the devices are made of steel.

Figure 2.37 shows the arrangement of a shell-and-tube surface heater of the low-pressure feed water. The steam enters the preheater through connecting pipe *1*, the condensate leaves through connecting pipe *4*. The feed water enters through connecting pipe *3* and, after making four passes through the heater tubes, leaves through connecting pipe *2*. The heater is covered with thermal insulation *6*. To improve the steam washing of the heater tubes, the body is fitted with deflecting baffles *5*.

Figure 2.38 shows a preheater of the high-pressure feed water with flat spiral tubes *1* that are welded to feed *3* and collecting *2* receivers. The heater consists of three

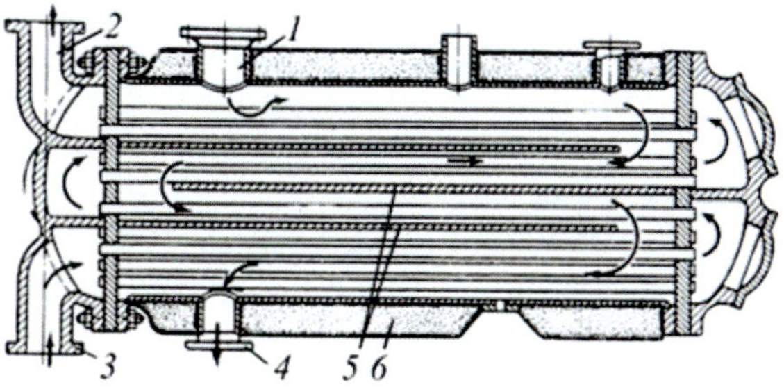

Fig. 2.37 Surface shell-and-tube heater of the feed water

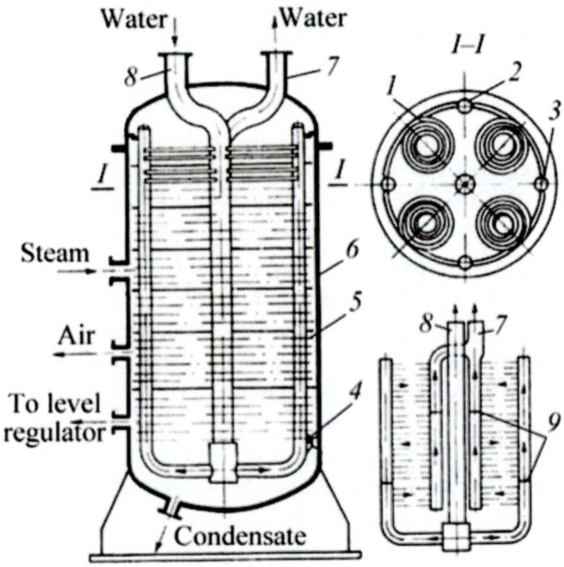

Fig. 2.38 Three-section heater of the high-pressure feed water

parts: main heater located in the central part of body *6*, steam cooler in the upper part, and condenser. The sections are formed by means of cross baffles *5*, which ensure the steam cross-flow in the intertubular space of the heater.

Sliding support *4* ensures compensation of the thermal expansions of the heater tube bundle. Change in the direction of the feed water movement in serpentine pipes is carried out by baffles *9* of collector pipes. Inlet *8* and outlet *7* connecting pipes of the feed water are welded to the heater cover.

The fuel heater (Fig. 2.39) is made in the form of a heat exchanger with loop pipes along which the fuel moves. Heater body *2* in the upper part is connected by the flange to tube board *5* and cover *7*. Flanged tubes *3* in the tube board are made in the form of the loops. Inlet *6* and outlet *8* connecting pipes and baffle *9* in the cover form a series of cavities, which ensures consistent passage of fuel through all the loops. The heating steam flows through connecting pipe *4* and makes several moves in the heater body due to diaphragms *1*. The level of the formed condensate is shown on water-indicative glass *10*; it is discharged through valve *11*.

The heaters of the considered type are reliable heat exchangers, since their design provides for free expansion of the tube bundle during heating.

General arrangement of the oil, fresh water and air coolers is not fundamentally different from the structure of the shell-and-tube water heaters. Figure 2.40 shows the construction of the oil cooler with the brass tubes 11×1 mm in diameter, which is used in marine diesel engines.

Design of the fresh water coolers is similar to that of the oil coolers. One of the tube boards is mobile, seawater is pumped inside the pipes, and fresh water washes the pipes from the outside.

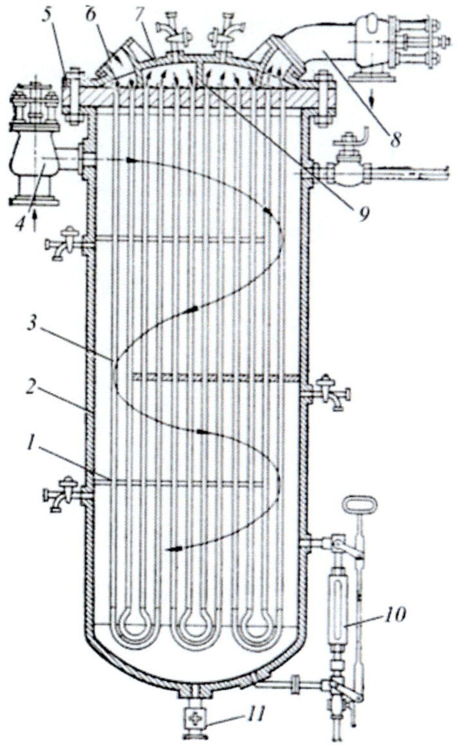

Fig. 2.39 Fuel heater

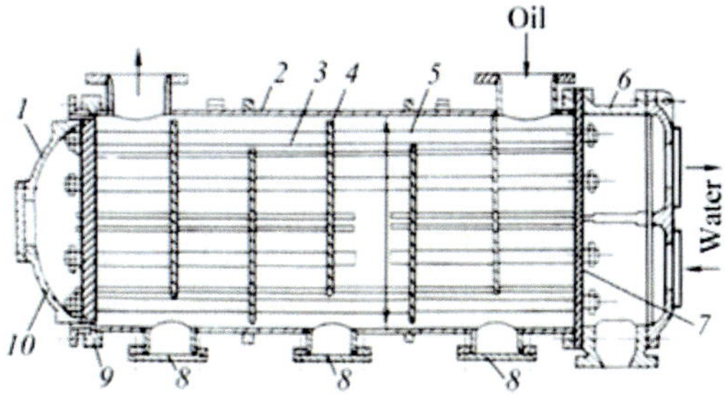

Fig. 2.40 Shell-and-tube oil cooler: 1, 6—covers of water chambers; 2—body; 3—tubes; 4—segment baffles; 5—remote tubes with longitudinal connections; 7, 10—tube boards; 8—stubs; 9—stuffing box seals

The charge air coolers of diesel engines (Fig. 2.41) employ the pipes ribbed from the outside; seawater is flowing inside the pipes.

Pipes *2* are flanged in boards *1*, the lower one of which is movable; it is sealed with rubber ring *4*. Seawater is supplied to cover *3*, makes two strokes and is discharged from the cover along connecting pipes *5*. The cooled air flows crosswise around the ribbed bundle from the outside.

Recently, the "Alfa Laval" plate heat exchangers have become widespread on ships. They are used to heat and cool the fuel, oil, and fresh water. The heat transfer

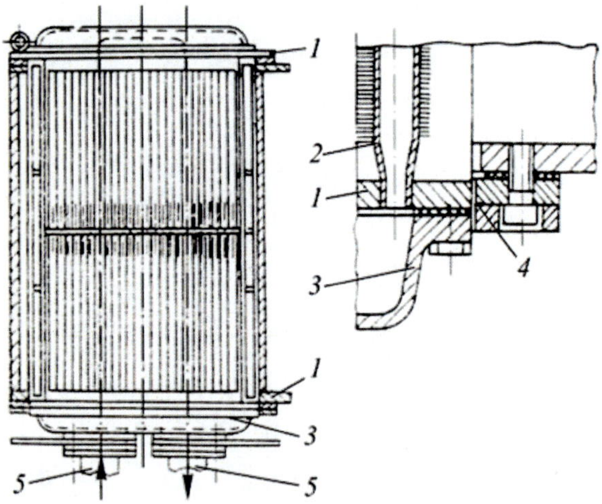

Fig. 2.41 Charge air cooler for diesel engines

surface in these devices (Fig. 2.42) is formed by the stamped corrugated plates 1.0–1.3 mm thick, made of titanium, stainless steel or aluminum bronze. Each plate has the ports for supply and discharge of the heating and heated or cooling and cooled media through connecting pipes *1, 2, 3, 4*.

The oil cooler consists of identical profiled plates *6*, which are assembled in a bag and pressed by the tightening bolts to the engine bed plate *5*.

Among the main advantages of the plate heat exchangers are their compactness, the possibility of easy changing of the surface area of the heat exchange within a

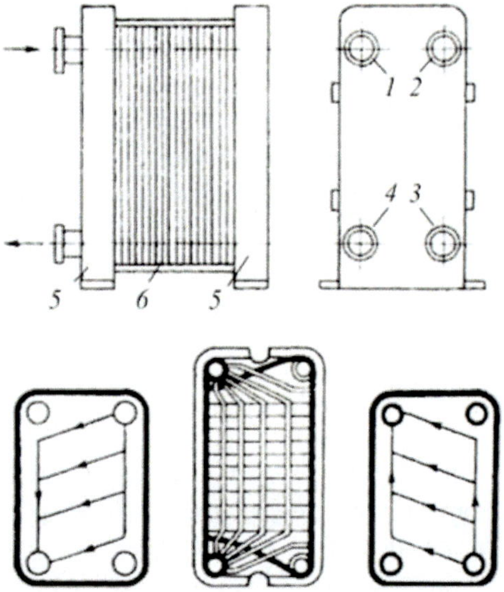

Fig. 2.42 "Alfa Laval" plate oil cooler: **a**—general view; **b**—stamped plate; **c**—scheme of the cooling water movement from one side of the plate; **d**—scheme of the oil movement from the other side of the plate; 1, 3—connecting pipes of oil and water discharge, respectively; 2, 4—connecting pipes of oil and water supply, respectively; 5—engine bed; 6—plate

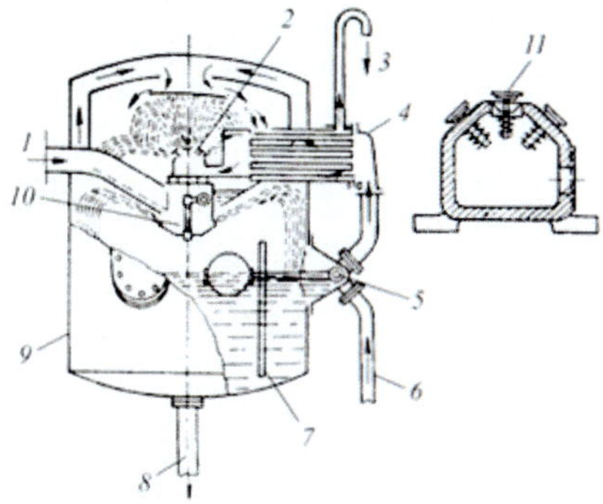

Fig. 2.43 Structural scheme of the ship deaerator: 1—steam supply pipeline; 2—spray head; 3—air discharge pipe; 4—vented steam condenser; 5—regulating valve; 6—feed water supply; 7—indicator of the level of feed water; 8—water supply to the feed pump; 9—deaerator body; 10—switchgear; 11—feed water sprayer

wide range by means of a set of plates, the possibility of rapid disassembly, and the convenience of cleaning.

Deaerators are designed to remove the dissolved oxygen and gases from the feed water, as they enter the condensate movement path and cause intensive corrosion of the whole path of the condensate-feed system. In addition, deaerators are the feed-water heaters of the mixing type.

The structural scheme of a ship deaerator is shown in Fig. 2.43.

The purpose of *evaporation-desalination plants* is to obtain fresh water from the sea for the provision of ship needs. There are other methods of water desalination, but the most widespread one in the marine fleet is application of the evaporation plants.

Fresh water is used on the ship not only for operation of the power plant, but also for the provision of household needs of the crew and the passengers at the rate of 200–250 L per person per day. Under substantial daily water needs in long-term voyages, it is not practical to take the full supply of fresh water on board because of the decrease in the net carrying capacity of the ship. In these cases, the required amount of water for the supply to boilers and household needs is obtained by evaporating seawater and its further condensing.

As a rule, modern evaporation plants are of the vacuum type. This means that the secondary steam (the one from which the fresh water condenses) is formed at a pressure below the atmospheric pressure. Production of the secondary steam in vacuum allows using the working fluid of a relatively low temperature for heating the evaporators:it is the water from the cooling system of the main diesel engines or the low-pressure steam in the steam turbine plant.

The constructive scheme of the recovery vacuum surface desalinator is shown in Fig. 2.44. The desalinators of this type with the capacity of 1–25 t/day are widely implemented on ships with the main ICEs. The basis of the construction is cylindrical flash drum *1*. The hot water is supplied to straight-pipe heater *5*, which is installed in the lower part of it, and there is horizontal separator *3* of the louvre type and two-way horizontal straight-pipe condenser *2* in the upper expanded part [1].

Fig. 2.44 Scheme of general arrangement of the recycling vacuum desalinator

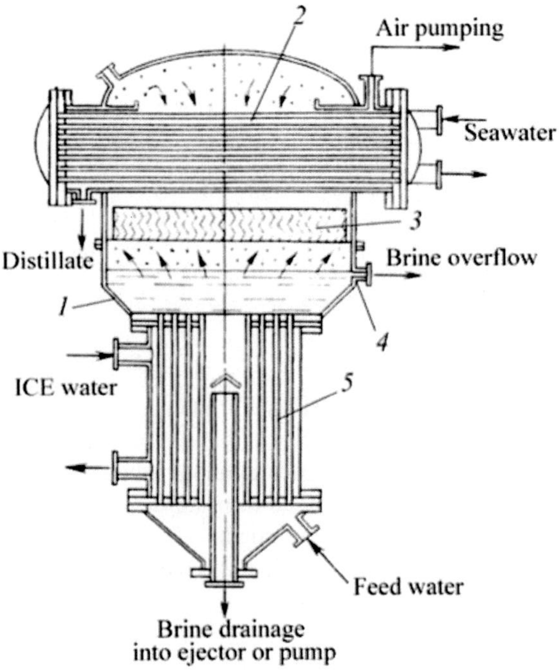

In the center of heating battery *5* there is a cylindrical shaft for the brine circulation. Through its central pipe, the brine is supplied to the ejector or the pump. The brine level in the casing is set at the height of the upper section of drain pipe *4*.

The desalination plant allows obtaining a distillate with the salt content of no more than 8 mg/l with the brine salt content of 50 g/l.

The vacuum desalination plants of the plate type have been widely used recently since they are highly efficient. By its design, the plant is a compact device with an evaporative part in the lower part of the body and a condenser in the upper part. The heat exchange surfaces of the evaporator and the condenser have a similar design. They consist of corrugated plates about 0.6 mm thick, which are gathered in a bag.

The plates are made of titanium alloys and materials with high anticorrosion properties, they have a special shape that provides a large heat exchange area and high efficiency.

The capacity of such plants reaches 60 tons per day, the salt content of the distillate does not exceed 5 mg/l, and the vacuum level is 90–95%. Seawater evaporates at the temperature of 40–50 °C.

2.4.3 Apparatuses and Devices for Cleaning Fuels and Oils

The fuel received by the ship during bunkering for use in the main and auxiliary engines must be brought to a state that meets the requirements of the manufacturer of appropriate power equipment. These functions are performed by the fuel preparation systems. Along with other tasks, they provide reduction in the content of water and mechanical impurities in the fuel. Purification of fuels from water and mechanical impurities occurs in apparatuses and devices that use the gravity forces (*sedimentation*), centrifugal forces (*separation*) and semipermeable materials (*filtration*).

Sedimentation of fuel. During sedimentation, the mechanical particles and water droplets of the fuel are influenced by the gravity force and the Archimedes buoyant force, which is directed upwards, opposite to the impact of gravity. In case of gravity excess, the particles precipitate, which ensures purification of the fuel from contaminants.

Under the shipboard conditions, sedimentation of low- and medium-viscosity fuels is most effective. To accelerate the sedimentation process, it is advisable to reduce viscosity and density of the fuel by raising the temperature in the tank, but not above the temperature that is 15 °C below the flash point and not higher than 90 °C.

The fuel sedimentation in the settling tank takes place for 20–22 h. During this time, the heavy particles and water settle on the bottom of the tank and are subsequently removed through the drain valve. The capacity of the settling tank should be sufficient to provide an eight-hour operation of the main and auxiliary engines and boilers at the maximum operating load. Considering the fact that the time of sedimentation also depends on the thickness of the fuel layer, the height of the settling tanks should be as low as possible [9, 17].

Separation. The principle of operation of centrifugal separators is based on the difference in the densities of the fuel and the impurities that contaminate it.

The increase in the efficiency of fuel purification by means of separation in comparison with the sedimentation system is explained by the fact that the centrifugal forces which separate the foreign particles from the fuel are almost by 1500 times higher than the gravitational forces.

Figure 2.45 shows the principle of operation of the centrifugal separator. The uncleaned fuel is continuously fed to rotary drum *4* through central channel *3*, then it flows to the periphery of the drum, passes between plates *1* and is discharged through annular channel *2* in the upper part.

Centrifugal separators can be used for processing both fuels and oils. The centrifugal separators with plate drums from such companies as Alfa Laval (Sweden), Shurpless (Great Britain), Vestfalia (Germany), Titan (Denmark), Mitsubishi (Japan) and others have become the most widespread in the fuel and oil systems of power plants of modern ships.

By the principle of cleaning of the drum from dirt, separators are divided into two groups:

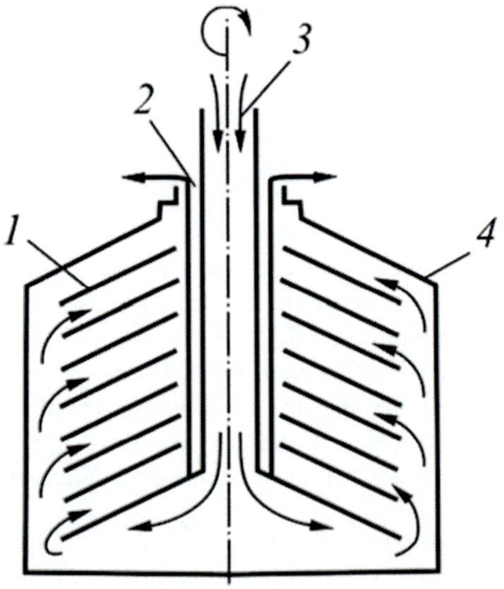

Fig. 2.45 Scheme of the flow of liquid in the drum of the centrifugal separator

- *non-self-cleaning (non-self-unloading)*, where the dirt accumulated in the drum is removed manually during the periodic stop and disassembly of the separator;
- *self-cleaning (self-unloading)*, where the dirt is removed by the washing water automatically under the action of the centrifugal forces during operation of the separator.

Modern ships almost exclusively employ the self-cleaning separators.

All separators can be set up for the *purification mode*, when water and mechanical impurities are separated simultaneously, or for the *clarification mode*, which only separates mechanical impurities (Fig. 2.48). In the purification mode, separators work at a considerable watering of fuel. Separation of mechanical impurities in this mode is less effective than in the mode of clarification.

The general view of the Alfa Laval separator is shown in Fig. 2.46 [9].

The fuel is supplied to the separator from above through connecting pipe *1*, under pressure it enters the lower part of drum *10* through the central channel. The drum is driven by an electric motor through shaft *9*. The fuel is rotated there, moving from the periphery to the axis of rotation through the narrow gaps formed by plates *6*. The speed of its movement increases to the speed of rotation of the drum. The fuel fills the gaps between the plates through the holes in distribution disk *8* and similar holes in the plates. The number of plates, depending on the size of the separator, is 50–150 pieces. The gap between them is 0.5–0.6 mm.

The fuel that has passed separation rises up inside the stack of plates and exits the separator through connecting pipe *2*. In the purification mode, the separated water that is thrown to the periphery of the drum creates water seal *7*, its excess rises above the vane packet and, passing pressure *4* and gravitational *5* disks, exits the separator through connecting pipe *3*. The boundary of water gate with fuel *11* should be located near the outer edge of the plates.

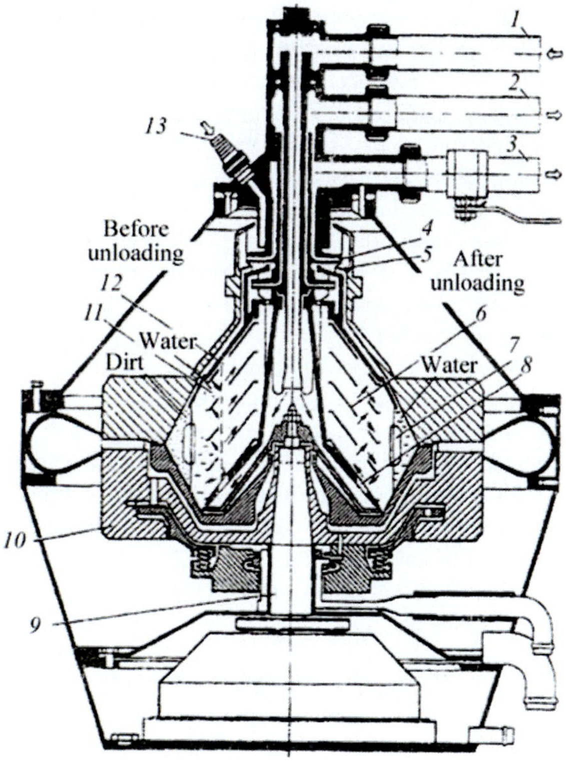

Fig. 2.46 “Alfa Laval” centrifugal fuel separator

To create the water seal, water is fed through connecting branch *13* through channel *12*. This water also provides washing of the drum during unloading. Usually, replenishment of the water seal is provided by the water separated from the fuel.

Selecting the dimensions of gravity disk *5* ensuresmaintaining of the equilibrium state between the amount of water that is separated from the fuel and enters the hydraulic seal zone and the water that leaves it.

In the clarification mode, the separator under consideration operates without the water seal; the water outlet from the fuel is blocked and the fuel is cleaned only of mechanical impurities.

In the practice of preparation of heavy fuels, it is recommended to perform separation in the following sequence: with two separators that operate in parallel in the low-productivity purification mode, and then with a sequentially activated separator configured as a clarifier [6].

Automatic separation systems, such as ALCAP, have been widely used on ships.

The ALCAP system consists of the following elements: the “Alfa Laval” self-cleaning separator of the FOPX series, the EPC electronic control system, the MARST-1 microprocessor, the water content sensor in the fuel, and the valve removing the separated water, and the other equipment [6].

The electronic control system provides unattended maintenance of the plant. All the operations are performed automatically according to the prescribed program, which takes into account the separator's type, size, and operating mode, the time between the unloadings, the intervals between operations during unloading, between the start and stop of the separator. It also provides pressure and temperature monitoring, emergency alarm and protection. The EPC system consists of the software unit and actuation devices.

Homogenization. Up to 3–6% of waste and sludge is formed during separation of heavy fuels. The sludge composition includes a considerable amount of asphaltic-resinous compounds that could be used in the boiler furnaces, or even in the cylinders of diesel engines.

Reducing the loss of combustible components with the sludge can be achieved by treating the fuel in a homogenizer, where gelled clots and solid agglomerates are destroyed. As a consequence, the fuel becomes homogeneous. It is separated and filtered with the minimum losses of the combustible part.

At the same time, the water with the fuel is transformed into the water-fuel emulsion (WFE), which consists of water microdroplets surrounded by the shell of heavy fuel constituents.

The use of WFE reduces the temperature of the fuel combustion in the cylinder, which helps to reduce formation of solid soot particles. At the same time, it decreases the content of nitrogen oxides in the exhaust gases and the precipitates on the surfaces of the combustion chamber and the cylinder of resinous constituents.

For homogenization of the fuel on ships, apparatuses and devices of rotational (Fig. 2.47), valve (Fig. 2.48), vibrational, hydrodynamic and electrophysical types are implemented.

Filtration. Filters have been widely used in MPP to clean fuels and oils that circulate in the appropriate systems. They are characterized by such indicators as the fineness of screening (the largest size of the contaminating particles that passes through the filter) and the degree of filtration (the ratio of the weight of the removed impurities to its initial value).

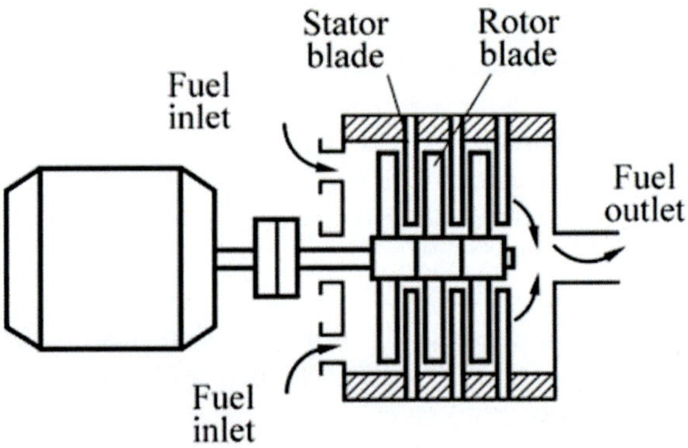

Fig. 2.47 General arrangement of the homogenizer of the rotary type with vanned working tools

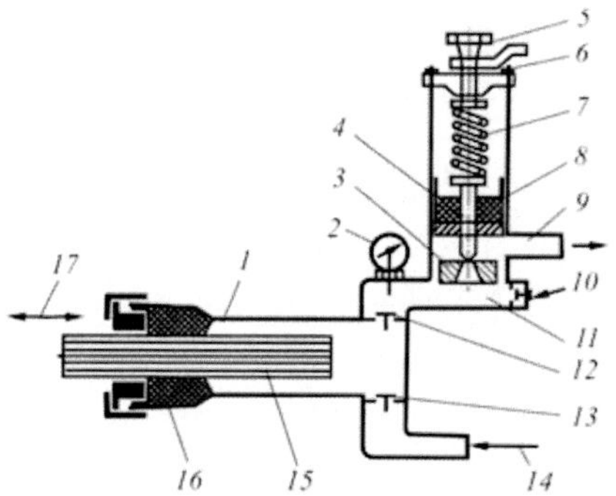

Fig. 2.48 Principal scheme of the valve homogenizer: 1—plunger pump unit; 2—manometer; 3—seat of the homogenization valve; 4, 10, 13—homogenization, safety and suction valves, respectively; 5—pressure regulator; 6—stopper; 7—spring; 8—gland seal; 9, 14—discharge and supply of fuel, respectively; 11, 12—discharging chamber and valve, respectively; 15—plunger; 16—sealing of the plunger; 17—direction of the plunger movement

By the fineness of screening, filters are divided into the following groups:

input filters prevent accidental large contaminating particles from entering the fuel system (the filters in front of the fuel transfer pumps);
coarse mesh filters (CMF) remove particles of more than 40 μm from fuel;
final stage filters (FSF) remove impurities larger than 6 μm from fuel (using paper filters, if the impurities are larger than 4 μm).

Depending on the principle of operation, filters can be *surface* or *volumetric* (*capacitive*).

During cleaning of the fuel in *the surface filter*, impurities settle on the surfaces of the elements, the edges of the cells or the gaps. A mesh with 10–2000 μm cells, paper, and cloth are used as the filtering material. Sometimes the filtering elements are formed by plates, turns of wire or tape (gap filters).

In *the volumetric filter*, fuel or oil passes through the filtering material that contains a large number of channels and pores. Such filters are characterized by deposition of impurities inside the filter element (in the thickness, pores and on its surface). The filter elements of volumetric filters are made of felt, wood-fibrous materials, metal-ceramics, or porous bronze.

Theinput filters often have the form of flat metal sheets or intensively perforated hollow cylinders. For example, a mud box or a portable deck filter with a mesh filtering element can be used as the input filter.

The coarse mesh filters are placed in the fuel and oil systems in front of separators or FSF. Such filters with the fineness of screening of 40–150 μm are often placed in front of engines instead of FSF. CMF is made in the mesh, plate-tape and wire-slotted form. Purification of the contaminated elements of these filters is carried out by crossflow of the filtered liquid or the compressed air. The frequency of cleaning filters depends on their contamination, characterized by the increase in the difference

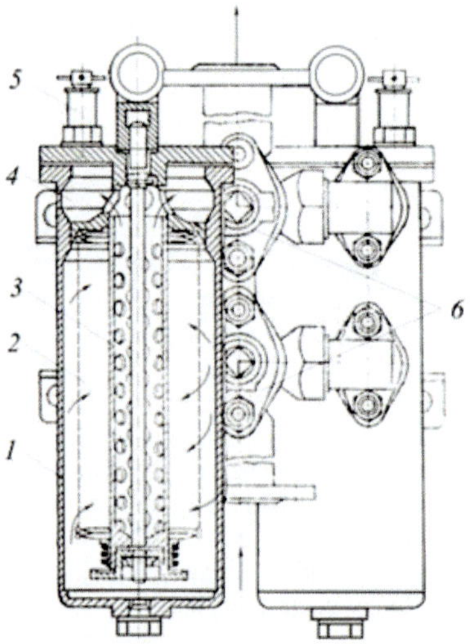

Fig. 2.49 Two-section coarse mesh filter: 1—body; 2—filtering connecting pipe; 3—rod; 4—chamber; 5—air discharge valve; 6—taps for switching filter sections

in pressure in front of and behind the filter. On pure CMF, the pressure difference is 0.02–0.04 MPa, on the contaminated one—0.08–0.10 MPa.

Figure 2.49 shows a two-section CMF of the mesh type designed to separate solid particles larger than 100 μm [2]. The filter cartridge consists of lenticular elastic meshes, which are compressed by the spring. The liquid is supplied from the outside through the inlet connecting pipe, passes through the meshes into the inner rod with the holes, through which it rises to the upper chamber, after which it leaves the filter the outlet connecting pipe.

The final-stage filters provide the fineness of screening of 2–40 μm. They operate with substantial pressure falls, so their filter elements must have a sufficient strength. Textile woven and non-woven materials, paper, cardboard, and metal-ceramics are used to produce such filtering elements.

Figure 2.50 shows a filter with the fineness of screening of 6 μm, which is used for fine filtration of oils. The filtering candles are collected in a bag that is protected by a mesh covered with the filter cloth.

The magnetic insert, which is the part of the filter bag, provides removal of ferromagnetic particles from the oil product subject to purification. The company Ball &Kerh also produces final-stage filters with the counterflow washing and filtering element which constantly rotates.

The "Winslow" volumetric FSF (UK) has been designed to filter high-viscosity fuels [2]. It is shown in Fig. 2.51.

Clogging of filter elements with asphaltic-resinous compounds during the heavy fuels filtering, the difficulty of separating water, and the need for frequent cleaning become a serious obstacle in the use of traditional filters on modern ships, especially taking into account the decrease in the level of qualification of the ship staff serving

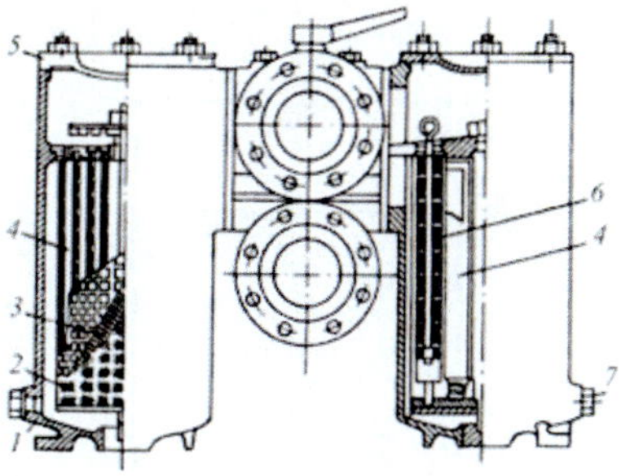

Fig. 2.50 Double Ball & Kerh filter: 1—body; 2, 3 and 4—filtering cloth, mesh and candles, respectively; 5—body cover; 6—magnetic insert; 7—drain plug

Fig. 2.51 Winslow volumetric filter: 1—filter element; 2—body; 3—cover; 4, 6—holes for air and residual discharge, respectively; 5, 7—supply and discharge connecting pipes, respectively; 8—drainage

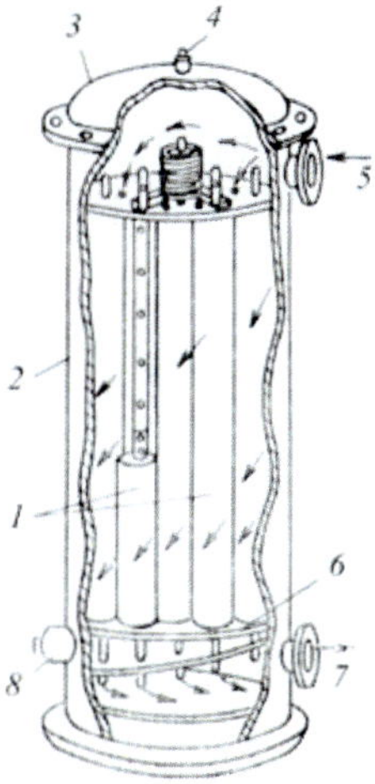

the MPP. Thus, self-cleaning filtering plants are becoming more and more widespread nowadays (Fig. 2.52).

The Sofrance high-efficiency filtration plants of the surface type (France) have received a positive feedback from experts. They are used on ships for cleaning fuels and oils (Fig. 2.53). The productivity of these plants is 1.5–6.0 m^3/h, the fineness of the purification is 3–20 μm.

The principle of operation of such plants is typical for modern self-cleaning filtration devices. The filter elements consist of metal lens-shaped disks with the two-layer mesh surface (inner brass mesh with the cells of 50 μm, external ones—3–10 μm), packaged in a bag on the central perforated rod.

The heated fuel is pumped to the filter elements by the pump and passes through them. The purified fuel is taken to the fuel system along connecting pipe *8*. Solid depositions and droplets of water accumulate on the surface of the filter, and then they flow into the lower part of the filter as they enlarge.

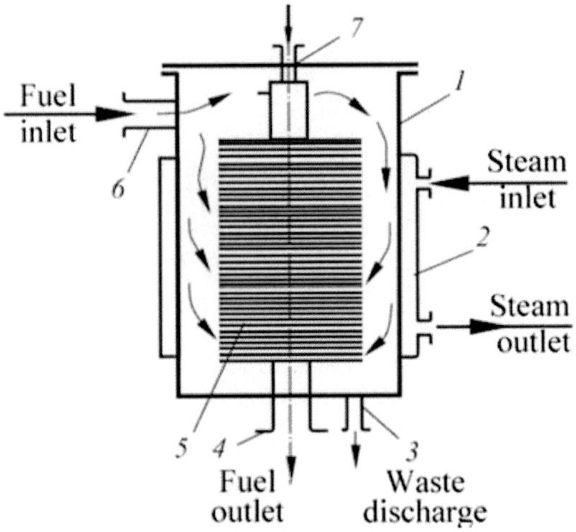

Fig. 2.52 Scheme of fuel cleaning in the Scamatic filtration plant: 1—body; 2—steam shell; 3, 4—connecting pipes; 5—filter element; 6—inlet connecting pipe

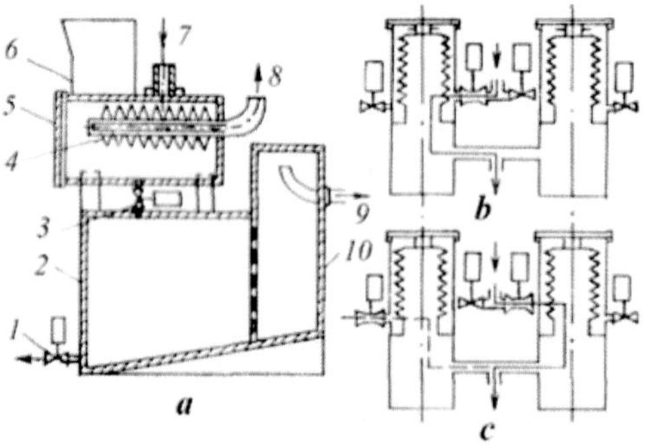

Fig. 2.53 Sofrance filtration plant: **a**—general view; **b**—operation of the left filter element; **c**—operation of the right filter element and cleaning of the left element by the reverse flow; 1, 3—drainage and bypass valves, respectively; 2, 10—sedimentation and overflow tanks, respectively; 4—filter element; 5—body; 6—software device; 7, 9—fuel movement; *8*—connecting pipe

Along with the counter-current fuel cleaning of the filter elements, other engineering solutions are also used. For example, design of the widespread models of the automatic filters manufactured by Ball &Kerh makes provisions for the use of compressed air for such purposes [2].

2.5 Atmospheric Pollution of the World Ocean by Marine Power Plants

The sources of atmospheric pollution during marine power plants operation are as follows [6]:

- exhaust gases of heat engines, boilers, incinerators;
- ventilation vapors, coolant leaks of the air conditioning and refrigeration plants;
- steam discharge into the atmosphere during operation of boilers;
- leakage and evaporation of the gas working medium of ship systems.

The main part of the atmospheric pollution by the marine power plant is the exhaust gases of the heat engines and boilers (Table 2.6) [4, 6]. The chemical composition of the emissions depends on the type and quality of the fuels, the production technology, the method of their combustion, and the technical state of the engines and boilers.

Details on the composition of the exhaust gases of the ship ICE is given in the data below [4], which correspond to the amount of fuel, air and oil per 1 kW·h.

Exhaust gas components of LSE of MC (MAN B&W) series

Air—8.5 kg/(kW·h): O_2–21%; N_2–79%Fuel—175 g/(kW·h):S—3%; HC—97%

Oil—1 g/(kW·h): S—0.5%; Ca—2.5%; C_xH_y—97.0%

Exhaust gases: CO_2–5.20%; H_2O—5.35%; O_2—13.00%; N_2–75.80%; CO—0.80 g/(kW·h); C_xH_y—0.90 g/(kW·h); PM—1.20 g/(kW·h); SO_x—12.00 g/(kW·h); NO_x—17.00 g/(kW·h)

What follows is the qualitative composition of the cyclic air (taking into account the content of seawater aerosols), fuel (medium-viscosity TG gas turbine fuel with the sulfur content of 2% and the additive of Mg and Cr, which binds Na and V) and combustion products for GTE with the capacity of 10 MW. This data corresponds to the combustion of 1000 kg of fuel [19].

Components of air, fuel and exhaust gases of GTE

Air, kg: N_2–53900; O_2—16100; Ar—630.00; CO_2–21.000; $MgCl_2$—0.0004; NaCl—0.0025; $CaSO_4$—0.0002

Table 2.6 Emissions of harmful substances with the exhaust gases of engines and boilers of marine power plants, kg/ton of fuel

Engines, boilers	Fuel	NO_2	CO_2	SO_x	CO	Solid particles	Volatile organic compounds
Steam boiler	Fueloil	7.0	3200	60	0.43	2.5	0.5
LSE	Fueloil	87.0	3200	60	7.40	1.2	2.4
MSE	Fueloil	57.0	3200	60	7.40	1.2	8.4
LSE	Diesel	87.0	3200	20	7.40	1.2	2.4
MSE	Diesel	57.0	3200	20	7.40	1.2	2.4
HSE	Diesel	70.0	3200	20	9.00	1.5	3.0
GTE	Diesel	16.0	3200	20	0.50	1.1	0.2

Fuel with the additive, kg: C—858.000; H—120.000; H_2O—1.992; S—20.000; V—0.005; Na—0.003; Mg—0.015; Cr—0.006

Combustion products, kg: N_2–$_5$3900; O_2—12809; CO_2–$_3$167; H_2O—1082; Ar—630; SO_4–$_4$0; NO_x—33; CO—30; C—5

The exhaust gases of heat engines contain a substantial number of components; there are about 280 of them in the internal combustion engines. By the nature of their effect on the human body and the environment, by chemical structure and properties, they are divided into seven groups [4, 6]. The first group is non-toxic substances: nitrogen N_2, oxygen O_2, hydrogen H_2, water steam H_2O, and carbon dioxide CO_2.

The second group is *carbon monoxide CO*. It is a colorless gas without taste and smell, lighter than air, practically insoluble in water. Carbon monoxide is formed in heat engines during combustion of fuel-air mixtures with some oxygen deficiency or due to dissociation of carbon dioxide at high temperatures.

The volume concentration of carbon monoxide in the atmosphere determines the level of its influence on the human body: $0.1 \cdot 10^{-3\,\%}$ is harmless; $0.1 \cdot 10^{-2\,\%}$ causes chronic poisoning at prolonged exposure; $5 \cdot 10^{-2\,\%}$ causes mild poisoning in 1 h; 1.0% causes loss of consciousness after several breaths.

The third group includes nitrogen oxides. Nitric oxide NO is a colorless gas. The dioxide NO_2 is a reddish-brown gas with a peculiar smell, it is heavier than air. Nitrogen oxides in the exhaust gases of engines can be formed in one of three ways: thermal, fuel (with a high content of nitrogen in the fuel), or rapid formation.

Thermal formation is considered the main one. The nitric oxide NO is formed in the engine due to high temperatures in the cylinder during combustion, substantial excess of air and a long time of the gases staying in the reaction zones (especially in two-stroke ICEs). A part of NO is oxidized to NO_2. By the effect on the human body, nitrogen oxides are much more dangerous than carbon oxides. When ingested in the body and interacting with water, they form compounds of nitric and nitrous acids, which irritate the mucous membranes of the eyes and the respiratory tract.

The volume concentration of nitrogen oxide in the atmosphere above 0.008% is fatal for humans.

The fourth group is the most numerous one (about 160 names). It includes various *hydrocarbons* (compounds of the type C_xH_y): paraffin (alkanes), naphthenic (cyclanes), aromatic (benzenes). Hydrocarbons are formed due to incomplete combustion of the fuel in heat engines and boilers of the MPP. They are toxic, and some of them (benzopyrene $C_{20}H_{12}$) have carcinogenic properties. Hydrocarbons interact with nitrogen oxides to form photo-oxidants, which are the basis of smog.

The fifth group includes aldehydes—organic compounds containing the aldehyde group bound to a carbohydrate radical (CH_3 or others). In the exhaust gases, there are mainly simple aldehydes, such as formaldehyde and acrolein.

The sixth group is soot and other dispersed particles. The main component of soot is solid carbon, which does not pose an immediate danger to the human body.

The consequences of deposition of the soot particles in the gas outlet and on the surfaces of waste heat boilers are the increase in the resistance of the tail surfaces of engines and boilers and the threat of fire in them. Soot has a polydisperse structure. Most of soot particles (85–95%) have the size of 0.004–0.500 μm; the size of individual particles can reach 1 μm in the gas path.

The seventh group includes sulfurous compounds, such as sulfurous anhydride SO_2 and hydrogen sulfide H_2S, which appear in the composition of exhaust gases during combustion of the fuels with an increased content of sulfur.Sulphurous anhydride is a colorless gas with a pungent smell. It is heavier than air and dissolves well in water, forming sulfuric acid.

The environment is threatened by the ship air conditioning and refrigeration systems and the gas fire extinguishing systems, which use ozone-depleting substances. These are chlorofluorocarbons, hydrochlorofluorocarbons and halogens, which are the substances containing methyl bromide, carbon tetrafluoride, methyl chloroform [4]. The listed substances evaporate into the atmosphere and decay under the impact of intense ultraviolet irradiation. During these processes, chlorine and bromine atoms are released, which then destroy the ozone layer.

Long-term pollution of the atmosphere by the above-mentioned substances can be disastrous for the World Ocean, as it leads to persistent, and often irreversible, consequences. Therefore, it is very relevant to develop a set of measures aimed at reduction of the harmful impact of MPP on the atmosphere of the World Ocean.

References

1. Artemov H. A., Horbov V. M. *Sudnovienerhetychniustanovky:navch. posib.* [Marine power plants: textbook]. Mykolaiv, UDMTU Publ., 2002. 356 p.
2. Artemov G. A., Voloshin V. P., Shkvar A. Ya., Shostak V. P. *Sistemysudovykhenergeticheskikhustanovok:ucheb. posobie*[Systems of marine power plants: textbook]. Leningrad, Sudostroenie Publ., 1990. 376 p.
3. Controlling emissions in two-stroke marine diesel. MER, 2008, November, pp. 16–21.
4. Horbov V. M., Ratushniak I. O., Trushliakov Ye. I., Cherednichenko O. K. *Sudnovaenerhetyka ta Svitovyiokean:pidruchnyk* [Marine power engineering and the World Ocean: textbook]. Mykolaiv, NUK Publ., 2007. 596 p.
5. Kornilov E. V., Boyko P. V., Smirnov V. P. *Deydvudnyeustroystvaivaloprovodmorskikhsudov. Konstruktsiya, ekspluatatsiya, remont* [Stern tube assemblies and the shafting of sea vessels. Construction, operation, repair]. Odessa, Feniks Publ., 2008. 200 p.
6. Horbov V. M. *Entsyklopediiasudnovoienerhetyky: pidruchnyk* [Encyclopedia of marine power engineering: textbook]. Mykolaiv, NUK Publ., 2010. 624 p.
7. Pakhomov Yu. A. *Sudovyeenergeticheskieustanovki s dvigatelyamivnut-rennegosgoraniya: uchebnik* [Marine power plants with internal combustion engines: textbook]. Moscow, TransLit Publ., 2007. 528 p.
8. *Nikolaevskiegazoturbinnyedvigateliiustanovki. Istoriyasozdaniya/ GP NPKG "Zorya"–"Mashproekt", Tsentr NIOKR "Mashproekt"* [Nikolaev gas turbine engines and plants. History of establishment/ The State Enterprise of Scientific and Production Complex of Gas Turbine Building Industry Zorya-Mashproekt, Research and Development Center Mashproekt]. Nikolaev, Yug—Inform Publ., 2005. 304 p.
9. Voznitskiy I. V. *Praktikaispolzovaniyamorskikhtoplivnasudakh: ucheb. posobie* [The practice of using marine fuels on ships: textbook]. Saint Petersburg, Morkniga Publ., 2006. 122 p.
10. Horbov V. M. *Enerhetychnipalyva:navch. posib.* [Energy fuels: textbook]. Mykolaiv, UDMTU Publ., 2003. 328 p.

11. *RehistrsudnoplavstvaUkrainy. Pravylaklasyfikatsii ta pobudovymorskykhsuden* [The Register of Shipping of Ukraine. Rules for classification and construction of sea vessels]. Kyiv, RehistrsudnoplavstvaUkrainy Publ., 2002, ch. 1, 382 p.
12. Gorbov V. M., Mitenkova V. S. *Alternativnyetopliva v sudovykhenergeticheskikhustanovkakh* [Alternative fuels in marine power plants]. *Sudokhodstvo* [ShipNavigation], 2007, no. 1–2 (127), pp. 64–65.
13. Gorbov V. M. *Akusticheskayaobrabotkatoplivadlyasudovogoenergetichesko-gooborudovaniya:ucheb. posobie* [Acoustic fuel treatment for marine power equipment: textbook]. Nikolaev, NKI Publ., 1989. 51 p.
14. Gorbov V. M. *Primenenievodotoplivnykhemulsiy v sudovoyenergetike: ucheb. posobie* [Application of water fuel emulsions in marine power engineering: textbook]. Nikolaev, NKI Publ., 1991. 54 p.
15. Devyanin S. N., Markov V. A., Semenov V. G. *Rastitelnyemaslaitoplivanaikhosnovedlyadizel-nykhdvigateley*[Vegetable oils and fuels on their basis for diesel engines]. Kharkov, Novoeslovo Publ., 2007. 452 p.
16. Voznitskiy I. V. *Prakticheskierekomendatsiiposmazkesudovykhdizeley: ucheb. posobie* [Practical recommendations for the lubrication of marine diesels: textbook]. Saint Petersburg, Morkniga Publ., 2007. 129 p.
17. Voznitskiy I. V. *Sudovyedvigatelivnutrennegosgoraniya: uchebnik: v 2 t. T. 1* [Ship internal combustion engines: textbook. Ch. 1]. Moscow, Morkniga Publ., 2008. 283 p.
18. Gorbov V. M. *Osnovytekhnicheskoyekspluatatsiisudovykhgazoturbinnykhustanovok:ucheb. posobie* [Fundamentals of technical operation of marine gas turbine plants: textbook]. Nikolaev, UGMTU Publ., 1996. 139 p.

Chapter 3
Marine Diesel Power Plants

3.1 Thermal and Structural Schemes of Marine Diesel Power Plants

Diesel power plants with low-speed engines. Along with the main diesel engine, a diesel power plant includes auxiliary and recovery boilers, marine power station, electric engines for driving pumps, fans, separators, heat exchangers (heaters, evaporators, coolers, condensers), and oil, fuel, cooling, air-gas, steam systems. All this equipment is a single energy complex. To visualize, study and calculate the efficiency of the power plant, a thermal scheme is used. *The thermal scheme* is a conditional model of a real plant, which shows the interconnections between its parts needed to ensure the functioning of the marine power plant (MPP).

Depending on the set goals, the thermal scheme is made in the principal or extended form. The principal schemes which display sequential arrangement of the elements of diesel power plants have become widely spread. They also show how the ship electrical network and general consumers are provided with power and what the essentially important armature is.

Diesel power plants with LSE with the direct power drive have acquired the most widespread application on ships. The main engine is rigidly connected with the ship shafting in this case. The propeller thrust is perceived by a special thrust bearing, which is located on the shaft line closer to the main engine, or, as a rule, fixed in the stern of the engine.

All the elements of a typical diesel plant with LSE are connected into a thermal scheme, the simplified image of which is shown in Fig. 3.1.

The propulsion part of the plant consists of the main engine, thrust bearing, shafting with shaft bearings, and propeller. It should be noted that almost all the MPPs with LSE installed on modern ships include only one main engine, which is located in the center of the ship; the use of two LSEs in the power plant is an exception.

The gases that leave the main engine pass through the gas turbocharging unit of the main engine and are fed to the recovery boiler (RB). Steam from the recovery

Z. Yang et al., *Marine Power Plant*,
https://doi.org/10.1007/978-981-33-4935-3_3

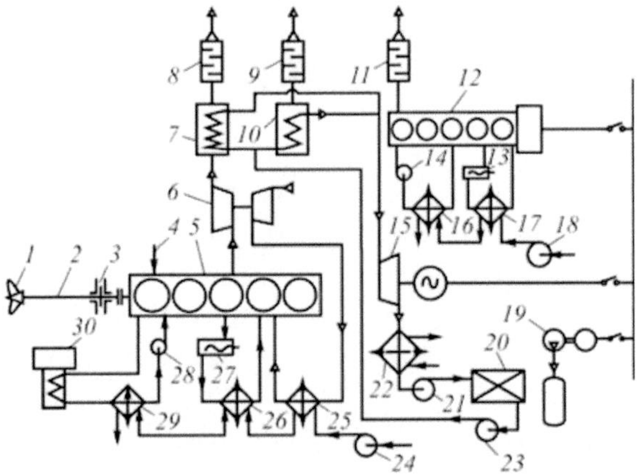

Fig. 3.1 Principal thermal scheme of a diesel plant with the direct drive: *1*—propeller; *2*—shafting; *3*—main thrust bearing; *4*—fuel supply; *5*—main engine; *6*—turbocharging unit; *7*—recovery boiler; *8, 9, 11*—muffler; *10*—auxiliary boiler; *12*—diesel generator; *13, 27*—oil pumps; *14, 28* and *18, 24*—pumps of cooling fresh water and seawater, respectively; *15*—turbine generator; *16, 29*—water-to-water fresh water cooler; *17, 26* and *25*—oil and charge air coolers, respectively; *19*—air compressor; *20*—warm box; *21, 23*—condensing and feed pumps, respectively; *22*—condenser; *30*—vacuum desalinator

boiler is supplied to the steam turbine generator (TG), which transfers the generated electric current to the ship network and general consumers.

The gases coming to the RB have the temperature of 220–260 °C; the higher the efficiency of the main engine is, the lower the temperature is. This allows the recovery boiler to generate the saturated steam at the pressure of 0.5–1.0 MPa and the temperature of 190–210 °C. Gases leave the RB with the temperature close to 180 °C. This temperature can be reduced, which will allow receiving the additional amount of steam and increase the parameters in the RB. However, further reduction of the temperature of the exhaust gases will contribute to the development of corrosion of the tail surfaces of the boiler due to the impact of sulfur compounds formed during the fuel combustion. In case of using corrosion-proof materials and anticorrosion coatings for the gas removal device of the RB, the temperature of the gases that leave the RB can be reduced. The use of liquefied gas as the main engine fuel eliminates these restrictions on the temperature level of the exhaust gases.

The work of the main engine is ensured by a number of systems, some of which are presented in the principal scheme.

The cooling system of the main engine has two circuits: fresh water is cooled directly by the engine (pump *28*), and heated fresh water is cooled by seawater (cooler *29*). The energy of the heated fresh water that leaves the engine provides operation of the vacuum desalination plant *30*. The oil that lubricates the friction surface of the main engine and provides heat removal is cooled by seawater pumped through cooler *26*. The air compressed in the turbocharging unit compressor passes through the charge air cooler before being supplied to the main engine.

The marine power plant consists of diesel generators (only one is shown in the thermal scheme), recovery steam turbine generator and emergency diesel generator (located outside the engine room). The operation of the recovery TG is provided by the condensate-feed system, the elements of which (condenser *22*, pumps *21*, *23*, warm box *20*) are shown in the scheme.

The auxiliary boiler plant of the marine diesel plant under consideration includes an auxiliary boiler and a recovery boiler of the main engine.

Along with the equipment considered in the thermal scheme, there is other equipment: compressor station *19*, SPS units, heat exchangers and system mechanisms that provide operation of the installation.

Diesel power plants with medium-speed engines as the main ones take the second place after LSE as for their distribution on ships (transport, technical and fishing fleet, mixed navigation, river and specialized ships).

MSEs include diesel engines with the rotation frequency of 250–750 rpm. This significantly exceeds the necessary propeller speed; therefore, it is necessary to introduce the gear in the structure of diesel plant with MSE. Combination of the main engines mounted on a common foundation frame with mechanical or hydraulic gears and the connecting and disconnecting elastic couplings is called a diesel-gear unit (DGU). One or two shaft generators are conveniently assembled with the gear of DGU, which somewhat complicates the kinematic scheme of the plant, but gives the opportunity to save fuel for the electric power production in the operation of the main engine and provides additional possibilities for increasing the survivability of the MPP in case of failure of the ME. In addition, such a solution allows reducing the number of diesel generators of SPS and saving its resource.

Reducers and couplings increase the mass and the size of DGU by 25–60% and 30–50%, respectively, but in general they are smaller than those for the plant with LSE. The length and width of the DGU and LSE of the same power are almost the same, however, LSE is almost twice as high. The moderate height of MSE allows them to be used on ships loaded through the aft part, for example on ships with horizontal cargo handling.

The main power plants with MSE and mechanical gears are performed in the single-, double-, three- and four-machine form; they include one or two reducers. The structural schemes of such plants with one common reducer are shown in Fig. 3.2.

Plants with MSE transmitting power to the propeller shaft via the reducer have certain advantages if compared to the plants employing LSE with direct power transmission to the propeller shaft. Thus, the presence of a reducer allows using the engines at the rotation speeds that correspond to the most rational efficiency coefficient of the propeller. At the same time, the operational characteristics of the plant are improved, since some engines can be switched off with the decrease in the speed of the ship, and the remaining ones will operate at full power. Failure of one of the engines does not lead to a stop in the ship movement, and the ability to switch off any engine allows it to be repaired during the voyage.

At the same time, the installations with MSE have certain disadvantages in comparison to those with LSE. Thus, the MSE motor resource is much smaller than the LSE motor resource, and SPPs with MSE have a lower efficiency coefficient

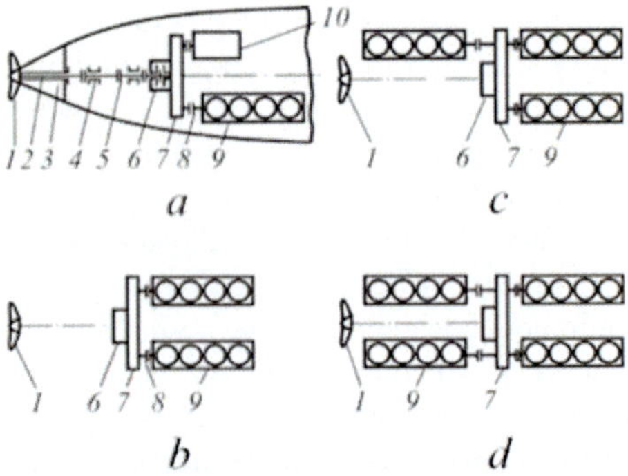

Fig. 3.2 Structural schemes of diesel-reducer plants: **a**—single-machine plant; **b**—double-machine plant; **c**—three-machine plant; **d**—four-machine plant; **1**—fixed pitch propeller; *2, 5*—rowing and intermediate shafts, respectively; *3*—stern tube device; *4*—shaft bearing; *6*– main thrust bearing; *7*—reducer; *8*—flexible disconnector; *9*—main engine; *10*—shaft generator

because of the presence of additional energy losses in the reducer and couplings. Due to the increase in the total number of cylinders, operation of plants with MSE is complicated; such plants have an increased noise level, which leads to the need for additional measures to reduce it. The thermal scheme of the MPP with MSE is completely identical to that of the MPP with LSE, but there is a gear between the ME and the shafting.

Diesel power plants with high-speed engines. HSEs are the engines with the rotation speed higher than 750 rpm; therefore, the propulsive plants necessarily include a downshift to the propulsor. The principal structural schemes of DPP with HSE do not differ from the schemes of DPP with MSE.

High-speed engines have smaller mass and dimensions, lower cost and better maintainability in comparison with MSE. At the same time, they are inferior to MSE by their economical efficiency, resource and require the use of light diesel fuel.

The use of engine-mounted auxiliary machinery (electric generators, air compressors, pumps) is typical for plants with HSE. It simplifies arrangement of the systems and reduces the load of the SPS. In addition, mounted mechanisms can reduce the reliability and maintainability of the plant.

The MPP with HSE as the main engine have acquired the most widespread application on fishing seiners, river fleet vessels, port tugboats, boats and hydrofoil ships. They are also used on all types of vessels as drives of auxiliary and emergency current generators.

3.2 Classification and Marking of Marine Diesels

Classification of marine diesel engines. According to their purpose, marine diesel engines are divided into *main* and *auxiliary* ones (Fig. 3.3). The main engines drive propeller shafts with propellers and ensure the movement of the ship. The auxiliary

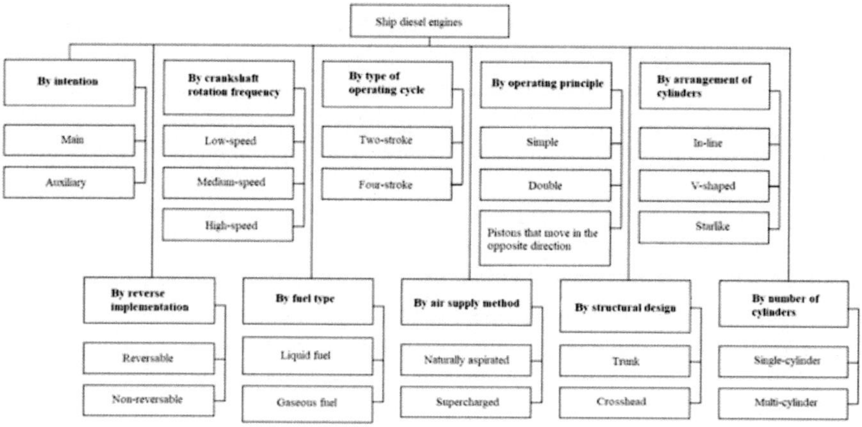

Fig. 3.3 Classification of marine diesels

ones are designed for driving the electric current generators of auxiliary SPS, pump drives and other mechanisms. In addition, diesels can be *reversible*, where the direction of rotation of the crankshaft can vary, and *non-reversible* with a constant direction of rotation. Depending on the rotation speed of the crankshaft, diesel engines are divided into LSE (above 250 rpm), MSE (up to 750 rpm) and HSE (above 750 rpm).

Marine diesels can operate on *liquid* and *gaseousfuels*. The former are the engines which can work on different fuels including gas without structural modifications. They are called bi-fuel engines. Gas engines are the diesel engines where the main fuel is gaseous, and liquid fuel in a small amount is used for ignition.

The whole variety of marine diesel engines is divided into *four-* and *two-stroke* engines. In the former, the entire operating cycle is carried out in four strokes of the piston, in the latter—in two strokes.

Diesel engines are also divided according to the nature of filling the cylinder with air into *naturally aspirated* and *supercharged* engines where the air is fed into the cylinders under pressure.

There are *simple-* and *double-acting* diesel engines. In the former, the processes of the working cycle take place in the cylinder on one side of the piston. Meanwhile, in the latter they take place alternately on both sides of the piston, i.e. the fuel is injected first into the top of the cylinder (the working stroke is directed from the top dead center (TDC) to the bottom dead center (BDC), and then into the bottom (the working stroke is directed from BDC to TDC). In each cylinder of the *diesel engines with pistons that move oppositely*, there are two pistons moving vertically in opposite directions and forming a common combustion chamber in the middle part of the cylinder. Each piston rotates the crankshaft through a connecting rod; the power from both shafts is transmitted to one propeller via reduction gear.

According to the design of the crank mechanism, all marine diesel engines are divided into *trunk* and *crosshead ones*. In the trunk engine, the normal pressure forces that occur when the connecting rod is tilted are transmitted to the cylinder by the lower piston guide part—the trunk.

In the crosshead engine, the normal force that occurs in the crosshead connection is transmitted by the sliders to the parallel guides fixed to the engine bed. All MSEs and HSEs are trunk engines, and modern LSEs are only performed in the crosshead form.

Depending on the number of cylinders, diesel engines can be single- or multi-cylinder. Cylinders of multi-cylinder diesel engines can be arranged in one row (*in-line*), in the form of the V letter (*V-shaped*) or a star (*star-shaped*). Single-cylinder engines on ships are used to drive propellers of the lifeboats and workboats. The maximum number of cylinders in modern marine diesel engines is limited to 20, and there are in-line diesel engines with 14 cylinders. The maximum diameter of the cylinder is 980–1080 mm.

There are other characteristics that marine diesels are classified by, but even the above mentioned classification indicates the diversity of types of diesel engines used on ships.

Marking of marine diesel engines. There is no single designation for diesel engines throughout the world, therefore each diesel engine company uses its own systems to mark its engines.

The following are the designations of marine diesel engines produced by major foreign companies.

Low-speed engines.

MAN Diesel and Turbo Company

Until 2013, the company had been called MAN B & W Diesel Group. The structure of the engine marking is shown in Fig. 3.4.

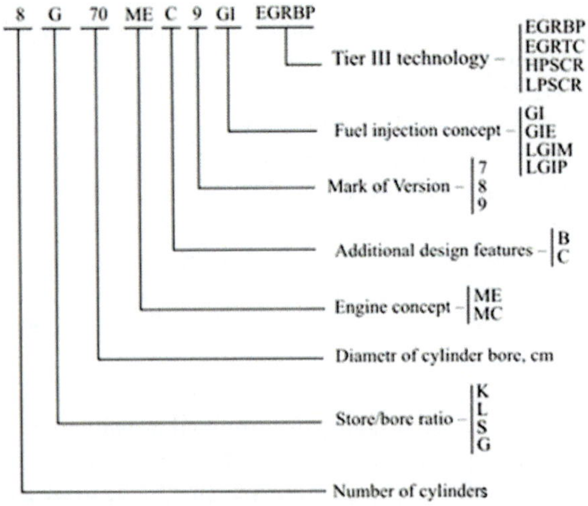

Fig. 3.4 Structure of marking of MAN Diesel and Turbo low-speed engines

The number of cylinders is given in the first block of the marking structure. It is 5–14 for the MAN Diesel and Turbo LSE.

In the second block of the marking structure, the letter designations correspond to the following values of the ratio of the stroke length to the cylinder diameter *S/D*: K—2.65–3.2; L—3.37; S—3.54–4.42; G—4.65.

In the fourth block, the concept of engines is given: MC stands for an engine with mechanically controlled fuel injection, lubrication of cylinders, exhaust valves, and starter valve. ME is an electronically controlled engine for these functions.

The fifth block of the structure contains additional information about the engine design: B—control of the engine exhaust valve with electronic control from the camshaft; C—compact engine.

In the sixth block of the structure, the numbers 7, 8, 9 indicate the engine version depending on the mean effective pressure in the cylinder p_e.

The seventh block of the marking structure contains information about the use of gaseous fuel. The absence of any designation indicates that the engine only works on liquid fuel. The designation GI (Gas injection methane) corresponds to the use of natural gas, methane; GIE stands for Gas injection ethane, LGIM—for Liquid gas injection methanol, LGIP—for Liquid gas injection LPG.

The final eighth block contains information about the use of technologies in the engine that ensure the compliance of its characteristics with the requirements of the Rules of the International Maritime Organization Tier III.

The absence of any data in this block indicates the compliance of the engine with the requirements of Tier II. Legend: EGRBP—EGR (Exhaust Gas Recirculation) with bypass matching; EGRTC—EGR with turbocharger cut-out matching; HPSCR—High-pressure SCR (Selective Catalytic Reduction); LPSCR—Low-pressure SCR.

Wärtsilä Corporation

Along with this, in recent years the company has been producing engines with the following designations: Wartsila X35, Wartsila X40, Wartsila X62, Wartsila X72, Wartsila X82, Wartsila X92, where the number corresponds to the cylinder diameter in cm. All these engines are controlled electronically; they have a higher efficiency coefficient if compared to the engines like Wartsila RT-flex. The number of cylinders is indicated after the brand mark: Wartsila 8X72 (Fig. 3.5).

Mitsubishi Heavy Industries Marine Machinery and Engine Co

See Fig. 3.6

Medium-speed engines.

MAN Diesel and Turbo

See Fig. 3.7.

WärtsiläCorporation

See Fig. 3.8.

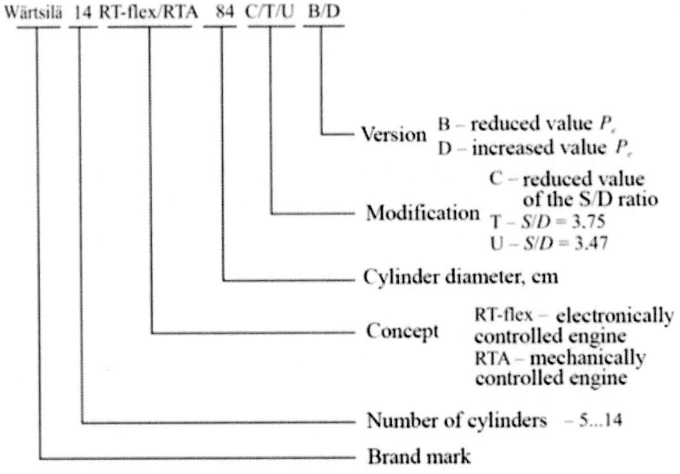

Fig. 3.5 Structure of marking of Wärtsilä Corporation low-speed diesel engines

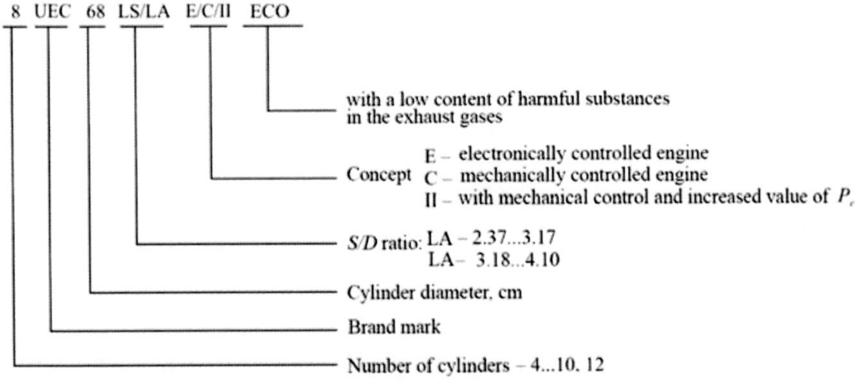

Fig. 3.6 Structure of marking of MNI low-speed diesel engines

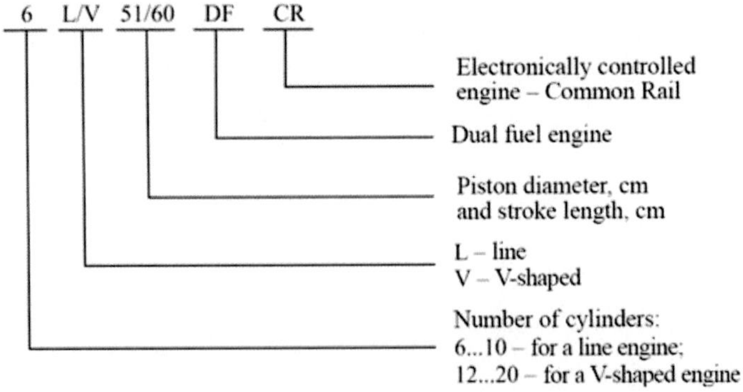

Fig. 3.7 Structure of marking of MAN Diesel and Turbo medium-speed engines

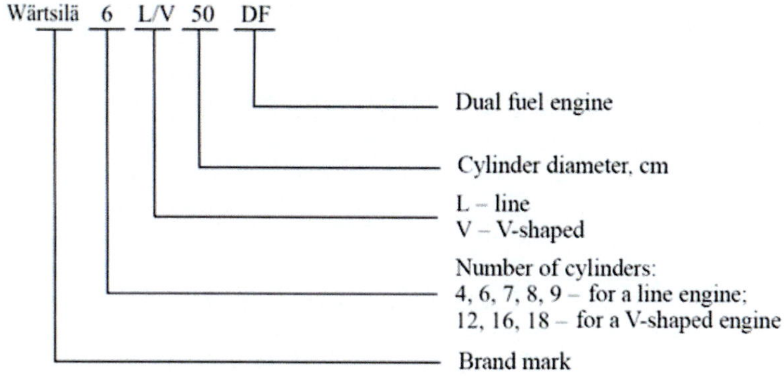

Fig. 3.8 Structure of marking ofWärtslilä Corporation medium-speed engines

Caterpillar Marine Power Systems

See Fig. 3.9.

3.3 The Operation Principle of Marine Diesels

Here are the general definitions that are needed to explain the operation principle of marine diesel engines.

Top and bottom dead points (TDP and BDP) are the points that correspond to the top and the bottom end positions of the piston in the cylinder.

Stroke of the piston is the distance that the piston travels from one extreme position to the other.

Cylinder capacity is the volume separated by the piston as it moves between dead points.

Working cycle is the set of different processes that take place in the ICE cylinder in a certain sequence. While the engine is running, they are periodically repeated.

Indicator diagram is a diagram of pressure change in the cylinder within the stroke of the piston per cycle. It is obtained during testing of the engine with the help of a special indicator device or built according to the results of the calculation of the working cycle. In such a diagram, the ordinates correspond to the gas pressure in the cylinder, and the abscissas correspond to the piston stroke and the appropriate cylinder volumes.

The horizontal thin line plotted on the diagram characterizes the pressure in the exhaust pipe; the vertical lines indicate the extreme points of the piston position (TDP and BDP).

Four-stroke diesel engines. Let us consider the scheme of operation of the four-stroke diesel engine.

The first stroke is inlet or filling. At the beginning of this stroke the piston is in a position close to TDP. The combustion chamber is still filled with combustion

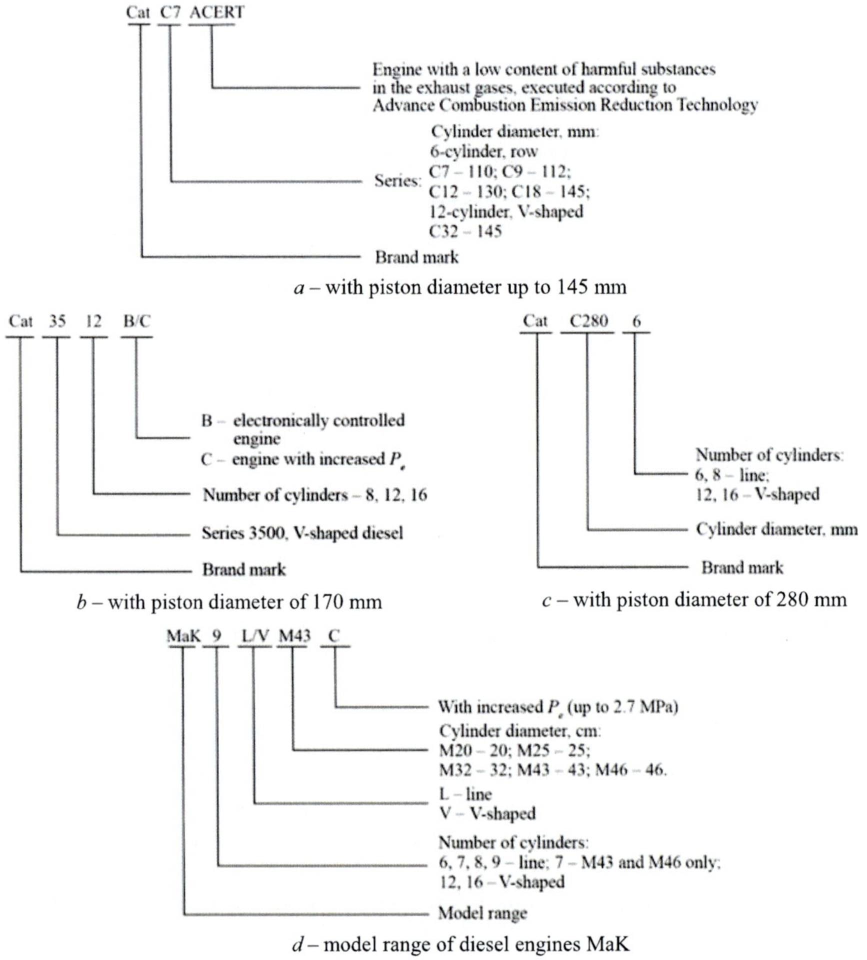

Fig. 3.9 Structure of marking of Caterpillar Marine Power Systems medium- and high-speed engines

products from the previous process with the pressure somewhat higher than the atmospheric pressure.

In the indicator diagram (Fig. 3.4), the initial position of piston *1* corresponds to the point *r* [1]. During the rotation of the crankshaft in the direction of the arrow (see Fig. 3.10), the piston moves to BDP, and the distributor opens inlet valve *2* and connects the over-piston space of the engine cylinder with the inlet pipeline through inlet pipe *3*, so that the cylinder is filled with air. In the indicator diagram, the inlet stroke corresponds to the line *ra*.

In the inlet pipeline, the pressure is close to atmospheric pressure (in the naturally aspirated engines) or above (p_c = 0.13–0.45 MPa), depending on the degree of

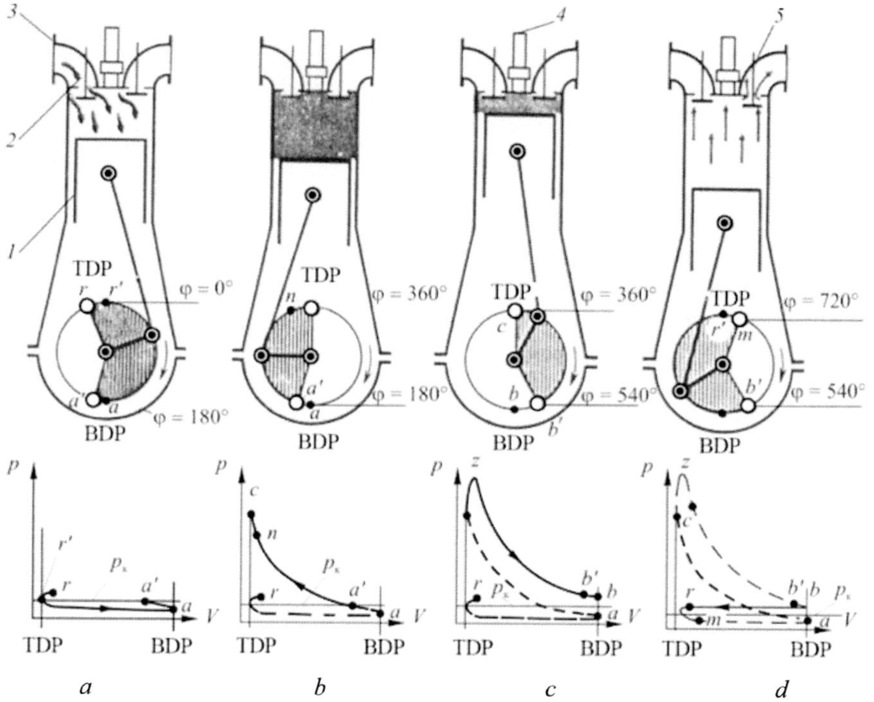

Fig. 3.10 Scheme of the four-stroke engine operation: **a**—first stroke (inlet); **b**—second stroke (compression); **c**—third stroke (combustion and expansion); **d**—fourth stroke (outlet)

pressurization. The pressurization increases the cylinder charge and engine power, but at the same time the pressure and temperature of the cycle also increase. The inlet valve opens a bit earlier (point *r*) than TDP (point *r'*) (i.e., with the advance angle $\varphi_{r\text{–}r'} \approx 20\text{–}50°$), which creates more favorable conditions for the air inlet at the beginning of the filling. The inlet valve closes after BDP (point *a*) with the retarding angle $\varphi_{a\text{–}a'} \approx 20\text{–}50°$.

The second stroke is compression. At the further piston movement to TDP from the moment of inlet valve closing, fresh air charge is compressed, so that its temperature rises to the level required for self-ignition. For the best use of heat released during combustion, it is necessary that the fuel combustion ends when the position of the piston is as close to TDP as possible. Therefore, fuel is injected into the cylinder with some advance (angle $\varphi_{n\text{–}c} = 1\text{–}16°$), which is necessary to prepare it for self-ignition at the time of the piston arrival in TDP. Thus, at the beginning of the second stroke, charging of the cylinder continues, the charge is compressed, and the fuel combustion begins at the end. In the indicator diagram, the second stroke corresponds to the *ac* line.

The third stroke is the working stroke (combustion and expansion). During the piston movement from TDP to BDP, the fuel sprayed by nozzle *4* and mixed with hot air ignites and burns, which results in a sharp increase in the pressure of the gases (point *z*) and their further expansion. It is in this stroke that gases, acting on the piston, carry out useful work, which is transmitted through the crank mechanism to the consumer of energy. The process of expansion is coming to an end at the time of the start of the opening of exhaust valve *5* (point *b*), which occurs with the advance angle $\varphi_{b\text{–}b'} = 20\text{–}40°$.

During expansion, gases perform useful work, so the third stroke is also called *the working stroke*. In the indicator diagram, the third stroke corresponds to the line *czb'*.

The fourth stroke is outlet. Only the combustion and expansion stroke is a working one; the other three strokes take place due to the kinetic energy of the crankshaft with the flywheel and the operation of other cylinders.

The better the cylinder is cleaned from exhaust gases and the more fresh charge comes to it, the more useful work you can get per cycle.

During the movement from BDP to TDP, the piston pushes exhaust gases out of the cylinder. For their complete removal, the exhaust valve closes after the piston passes TDP (the closing lag angle is $\varphi_{r'\text{–}m} \approx 10\text{–}60$ of the crankshaft revolution). Due to this, during the time that corresponds to the angle $\varphi_{r\text{–}m'} \approx 30\text{–}110$ of the crankshaft revolution, the inlet and exhaust valves are simultaneously open in the cylinder. This position of the valves is called shutting off of the valves. It contributes to the improvement of the filling of the cylinder due to the ejecting effect of the gas flow in the outlet pipeline. This improves the process of the combustion chamber cleaning from exhaust gases, so the charge air pressure is higher than the pressure of exhaust gases in a supercharged diesel during this period. The exhaust valve is open during 210–280° of the crankshaft revolution. In the indicator diagram, the fourth stroke corresponds to the line *b'rm*. The fourth stroke is the end of the working cycle. With the further movement of the piston, all the processes of the cycle are repeated in the same sequence.

The useful work received during the cycle is determined by the *aczba* area of the indicator diagram.

A graphic illustration of the opening and closing of the inlet and exhaust valves is a circular diagram of the phases of gas distribution (Fig. 3.11), which shows the opening and closing moments of the valves, as well as the angles corresponding to individual phases. When considering the scheme of the diesel engine, it has been found that the phases of gas distribution do not coincide with TDP and BDP of the piston. The correct choice of the phases of gas distribution largely affects the power and economic efficiency of the engine. The final setting of the gas distribution phases is carried out when the engine is mounted to the test stand and adjusted in accordance with the obtained indicator diagrams.

Two-stroke diesel engines. Thus, a four-stroke engine operates as a heat engine only half of the cycle duration (compression and expansion strokes). In two-stroke engines, the working cycle occurs in two strokes, i.e. in one revolution of the crankshaft. The time allotted to the working stroke is used more rationally. Unlike four-stroke engines, in this case cleaning of the working cylinder from combustion products and filling it with a fresh charge (gas exchange) occur only when the piston

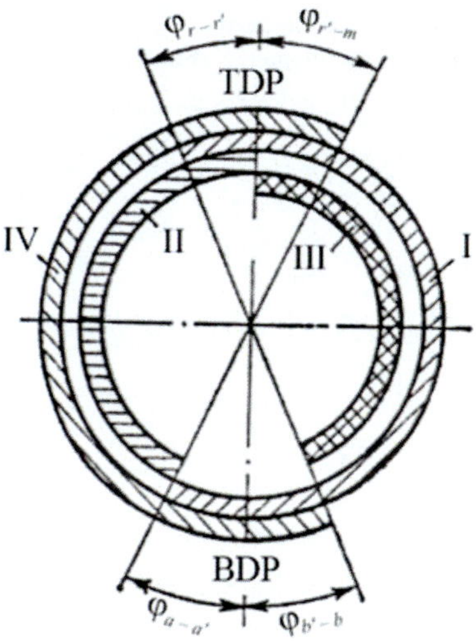

Fig. 3.11 Circular diagram of the phases of gas distribution of a four-stroke diesel engine: I—inlet; II—compression; III—combustion and expansion, IV—outlet

moves near BDP. Cleaning of cylinders from the exhaust gases is performed by displacing them with air previously compressed in the turbocharging unit. In the process of gas exchange in two-stroke engines, a certain part of the air is inevitably removed from the cylinder along with the exhaust gases through the exhaust bodies.

Figure 3.12 shows the operation scheme of a two-stroke engine with a direct-flow valve scavenging [1]. The structure of this engine is distinguished by the following features:

- inlet ports are located in the lower part of the cylinder; their height is 10–20% of the piston stroke, and their opening and closing are carried out by the piston when it moves in the cylinder;
- exhaust valve *3* (modern ICEs employ one exhaust valve) is located in the cylinder cover; it is driven from the camshaft, the rotation speed of which ensures one opening of the valve per revolution of the crankshaft;
- air compressor (driven by a gas turbine that operates on the diesel exhaust gases) injects air under pressure into the receiver to clean the cylinder from the combustion products and fill it with a fresh charge.

The working cycle in the engine is as follows.

The first stroke corresponds to the piston movement from BDP to TDP. At the beginning of the stroke of piston *4*, scavenging ports are overlapped. The end of cylinder scavenging and filling with a fresh charge (point k) is determined by the moment of closing of the inlet ports and the exhaust valve. The valve closes simultaneously with the inlet ports or somewhat earlier. The pressure in the cylinder at the end of gas exchange is above atmospheric and depends on the pressure in the receiver. After this, there starts the process of air compression in the cylinder, which ends when the

Fig. 3.12 Scheme of a two-stroke engine: **a**—first stroke; **b**—second stroke

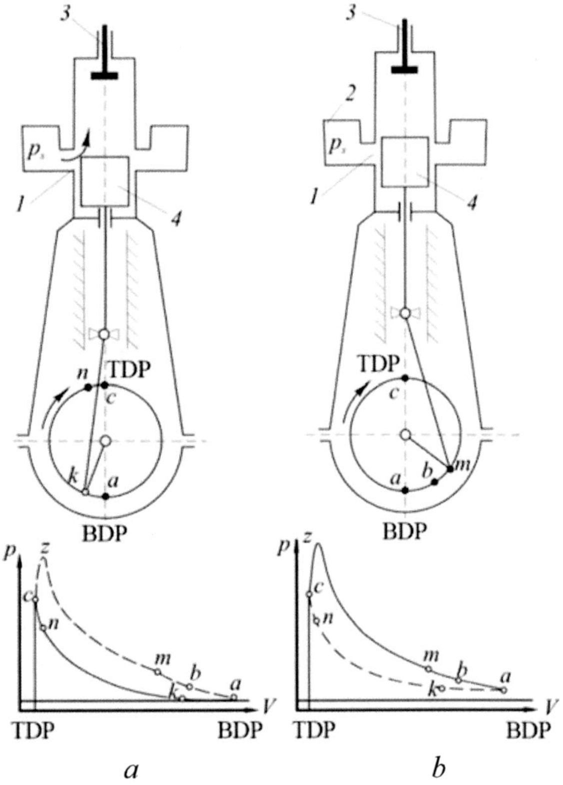

piston comes to TDP (point c). Point n, located at 10–30° before TDP, corresponds to the moment of the fuel injection by the nozzle (the latter is not shown in the scheme).

Thus, during the first stroke, the discharging, scavenging and filling of the cylinder end, the fresh charge is compressed, and fuel injection starts.

The second stroke is the piston movement from TDP to BDP. In the TDP area, injected fuel ignites and burns. The pressure of the gases acquires the maximum value (point z), and their expansion begins, i.e. the working stroke is carried out. A little before the moment when the piston approaches the inlet ports, the exhaust valve opens and the combustion products begin to flow out of the cylinder into the outlet pipe, with the pressure in the cylinder dropping sharply (section *mb* in the indicator diagram). Inlet ports *1* are opened by the piston when the pressure in the cylinder is equal to the pressure of the compressed air *ps* in receiver *2* or is a little above it. The air that enters the cylinder through the inlet ports displaces the combustion products that remain in the cylinder through the exhaust valve and fills the cylinder; i.e. gas exchange is carried out (section *ba* in the indicator diagram).

Thus, during the second stroke in the cylinder, fuel is burned, gases are expanded, exhaust gases are produced, scavenging and filling of the cylinder occurs.

The efficiency of two-stroke engines largely depends on the adopted gas exchange scheme. Direct-flow valve (discussed above) and contour valve have received widespread use in the modern two-stroke ICE.

Design of the vast majority of modern ship low-speed two-stroke ICE provides for the use of a direct-flow valve scavenging.

The indicator diagram of a two-stroke engine (see Fig. 3.12) is the same both for naturally aspirated and supercharged diesel engines. The useful work of the cycle is determined by the area of the diagram *aknczmba.*

3.4 The Main Structural Elements, Units and Systems of Ship Internal Combustion Engines

In the design of the ship ICE (both two-stroke and four-stroke), the following basic mechanisms, systems and groups of parts can be distinguished [1, 2].

The body of the engine includes fixed parts that support the movable elements of the crank mechanism and perceive all the efforts while the engine is running: foundation frame, cylinders, cylinder covers, anchorages, pins and bolts that pull these parts together.

The foundation frame is the base of the body on which the entire engine is assembled. Separate foundation frames are used mainly for LSE and MSE; they are not used in HSE.

The foundation frame contains the basic (frame) bearing shells constituting the support of the crankshaft.

The most important elements of the body are *cylinders*—the elements of the engine where the working cycle occurs. The cylinder consists of a *jacket* and an insertion liner of the working cylinder. In the liner, the piston moves and work processes take place. The jacket is the support for the liner and forms a cooling cavity for it. The inner walls of the cylinder liner—*cylinder mirror*—are the guides for the piston in its reciprocating motion.

In four- and two-stroke marine diesel engines, the cylinders are structurally designed in the form of box-like units with vertical bulkheads between individual cylinders, as well as cavities for cooling water, scavenging air supply and exhaust gas removal (only for two-stroke diesel engines).

The inner part of the cylinder, limited by the cylinder cover and the bottom of the piston, forms a combustion chamber.

During the operation of a diesel engine, the cover is exposed to significant gas pressures and temperatures. In most ship ICEs, the cover has a box-like structure, the shape of which is determined by the engine type, the design of the combustion chamber, the number of operating valves, the shape of the supply channels, and the arrangement of the nozzles. For HSE, the covers can be made block-wise for the entire engine or for a group of two or three cylinders; the LSE and MSE covers are always carried out individually for the convenience of installation.

Modern LSE covers are usually equipped with a single exhaust valve, while MSE covers are equipped with two or four of them. In addition to the operating valves, nozzles, an indicator valve (for LSE and MSE), and a start-up valve are placed in the covers. To reduce the temperature of the heated parts, the covers are cooled with fresh water.

The crank mechanism is designed to convert the reciprocating motion of the piston into the rotational motion of the crankshaft. When the engine is running, the driving force P_p impacts the piston, and one of its components Q (the projections of this force T and Z) is directed along the connecting rod, the other (N) is normal and perpendicular to the axis of the cylinder (Fig. 3.13) [2].

The composition of the crank mechanism and the place of application of the normal force depend on the design features of the trunk and crosshead engines. The crank mechanism of the trunk engines includes piston *3*, connecting rod *2* and crankshaft (see Fig. 3.13a). The crosshead engines include piston *2*, stem *3*, crosshead *4*, connecting rod *5* and crankshaft *1* (see Fig. 3.13b).

Translational motion of the piston in the trunk engine is converted into rotational motion of the crankshaft by means of a connecting rod linked by a hinged top head to a piston pin and by a lower head to a crankshaft neck.

In trunk engines, the piston is pressed by the normal force against the wall of cylinder *a*, which increases the thermal and mechanical loads of the piston-cylinder pair and accelerates the wear of its surfaces. In the crosshead crank mechanism, the

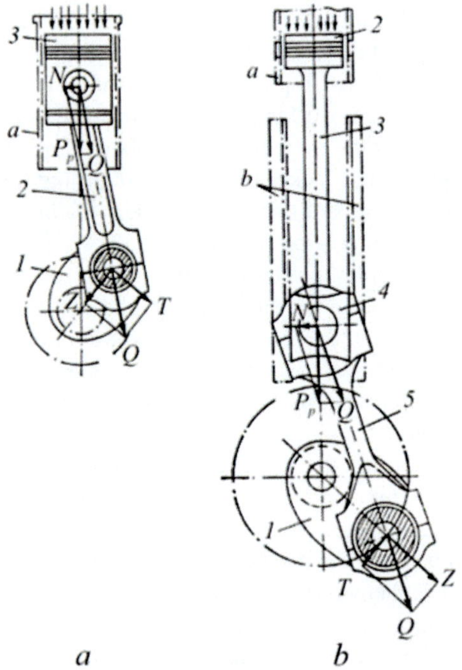

Fig. 3.13 Crank mechanism of trunk (**a**) and crosshead (**b**) engines

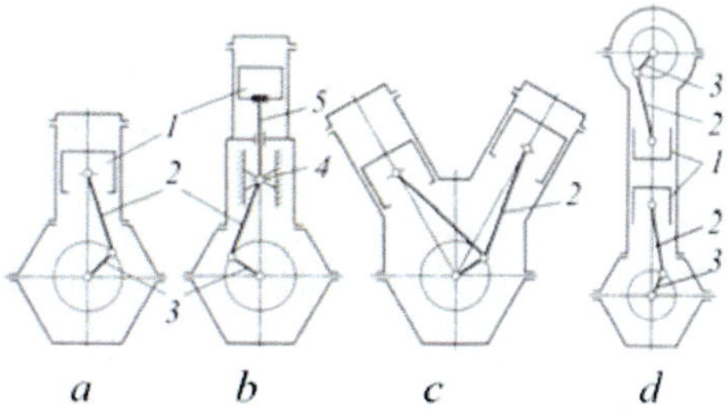

Fig. 3.14 Schemes of crank mechanisms of ship internal combustion engines: *1*—piston; *2*—connecting rod; *3*—crankshaft crank; *4*—crosshead; *5*—stem

piston is connected to the connecting rod by a stem and a crosshead that are rigidly connected to the piston and are characterized by translational motion.

The normal force in the crosshead engines is transferred from the connecting rod through the crosshead consisting of a cross member and a slider onto the parallel guides *b*. This ensures unloading of cylinder *a* from the impact of the normal force. The slider-parallel pair is located beyond the low temperature zone, which contributes to its good lubrication and cooling.

Modern four-stroke engines are made exclusively in the trunk form, and two-stroke low-speed engines are crosshead.

Figure 3.14 [1] shows the most common design schemes of the crank mechanisms of marine diesel engines:

- trunk engine (see Fig. 3.14a);
- crosshead engine (Fig. 3.14b);
- V-shaped engine (Fig. 3.14c);
- engine with pistons that move in the opposite direction and two crankshafts (Fig. 3.14d).

Piston group of crank mechanisms of marine diesels. The piston group includes the piston, piston pin, piston rings, and piston stem (in crosshead engines).

The piston is one of the most responsible and strenuous elements of the engine. It performs the following functions:

- perceives the force of the gases pressure and transmits it through the connecting rod to the crankshaft;
- provides the desired shape of the combustion chamber and the tightness of the intracylinder space;
- transmits the normal force to the cylinder liner (in trunk engines), controls the gas exchange (in two-stroke engines with a direct-flow slot scavenging).

Depending on the type of engine, the crank mechanism of the piston is made solid or compound, cooled or uncooled.

Figure 3.15a [2] shows the structure of a solid uncooled piston of the trunk engine, which consists of head part *2* (sealing) and trunk part *4* (guiding). The head is formed by bottom *1*, a cylindrical wall, and grooves *3* made for sealing rings in the upper part of the lateral surface of the wall. The grooves ensure sealing of the cylinder from the gases breakthrough and oil leakage from the crankcase into the combustion chamber.

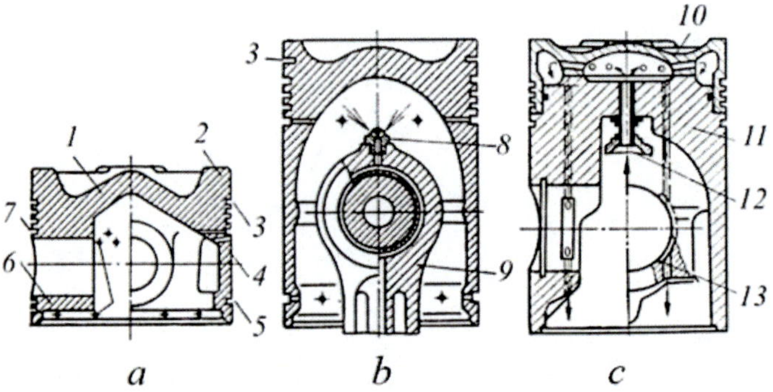

Fig. 3.15 Structures of trunk engine pistons

The trunk part provides direction of the piston movement in the cylinder and transfer of the normal (lateral) force to the engine frame. On both sides inside the trunk part, there are blockings *6* with holes for the piston pin. On the outer surface of the skirt, grooves *5* and *7* are made for the oil-removal rings. The oil which is scraped off by these rings is removed into the crankcase through the holes made in the piston.

Forced cooling of the piston is used to increase heat transfer from the bottom; trunk engines employ oil for that purpose. An example of a cooled piston is shown in Fig. 3.15b. Here, jet cooling is implemented, in which the inner surface of the piston head is washed with oil supplied from nozzle *8* located on connecting rod *9*.

High-capacity ship ICEs make use of cooled compound pistons (see Fig. 3.15c) with detachable head *10* made of heat-resistant steel and trunk *11* made of cast iron or aluminum alloy, connected by long pins. In the above mentioned design, oil is fed into the piston through the connecting rod and sealing cup *12* and is then drained into the crankcase through channels *13*.

The design of the crosshead engine piston is determined by the type of scavenging and the method of cooling.

The piston of the engine with a direct-valve scavenging (Fig. 3.16a) consists of head *6* fastened by the pins to upper flange *4*, piston stem*3* and short guide part *5*. Coolant (this can be either oil or water) is supplied to the piston along the annular channel between tube *2* and stem*3* and removed through pipe *2*.

In the engines with a contour scavenging (see Fig. 3.16b), the piston consists of head *6* and long guide part *7* that overlaps the windows. The coolant is fed into and removed from the piston by telescopic devices *8*. The material for manufacture of the pistons and their elements should have high mechanical characteristics, good thermal conductivity and a low coefficient of linear expansion, be heat-resistant and highly processable. These requirements are met by cast irons, alloyed heat-resistant steels and aluminum alloys.

The piston stem (only in crosshead engines) rigidly connects the piston to the crosshead cross member with flange *1* (see Fig. 3.16) and transmits the gas pressure force through it. It is a steel rod with an annular (see *3* in Fig. 3.16a) or round (see *9* in Fig. 3.16b) cross section.

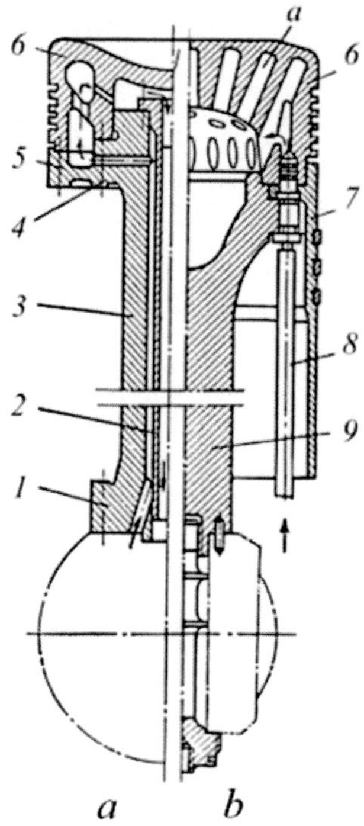

Fig. 3.16 Crosshead engine pistons

According to their purpose, *the piston rings* are divided into sealing (compressing) and oil-removing. The former seal the working gap, thus preventing the gas from bursting into the crankcase during compression and expansion, as well as removing heat from the piston. The ring is a split elastic element with a rectangular, conical or trapezoidal cross section. It should be tightly sealed against the inner surface of the cylinder. For this purpose, the ring is made split; its diameter in the free state is larger than that of the cylinder.

Oil-removing rings are used exclusively in four-stroke engines; the piston is fitted with 1-3 oil-removing rings placed on the head below the sealing rings in the lower part of the trunk. Production of rings employs gray cast iron with lamellar graphite or high-strength cast iron with spherical graphite, which are alloyed with chromium, nickel, molybdenum, copper, vanadium.

Parts of the crank group. They include connecting rods, upper and lower head shells, connecting rod caps, connecting rod bolts and nuts.

The purpose of the connecting rod, which links the piston or cross member of the crosshead to the crankshaft, is to ensure displacement of the piston at working and auxiliary strokes. The connecting rod is subject to the impact of the gas pressure force, the inertia forces of the masses characterized by translational motion, and the inertia forces that arise when the connecting rod is rocked.

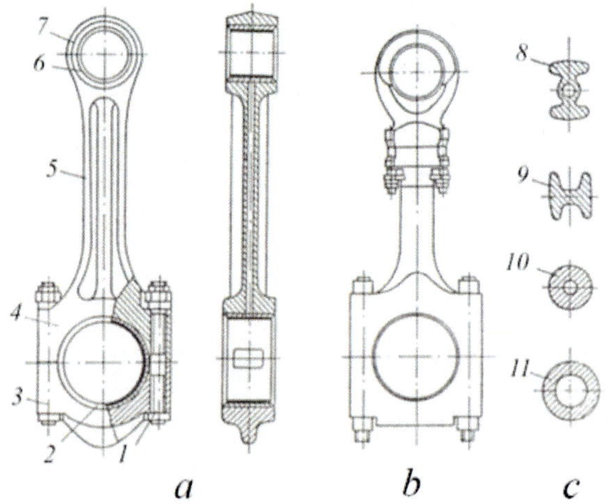

Fig. 3.17 Connecting rods of trunk engines: *a, b*—with solid and detachable rods; *c*—rod cross sections

The main elements of the connecting rod of the trunk engine (Fig. 3.17a) are upper piston head *7*, lower (crank) head *4* and rod *5* which connects them. Rod *5* can have a double-T (*8*), H-shaped (*9*) or round (*10, 11*) cross section (see Fig. 3.17c).

Bearing bush *6* made of antifriction bronze is pressed into the opening of the upper head (see Fig. 3.17a). The lower head carries out the pivot connection of the connecting rod with the crank pin of the crankshaft and forms the connecting rod bearing housing. It is performed with a straight or oblique connector. Detachable cover *3* of lower head *4* is secured by means of connecting rod bolts *1* (see Fig. 3.17a), studs or taper pins. The bearings of crank head *4* are made in the form of steel thin-walled liners *2*, the working surface of which is covered with a layer of babbitt, a lead-tin alloy. The connecting rods are made of carbon or alloyed steels.

Forged connecting rods of crosshead engines are linked to the cross member with upper head *1* (Fig. 3.18a) and to the crank pin of the crankshaft with bottom head *3*. The upper and lower heads are always made with connectors and can be detachable (see Fig. 3.18a) or forged together with connecting rod blade *2* (see Fig. 3.18b). The connecting rod blade has a circular cross section with a central hole for oil supply to the crank bearing. Gasket *4* is designed to regulate the length of the rod and the height of the combustion chamber.

Bearings of the connecting rods have a different arrangement. In the connecting rods with detachable heads, the antifriction alloy is most often poured into the body of the head (see Fig. 3.18a). Meanwhile, in the designs where the heads are non-detachable, the bearings have babbit thin-walled liners (see Fig. 3.18b).

The crosshead (Fig. 3.19) consists of cross member *1* and sliders *3*. The cross member connects piston stem2 and the sliders with a forked head of the connecting rod through trunnions *5* via nut *4* and transmits the force from the piston to the connecting rod. The stem, the slider, and the oil system pipes are fastened onto the treated surfaces of the cross member.

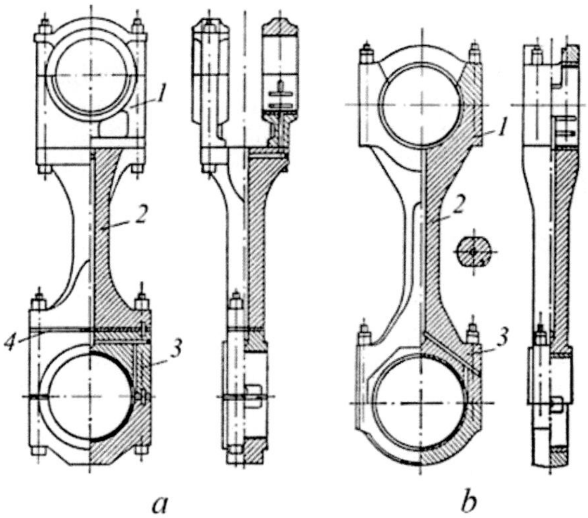

Fig. 3.18 Connecting rods of crosshead engines

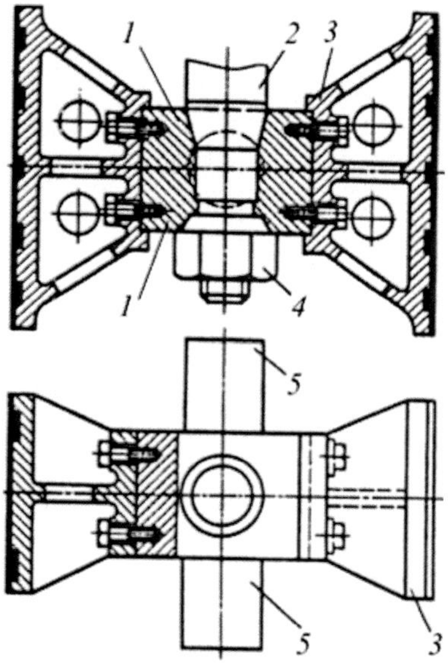

Fig. 3.19 Crosshead

The slider perceives the normal force from the connecting rod and transmits it to the parallels, ensuring rectilinear movements of the upper head of the connecting rod.

The working surfaces of the sliders are poured with Babbitt; they are provided with longitudinal and transverse grooves for distributing oil throughout the entire pouring plane.

The crankshaft is designed to perceive forces from the connecting rods, convert them into a torque and transmit it to the consumer, ensure the movement of the

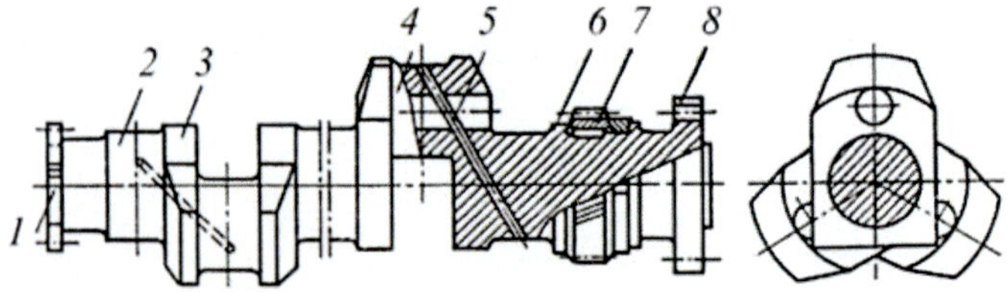

Fig. 3.20 Common crankshaft arrangement

pistons in the auxiliary strokes, transmit the motion to the camshaft, and also drive the auxiliary mechanisms. It perceives the gas pressure forces and the inertia forces of the masses of translational and rotational motion, as well as bending and torque moments from these forces. This leads to deformation in its elements, stress concentration, friction and wear of its stem and bearings.

Periodically alternating torques cause torsional oscillations of the shaft, which increase the stresses in its elements, thus leading to fatigue failure [2].

The main elements of the crankshaft (Fig. 3.20) are crankpins *4* and main journals *2*, cheeks *3* connecting them, the aft and the bow ends of the shaft. The aft end of the shaft ends with flange *8* to take power to the consumer. At this end, there is pinion *7* (an asterisk for driving a camshaft) and clamp *6* for an installed frame bearing. Pinion *1* secured to the nose end of the shaft is intended for driving the mounted auxiliary mechanisms. A flywheel can be located at the nose end of the shaft, providing the necessary uniformity of rotation of the crankshaft. In high-capacity MSE and LSE, the moments of inertia of the rotational masses provide the necessary uniformity, therefore, in these engines, there is a gear of a shaft-turning device instead of the flywheel on the nose end of the crankshaft.

The crankpins together with the cheeks form the cranks, the number of which corresponds to the number of cylinders. The main journals link the cranks located relatively to each other at the angle $\alpha = (360/i)z$, where i is the number of cylinders; $z = 1$ for two-stroke engines and $z = 0.5$ for four-stroke solid engines. This arrangement of the cranks provides a certain sequence of flashes in the cylinders, obtaining a uniform torque and balance of the engine.

Oil is fed on the frame necks from the engine's circulation system under pressure. It flows through the holes in the shaft elements from the frame bearings to the connecting rod. Sometimes brass tubes *5* are inserted into these holes.

The HSE and MSE crankshafts are made solid, while those of LSE are composed of several parts connected by flanges. Separate cranks of the shaft are made solid, semi-composite, composite and welded.

Crankshafts are manufactured by forging or casting from carbon-bearing or alloyed steels, sometimes from modified cast iron.

The gas distribution mechanism is designed to control the processes of air inlet and exhaust gas release in accordance with the accepted gas exchange phases. Valve gas distribution systems are the most widespread ones in the ship internal combustion engines.

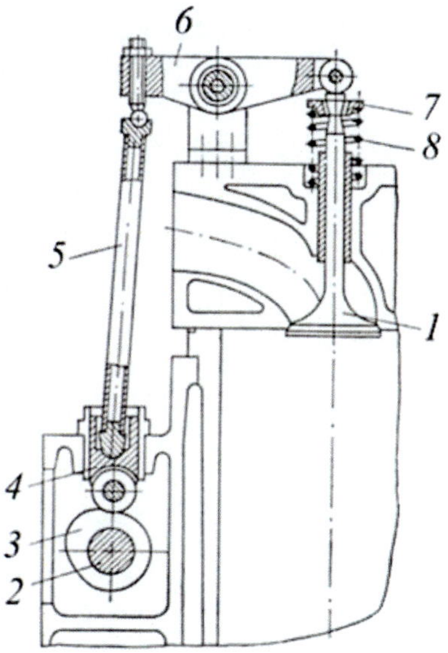

Fig. 3.21 Valve mechanism of gas distribution

At the valve gas distribution in four-stroke engines, the inlet and outlet openings in the covers are opened and closed by appropriate valves. The slide distribution is used in two-stroke engines with direct-flow slot scavenging, where the inlet and outlet ports are overlapped by a piston that functions as a slide. The combined valve-slide gas distribution is used in two-stroke ICEs with direct-valve scavenging.

The valve mechanism (Fig. 3.21) includes: valves *1* (inlet and outlet), their drive parts (pushers *4*, rods *5*, levers *6*), camshaft *2* with cam rings *3*, valve springs *8*, and plates *7*.

Two-stroke LSEs most often have one valve located in the center of the cylinder cover, while four-stroke MSEs employ two or four valves. The valves can be directly located in the cylinder cover (Fig. 3.22a) or installed in removable housing *9* (see Fig. 3.22b). The valve consists of plate *1*, mount rod *2*, guide bushing *3*, valve spring *4* and its plate *5*, spring plate lock *7* in the form of split conical bits, cap *6*, and valve seat *8*.

To increase the service life and equalize the temperature field, the exhaust valves are rotated by means of impeller *10* set into motion by the gas flow. The exhaust valves, which withstand temperatures up to 700–800 °C, are made of high-quality heat-resistant steels.

The camshaft controls the valves. It consists of a shaft with supporting necks, cam rings for valve drives, a high-pressure fuel pump, a gear, and the drive of the shaft itself. The shaft is located along the engine in special cavities outside the crankcase.

In modern LSEs, the mechanical drive of the exhaust valve has been replaced by a hydropneumatic (Fig. 3.23) [2] or hydraulic one. Valve *1* is opened by the oil pressure of about 30 MPa, which acts on servoengine piston *3* pressed onto the shank

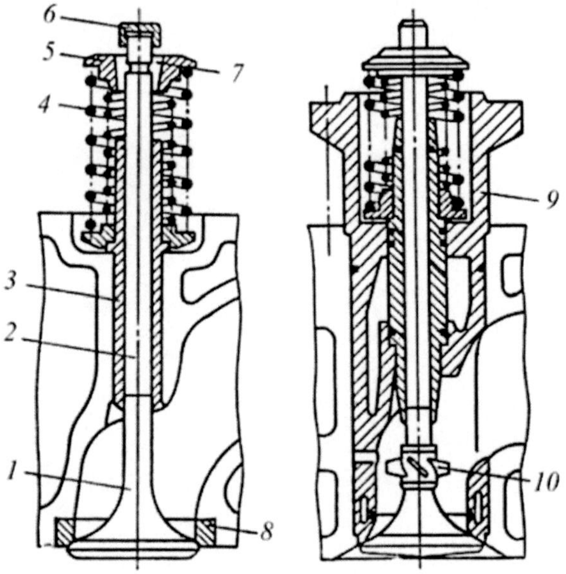

Fig. 3.22 Working valves

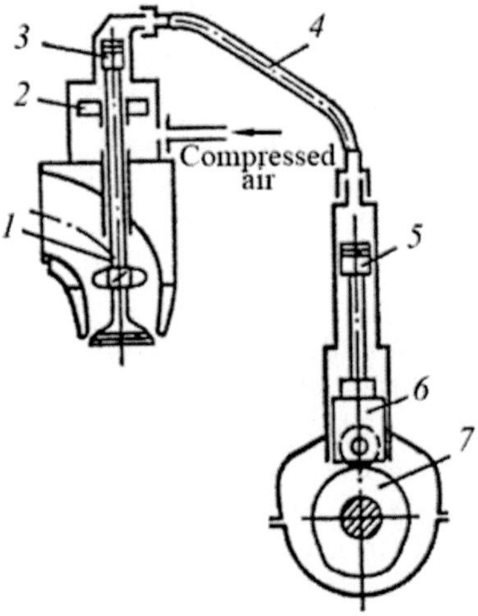

Fig. 3.23 Exhaust valve of a low-speed engine with hydropneumatic drive

of the mount rod. The oil is fed to the servoengine via pipeline *4* by piston *5*, which is driven through pusher *6* by cam plate *7*.

To close the valve, the pressurized air of about 2 MPa is fed to piston *2* connected to valve *1*. The use of the hydropneumatic drive improves the reliability of operation, reduces noise, wear, and increases the valve service life.

Charging units of diesel engines. Modern diesel engines are equipped with *turbocharger units* are used to provide scavenging and charging; they consist of a gas turbine and a centrifugal compressor located on the same shaft. The energy required to compress the air is supplied by the gas turbine. Turbochargers can be made according to different design schemes, the most common of which are shown in Fig. 3.24. Both sliding and rolling bearings are used as rotor bearings. In the modern designs of turbochargers, sliding bearings have an advantage because they have a higher durability.

The widest application for ICE charging was acquired by centrifugal compressors (Fig. 3.25). The operation of a centrifugal compressor is characterized by the air flow rate G_a,the degree of pressure increase π_c and the compressor efficiency η_c. In modern turbochargers, $\pi_c = 1.4$–4.5; $\eta_c = 0.78$–0.82.

The gas turbine of a turbocharger can be either axial or radial. Of the radial turbines, centripetal turbines are most often used in turbochargers; the gas there moves radially from the periphery to the center and, after making a 90° turn, leaves the turbine in the axial direction (Fig. 3.26).

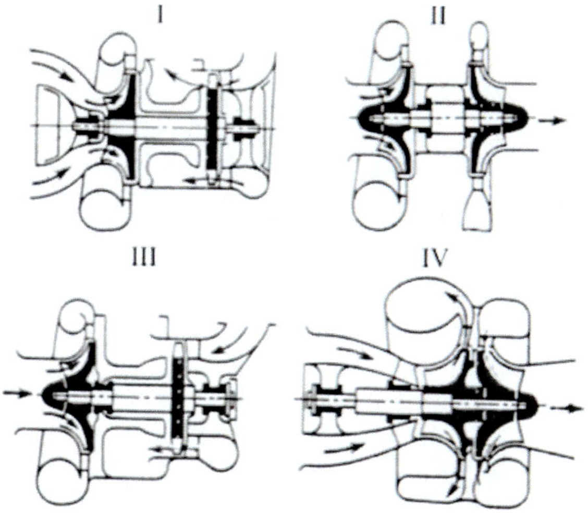

Fig. 3.24 Design schemes of turbochargers of ship internal combustion engines: I—non-console scheme (centrifugal compressor, axial turbine); II—two-console scheme (centrifugal compressor, centripetal turbine); III—one-console scheme (centrifugal compressor, axial turbine); IV—monorotor scheme (centrifugal compressor, centripetal turbine)

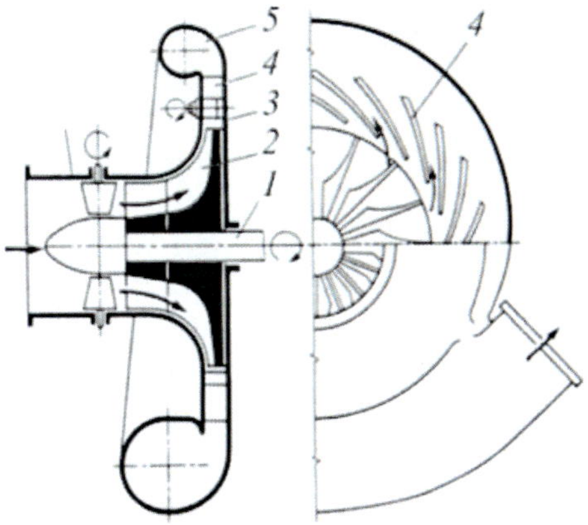

Fig. 3.25 Scheme of a centrifugal compressor: *1*—shaft; *2*—impeller; *3*—compressor case; *4*—directing device; *5*—snail

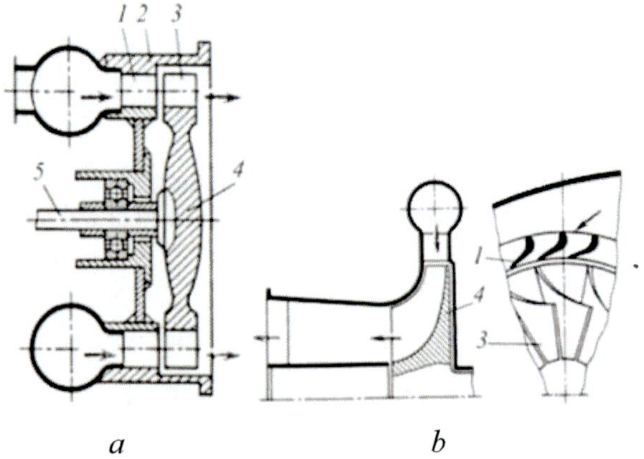

Fig. 3.26 Schemes of axial (**a**) and centripetal (**b**) single-stage gas turbines: *1, 3*—nozzle and working blades; *2, 5*—body and shaft of the turbine; *4*—impeller

Figure 3.27 shows the general view of the turbocharger of the ship ICE [3]. One, two or three turbochargers can be installed on the internal combustion engine, depending on its power.

As a result of air compression, the air temperature rises in the compressor of the charging unit, the mass filling of the cylinders with a fresh charge decreases, the heat stress of the engine increases. In order to avoid the negative consequences, the ship ICE cools air before feeding it into the cylinder.

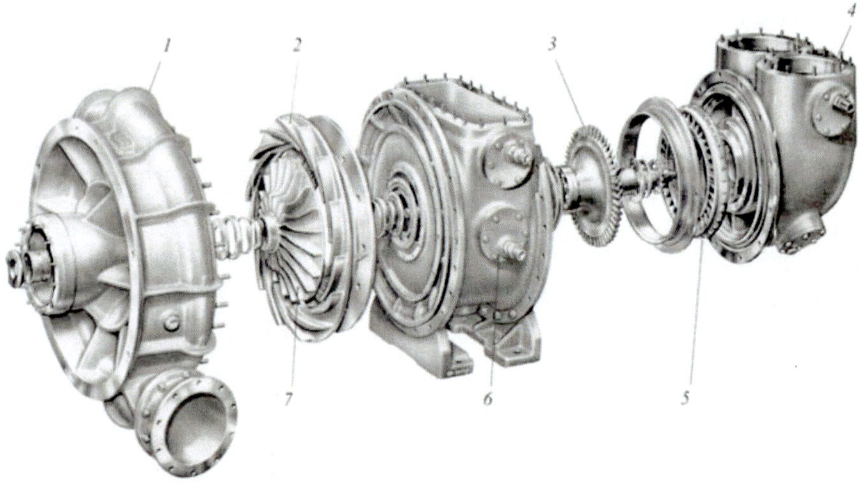

Fig. 3.27 General view of the turbocharger of the ship internal combustion engine: *1, 2*—body and stator of the compressor; *3, 7*—impellers of the turbine and compressor; *4, 5*—body and nozzle apparatus of the turbine; *6*—oil supply

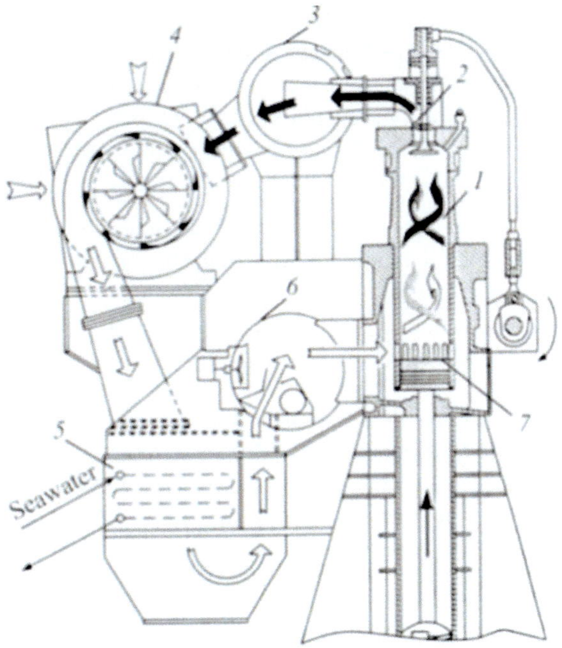

Fig. 3.28 Air and gases movement during scavenging of the cylinder of a working low-speed engine: *1*—engine cylinder; *2*—exhaust valve; *3*—gas exhaust drum; *4*—turbocharger; *5*—charge air cooler; *6*—air receiver; *7*—scavenging port

Ship ICEs implement only surface coolers, where heat transfer from the air to the cooling medium occurs through the interface formed by tubes or plates of a different configuration. The coolers of ship engines use seawater as a cooling medium.

Increase in the efficiency of air cooling can be achieved by evaporating water sprayed before or after the compressor [4]. Figure 3.28 shows the layout of the air supply units of the modern LSE and the directions of air and gas movement during the cylinder scavenging [3].

Systems that provide operation of the ship internal combustion engine

The fuel system is one of the most important systems for a diesel engine, since it supplies fuel to the working cylinders. It consists of low- and high-pressure systems.

The purpose of the low-pressure system is to prepare and supply fuel to the high-pressure system. It includes tanks, filters, pumps, separators, heaters and fuel lines.

Heavy fuel (the vast majority of ship engines operates on it) has a high viscosity and low pour point, so the reserve tanks undergo steam heating, and the fuel lines are isolated. The main method for fuel preparation on sea ships is separation.

The high-pressure system ensures the use of fuel in the combustion chamber of a diesel engine; it consists of a high-pressure fuel pump and nozzles connected by a high-pressure fuel line.

The high-pressure fuel system of an internal combustion engine should provide [2]:

- dosed cyclic fuel supply;
- predetermined phases of gas distribution (i.e. fuel supply at certain moments of the working cycle and for the same interval of the crankshaft rotation angle);
- high-quality fuel spraying (setting its high pressure before the spray holes) in all modes of operation of the diesel engine, taking into account small loads and idle stroke.

Fuel pumps in MSE and LSE are made autonomous for each cylinder. The most widely used pumps are those of the plunger type. The plunger is driven by a camring located on the engine's camshaft.

The nozzles of marine diesel engines have a needle which acts as a valve. After the end of fuel injection, the nozzle needle descends on the saddle under the impact of the spring, thereby excluding the possibility of fuel leakage into the working cylinder.

Recently, nozzles with electronic regulation of fuel injection have become popular.

Oil system. Oil is fed into the internal combustion engine to rubbing surfaces to reduce friction, remove the accompanying heat, and clean the friction surfaces from wear products, carbon deposits and other foreign particles. Along with this, oil is used as a coolant to remove heat from heated parts. Cylinder liners, bearings of crankshaft, camshaft, turbocharger, and pumps, drive gears, pushers and guiding surfaces of the gas distribution mechanism are all to be lubricated.

According to the location of oil in the system, the *"dry"* and *"wet" crankcase* systems are distinguished. In the first case, the waste oil that flows from the friction units to the engine crankcase does not accumulate in it, but is fed to a special waste-circulation tank. In the lubrication systems with "wet" crankcase, it is used as an oil tank. In some engines, part of the oil is in the crankcase, and the other part is in a special tank.

According to the method of providing pressure, the oil systems are divided into *gravity* and *pressure* systems. In the former, oil is fed from a tank located above the lubrication object to the discharge pipeline with the help of gravity; in the latter, pumps are used to supply oil to the consumer.

According to the oil flow, *circulating (closed) systems* and *linear (lubricating)* are distinguished. In the former, oil flows through a closed cycle repeatedly, while in the latter oil is supplied to the lubricant surface once and does not return to the system.

The main elements of the oil system are oil pumps, oil filters, oil coolers, oil tanks and cisterns, oil pipelines. Oil pumps are designed to continuously or periodically supply a portion of oil in the oil supply line. Pumps that supply oil continuously are called circulating pumps. They can be of the gear or screw type. In large ship engines, such pumps have an autonomous drive (most often electric), while elsewhere they are attached to and driven from the engine motion parts.

Lubrication of cylinders in large ship engines is carried out with the help of *lubricators*—multi-plunger pumps, in which each plunger provides periodic supply of small portions of oil to one opening on the cylinder liner. Lubricator drive is carried out from the engine crankshaft. Lubricators have a dosing device that allows changing the moment and extent of oil supply to the inner surface of the cylinder. In modern engines with electronic control, oil dosing is provided by electronic systems.

During the operation of the diesel engine, the oil that circulates in the system should be continuously cleaned from mechanical impurities and asphalt-resinous substances by means of oil filters. To cool the oil, the surface coolers with tubular or plate heat exchangers are used.

Water cooling systems are designed to cool parts that are heated from the combustion of fuel or during friction, heat removal from oil, water, or charge air.

The engine parts subject to cooling:

- cylinder cover—to reduce the firewall temperature to the values that ensure preservation of the mechanical properties of parts;
- cylinder liner—to reduce the mirror temperature to the values that ensure preservation of the oil film;
- piston—to reduce the temperature stresses and ensure the workability of the piston rings;
- turbocharger turbine housing—to reduce air heating in the compressor;
- exhaust drum—to reduce heat production in the engine room of the ship.

The cooling system includes pumps, coolers, an expansion tank, thermostats, and pipelines. The coolers are designed to remove excess heat from the cooled liquids and charge air into water. The expansion tank compensates for the change in the volume of water in the system when its temperature changes. The thermostat should automatically maintain the temperature of water and cooled liquids within the specified limits.

3.5 General Arrangement of Marine Diesels

3.5.1 Low-Speed Engines

High reliability, considerable motor resource, simplicity of structure and high environmental friendliness are characteristic of such engines. It is these features, as well as high aggregate power (up to 88000 kW), that determine their predominant use in propulsion plants of sea ships.

Only three companies in the world produce low-speed internal combustion engines: MAN Diesel and Turbo (almost 65% of the world production), Wärtsilä Corporation (about 25%), and Mitsubishi Heavy Industries Ltd (about 10%).

Let us consider the general arrangement of LSEs of these manufacturers.

The MAN Diesel and Turbo low-speed engines of the MC class. MAN Diesel and Turbo has been implementing the program for the production of direct-valve scavenging engines of the MC class since the mid-1980s.

It suggests manufacturing of high-efficiency long-stroke low-speed diesel engines that can successfully operate on low-quality heavy fuels with the help of modern technologies and materials.

Since 1996, the MC program has been expanded by including the S-MC-C, L-MC-C, and K-MC-C diesels, where C designates a compact engine that is smaller than the MC class. The use of such engines on ships greatly simplifies their placement in the engine room due to their compactness.

The power range of the engines includes 22 standard sizes with the power of 5.1-82.4 MW and the cylinder diameters of 30, 35, 35, 50, 60, 65, 70, 80, 90, 95 cm.

All engines of the MC class are based on the same design principles, which ensure the simplicity and reliability of their components.

A rigid foundation frame for the engines with a large cylinder diameter is assembled from longitudinal side beams with cast steel bearing supports. For small engines, the foundation frame is made cast. The main bearings are filled with babbitt. The thrust bearings are installed at the aft end of the frame, which has high rigidity.

Engine column is a welded structure for high-power engines and a cast structure for smaller engines. The cast steel cylinder block increases the rigidity of the engine design, and anchor bolts that are tightened using a hydraulic tool join the foundation frame, the engine column and the cylinder block into a rigid structure.

Crankshaft is standard, semi-assembled, with the chain-driven gear to the camshaft. Axial vibration dampers are located on the free ends of the crankshaft. In order to limit the height of the engine, the relatively short connecting rods are used.

Cylinder liners have a simple symmetrical structure that ensures low oil consumption and wear. Liners of the engines with a large piston diameter have channel cooling (absent in the engines with a small piston diameter). Steel cylinder covers are cast and have cooling water passages, a central channel for the exhaust valve, channels for the fuel nozzles and indicator valve.The oil-cooled piston head is made of heat-resistant chrome-molybdenum steel. It is rigidly secured onto the piston stem in order to ensure free transfer of forces from the hot gas pressure. The piston has four grooves for the piston rings, which are strengthened with a chrome coating on the upper and lower surfaces of the grooves; the steel cast trunk is attached to the bottom of the piston head.

The piston stem is subject to surface treatment to minimize friction in the stem seals and provide increased contact pressure in the sealing ring. The stem seal provides effective sealing between the clean compartment of the crankshaft and the combustion zone, which ensures the low level of cylinder oil consumption.The camshaft controls the fuel pump and the hydraulic drive of the exhaust valves. In the engines of the MC series, a chain drive is used to drive the camshaft.Exhaust valves are opened hydraulically by the oil received from the drive mechanism, and the closing force is provided by a "pneumatic spring", which allows the valve pin to rotate freely. The valve closure is damped by the oil cushion on the top of the valve pin.The engines make use of a close fuel system under pressure; the maximum temperature of the fuel heating to create the required viscosity at injection reaches 150 °C. Fuel nozzles are not cooled.The engines are reversed by a simple and reliable mechanism. The engine remains operable even when one of the cylinders fails, with the corresponding fuel pump being set to zero position.

The idea of the structure of the considered engines is given in Fig. 3.29, which provides a general view of the S60MC LSE [3]. The engine's foundation frame *1* consists of high longitudinal beams connected to welded-cast transverse shell structures where the frame support bearings are located. Column *2* is welded; cylinder block *7* is cast-iron. The foundation frame, the column and the cylinder block are tied together by anchor ties *3*. Cylinder liner *6* rests on cylinder block *7*, with its upper

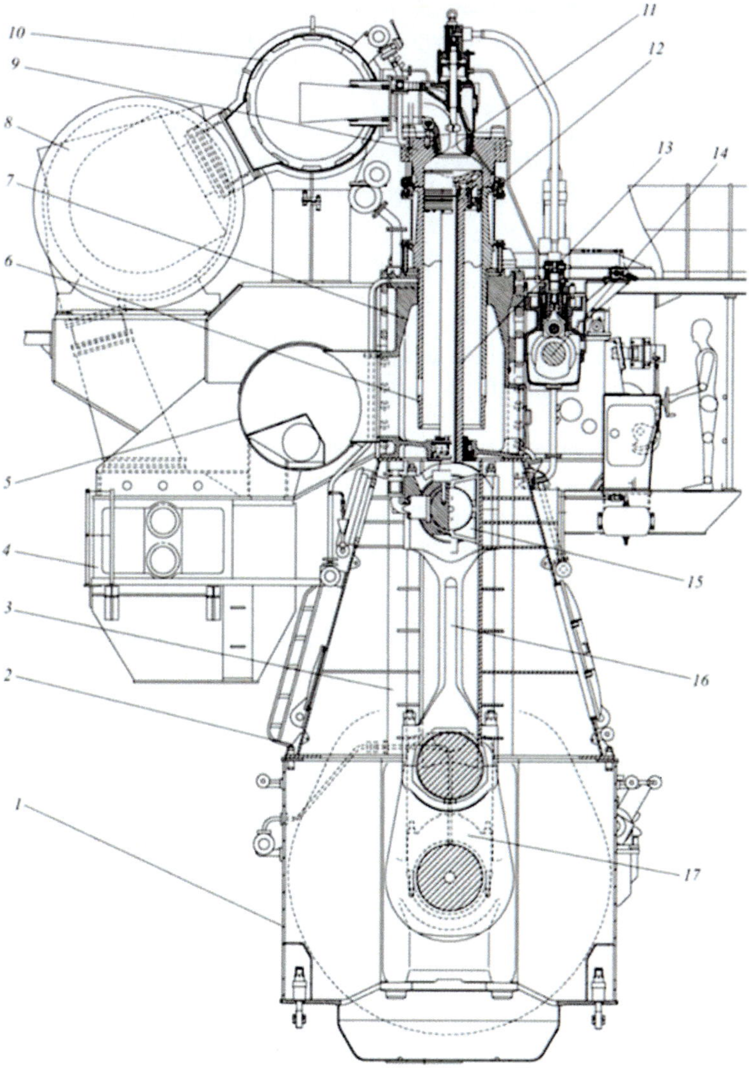

Fig. 3.29 S60MC low-speed engine

part withdrawn from the unit block and covered with a thin jacket creating a cooling cavity. The fittings for the cylinder oil supply are located at the top of the liner. Cylinder cover *9* is forged with drillings for cooling water. The cover accommodates a central exhaust valve, through which the gases are released from the cylinder, two nozzles on the sides of the exhaust valve, and starting and safety valves. Exhaust valve *11* has a hydropneumatic actuator. The S60MC-C engine employs a cover of the cap type, so when the piston is in TDC, the piston head is located above the sealing surface of the cover and the cylinder liner. Piston *12* is made of heat-resistant chrome-molybdenum steel; it is cooled by the oil fed by means of a telescopic device to piston stem*13* in the area of crosshead connection *15*.

Connecting rod *16* has a relatively short pivot, which helps to reduce the overall height of the engine. Crankshaft *17* is welded. The thrust shaft is made in one piece with the crankshaft. Camshaft *14* rotates from the crankshaft by chain gear; it drives the exhaust valve, high-pressure plunger pumps and hydraulic actuator pistons of the exhaust valves. The exhaust gases from gas drum *10*are supplied to gas turbocharger (GTC) *8*. The cylinders are scavenged with air from receiver *5* into which it is pumped by the GTC through cooler *4*.

A deeper insight into the LSE arrangement is provided by the 3D image of the S60MC-C engine [3] presented below (Fig. 3.30).

MAN Diesel and Turbo low-speed engines of the ME class. At the beginning of the 21st century, further improvement of the engines of the MC series produced by MAN B&W was constrained by the capabilities of the applied gas exchange and fuel supply systems.

Creation of the first piston engines enabled using a camshaft with cams for adjusting the phases of fuel supply and gas exchange; it was driven from the engine shaft. Later, its design was repeatedly upgraded, but the cam with a pre-installed profile mounted on the shaft remained a definite obstacle to the improvement of engines.

The rapid development of electronics over the past decade has allowed applying microprocessor equipment first in low-power engines, and then in the ship LSE and MSE. It was originally intended for monitoring and remote control of the engine operation, which was extended by control of the processes of fuel supply, gas distribution, and cylinder lubrication.

The first engines with electronically controlled operation were produced by Caterpillar; the first electronically controlled ship engines were put into operation by Sulzer in 2001, and by MAN B&W in 2003.

The engine with electronic control of the ME type was created by MAN B&W on the basis of the MC model. The camshaft with the drive was removed from the design of this engine, and there was introduced electronic control of fuel supply, rotation frequency, the processes of starting and reversing, the exhaust valve and the lubrication of the cylinders.

The fuel injection and exhaust valves are controlled by hydraulic servodrives. For this purpose, the oil from the circulating lubrication system is passed through a self-cleaning fine filter and compressed to 20 MPa by the pumps driven by the engine or the electric drive (during start-up) (Fig. 3.31).

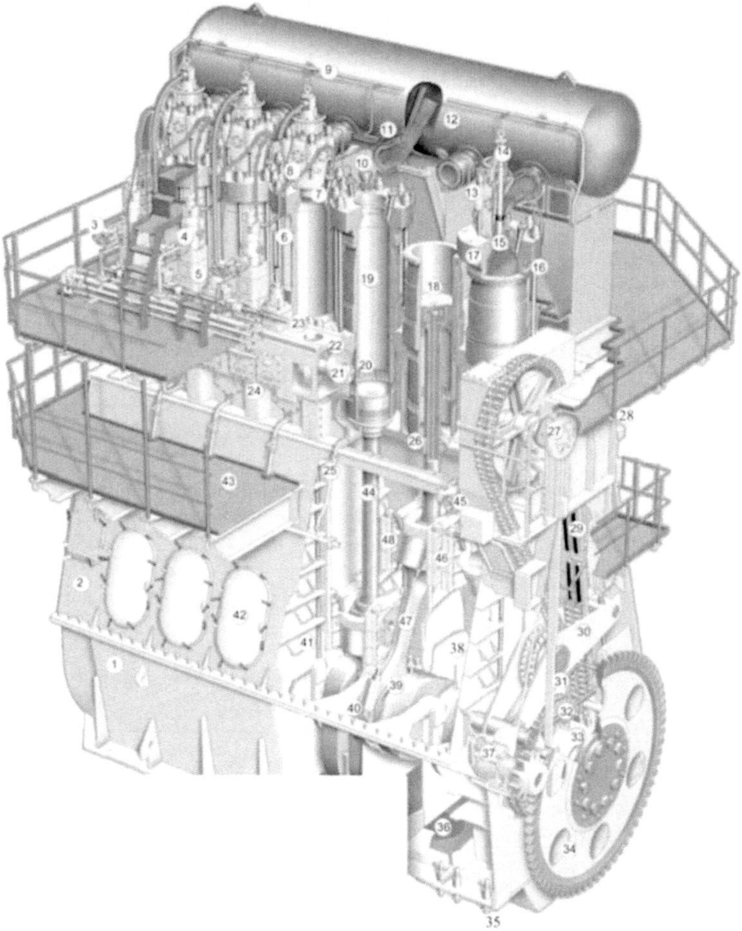

Fig. 3.30 General arrangement of the S60MC-C low-speed engine: *1*—foundation frame; *2*—crankcase; *3*—fuel pump regulator; *4, 8*—fuel pump and high-pressure pipes, respectively; *5, 13, 15*—hydraulic drive, the body and stem of the exhaust valve, respectively; *6*—cooling jacket; *7*—nozzle; *9*—cooling water outlet; *10, 27*—starting air valve and distributor, respectively; *11*—thermal expansion compensator; *12*—gas exhaust drum; *14*—pneumatic damper; *16*—cylinder cover studs; *17, 33, 40*—covers of the cylinder, frame and crank bearings, respectively; *18*—piston; *19*—cylinder liner; *20*—scavenging ports; *21, 39*—camshaft and crankshaft, respectively; *22*—gas distribution cam; *23*—drive of the fuel pump rails; *24*—cylinder block; *25*—drainage pipe of the airspace; *26*—seal of the valve pins; *28*—air receiver; *29, 30*—silencer and tensioner of the chain, respectively; *31, 34*—pinions of the crankshaft and shaft-turning device, respectively; *32*—thrust bearing; *35*—engine mounting bolts; *36*—oil draining in the pan; *37*—shaft-turning device; *38*—oil supply to the frame bearings; *39*—crankshaft; *41*—anchor bolts; *42*—crankcaseports; *43*—platform; *44*—piston stem; *45*—control unit of the control system; *46*—crosshead shoe; *47*—connecting rod; *48*—crosshead

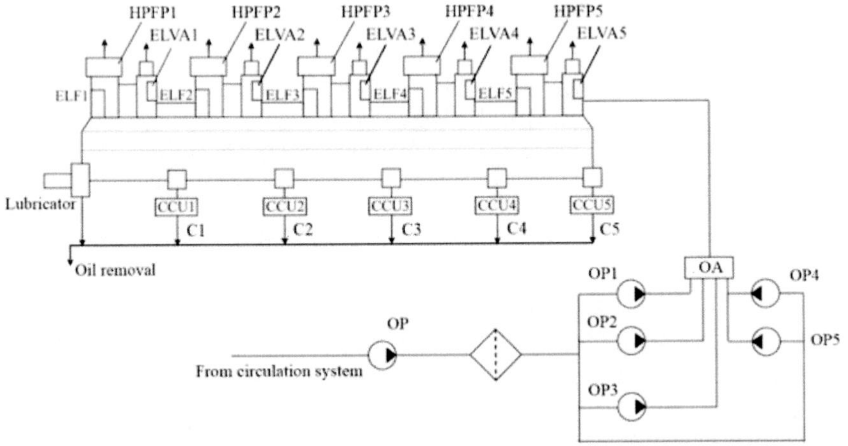

Fig. 3.31 Scheme of the hydraulic control system for fuel injection and exhaust valves: OP—circulating oil pump with electric drive; OP1, OP2, OP3—oil pumps with electric drives; OP4, OP5—oil pumps driven by the main engine; OA—oil accumulator; CCU1–CCU5—electronic units; ELF1–ELF5—electronically controlled fuel pump valves; ELVA1–ELVA5—electronically controlled gas outlet control valves; HPFP1–HPFP5—high-pressure fuel pumps; C1–C5—engine cylinders

The compressed oil is supplied to the accumulators, from which it flows through controlled valves into the hydraulic boosters of fuel injection pressure and the hydraulic drive pumps of the exhaust valves.

Electronically controlled valves ELF (fuel injection) and ELVA (opening of exhaust valves) are opened by a signal that comes from the electronic units (CCU) mounted on the cylinder. The hydraulic boosters of fuel injection pressure are piston servomotors. There, oil with the pressure of 20 MPa is applied to the piston of a larger diameter and fuel with the pressure of up to 100 MPa is applied when moving upwards to the plunger that is a continuation of this piston; the plunger's diameter is by 5 times smaller. The oil supply to the servomotor piston is determined by the control pulse from the CCU block [5].

The CCU program allows changing the phases of the exhaust valves opening and closing depending on the engine's operating modes. This provides a more economical long-term work at low loads.

The classic design of LSE provides for lubrication of the cylinder liners of multiple-plunger pump lubricators with a mechanical drive from the camshaft.

The electronic control system ensures synchronization of the delivery stroke of the plunger's lubricator with the movement of the working cylinder's piston by sending the electrical signal to the lubricator's solenoid valve from the controlled electronic CCU block at the necessary moment. The solenoid valve opens access for the servooil to the servomotor piston that moves the lubricator plungers. The economy in the cylinder oil consumption in the engines of the ME series is more than 0.3 g/(kW·h).

A number of changes have been introduced to the ME engines with electronic control in comparison with the MC engines [4]. The following elements were removed from the structure:

- chain drive of the camshaft;
- camshaft with cams of the HPFP's and exhaust valve's drive;
- previously applied HPFPs and the exhaust valve's drive;
- lubricators with a mechanical drive.

Instead, the following were installed:

- hydraulic system of high-pressure oil with pumps that are driven by the engine or the electric engines;
- electronic control system with shaft position sensors;
- FPHP and exhaust valves with a hydraulically controlled drive;
- regulator of the rotation speed and start-up of drive blowers (built into the electronic control system);
- electronically controlled lubricators.

Wärtsilä Corporation low-speed engines of the RTA series. In 1981, Sulzer (it joined Wärtsilä in the 1990s) launched production of a full model range of RTA engines with a direct-valve gas exchange system.

The power range of the Wärtsilä LSEs includes 26 standard sizes with a power of 6.9-80.8 kW and cylinder diameters of 35, 40, 48, 50, 58, 60, 68, 72, 82, 84, 90, 96 cm.

The framework of such an engine consists of a cast-iron foundation frame, welded columns connected by bolts, and individual cast-iron cylinder jackets bolted into a single rigid block. Anchor ties provide perception of tensile forces. Segment thrust bearing is integrated in the foundation frame.

The crankshaft includes the elements each consisting of crank necks and two cheeks into which the frame necks are pressed. Inserts of the frame bearings are filled with babbitt.

A steel one-piece forged cover is attached to the cylinder block with the use of eight pins. An exhaust valve is located in the center of the cover in a separate housing made of heat-resistant Nimonic8DA alloy with high corrosive properties.

A two-level cylinder lubrication system represents the ability provides an opportunity to form an oil film of the required thickness and ensures its renewal over the entire stroke length of the piston.

The heads of the pistons of the RTA48 and 50 engines are cooled by oil. It is fed through the piston stem into the nozzles installed in the head, from which the oil jets are directed to the openings made in the head. The oil is drained through the pipe installed in the stem. In larger engines, cooling takes place in water, which ensures a more efficient heat transfer. The water is fed to the crosshead unit and diverted from it through telescopic pipes.

Wärtsilä low-speed engines of the RT-flex series. The first engines with electronic control of the working process were put into operation by Sulzer in 2001.

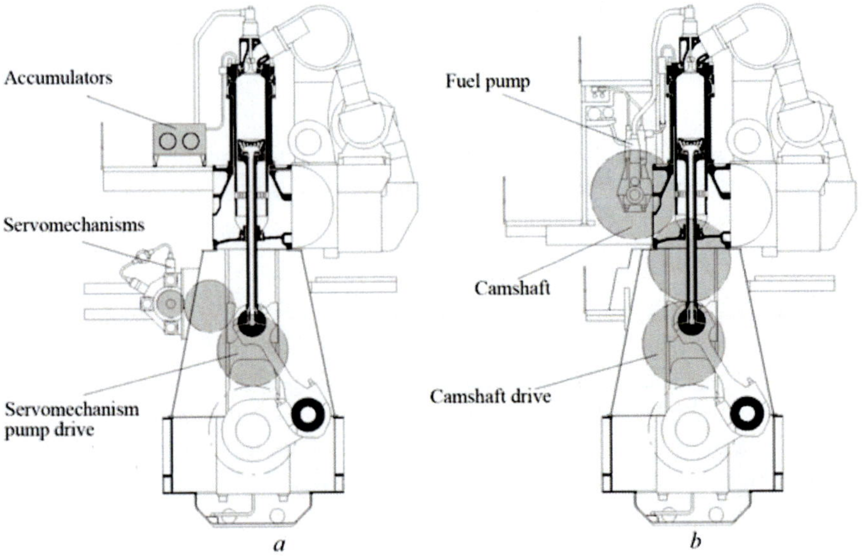

Fig. 3.32 Comparison of the RT-flex (**a**) and RTA (**b**) engines

In the electronically controlled engines of the RT-flex modification, the camshafts with their drives, traditional HPFPs and hydraulic drives of the exhaust valves are replaced with the accumulator system for fuel injection and control of the exhaust valves (see Fig. 3.32). This greatly simplified their design. The drive of hydraulic pumps of the servomechanisms is located in close proximity to the crankshaft. At the level of the cylinder caps there are oil and fuel pressure accumulators and servodrives of the HPFP and exhaust valves [2].

The use of electronic engine control has reduced the amount of harmful emissions (exhaust gases), reduced specific fuel consumption, and also increased flexibility in controlling the fuel supply pattern over the entire range of operating modes.

One of the main features of such engines is the accumulator fuel supply system, which consists of a HPFP providing the pressure of 100 MPa, a fuel accumulator, and electronically controlled valves that distribute fuel along the nozzles (Fig. 3.33).

The three nozzles in each cylinder are controlled independently, together (if necessary) or separately. The valve is controlled by the WECS9500 microprocessor electronic control system, which is made in a modular form with a separate microprocessor for each cylinder. It provides control of the entire engine.

Environmental parameters of the ship ICE are currently given considerable attention at the international level. Since 2000, all LSEs have been subject to compulsory environmental certification for compliance with the requirements of the International Convention MARPOL 73/78 on emission of exhaust gases containing nitrogen oxides [6].

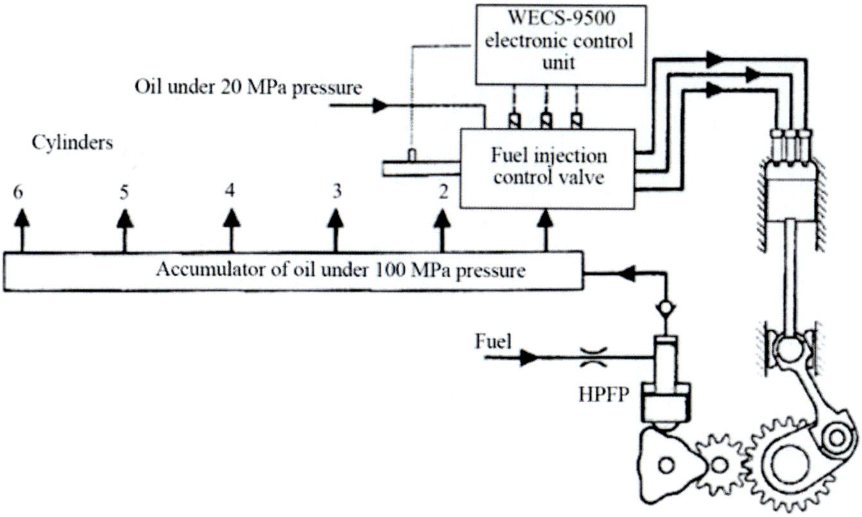

Fig. 3.33 Accumulator fuel supply system

The use of electronic control allows for individual control of each fuel nozzle installed in the cylinder cover. Its effectiveness ensures the improvement in the diesel performance at its operation in partial modes.

The scheme of the exhaust valve control in the electronically controlled diesel is shown in Fig. 3.34.

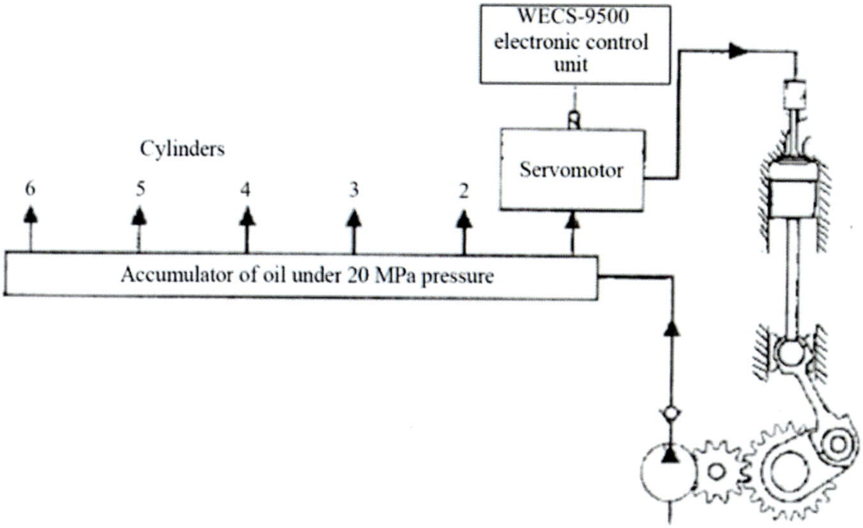

Fig. 3.34 Exhaust valve control system

Figure 3.35 is a cross-sectional view of the electronically controlled engine Wärtsilä RT-flex50.

Mitsubishi Heavy Industries Ltd low-speed engines. Mitsubishi began its activity with the production of two-stroke LSEs under the license of MAN B&W. In 1955, it switched to production of its own UEC engines using a proprietary technology. The first UEC75LS11 engine was released in 1987, then the company developed the

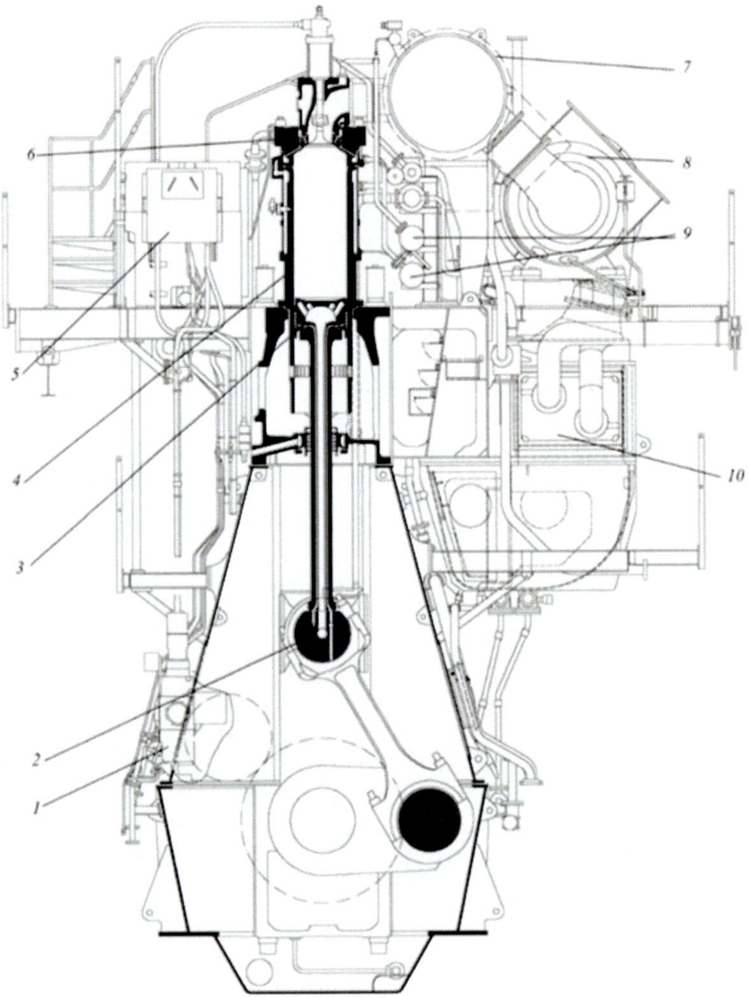

Fig. 3.35 Wärtsilä RT-flex50 engine with electronic control: *1*—drive of hydraulic pumps of servomechanisms; *2*—crosshead; *3*—piston head; *4*—cylinder liner; *5*—HPFP servodrives; *6, 7*—exhaust valve with impeller and drum, respectively; *8*—turbocharger; *9*—fuel and oil accumulator; *10*—air cooler

UEC-ECO engine and, finally, the UEC-LSE engine with electronic control system was released in 2001 [2].

The power range of the engines includes 19 standard sizes of 1–48 MW with the cylinder diameter of 33, 37, 43, 45, 50, 52, 60, 68, 75, 85 cm.

The construction principles of the UEC-LSE diesel control system are similar to the concept of the MAN B&W electronically controlled diesels. Fuel feeding, opening and closing of exhaust valves and cylinder oil feeding are carried out by means of hydraulic servomotors with electronic control via microprocessors. The main advantage of Mitsubishi electronic control is the reduction of NO_xemissions and the smokiness of exhaust gases. Flexible control allows reducing the specific fuel consumption at partial modes by 1–2% with the same amount of NO_xemissions as in conventional diesel engines or reducing NO_xemissions by 10–15% while maintaining fuel consumption unchanged.

The general arrangement of the UEC-LS engine is shown in Fig. 3.36.

Like in all UEC diesel engines, the engine frame consists of foundation frame *1* in the form of a base plate, frame *3* and cylinder block *4* made of cast iron, which are connected by anchor ties *11*. Crankshaft *2* is made in one piece; it has a device preventing longitudinal oscillations. Piston head *9* is made of molybdenum steel with internal radial and circular ribs; the head cavities are cooled by circulating oil. Insertion liners of the cylinders *5* have thick walls; horizontal reach-through openings are made in the flanges for better cooling. Cast-iron cover *10* of the engine cylinder has stiffeners. The cover bottom and the exhaust valve saddle *8* are intensively cooled. The exhaust gases from drum *7* are fed to turbocharger *6*.

3.5.2 Medium-Speed Engines

Modern MSEs differ significantly from the engines of earlier models, both in their design and principles of operation. Due to forcing of the operation process via gas turbine charging, the current level of the average effective pressure of the ship MSE makes up 2.1–2.9 MPa.

Wärtsilä has reached this level due to the use of the SPEX (Swirl-Puls-Exhaust) charging system, which combines the advantages of pulse charging and constant pressure charging. MAN B&W has adapted the constant pressure charging system (widely used in LSEs) for the four-stroke engines. Modern Caterpillar and Wärtsilä engines make use of the turbine-pump unit with controlled turbine nozzle devices. This allows to increase the maximum power of the ship MSE to 22 MW.

As the standards for emission of exhaust gases containing nitrogen oxides and smoke particles became more stringent, engine manufacturers had to look for ways to radically improve the processes of fuel combustion. Increased fuel combustion efficiency was promoted by electronic engine control systems, which allowed for optimization of the phases of fuel supply to the cylinders with load changes and control of the phases of opening and closing of the inlet valves.

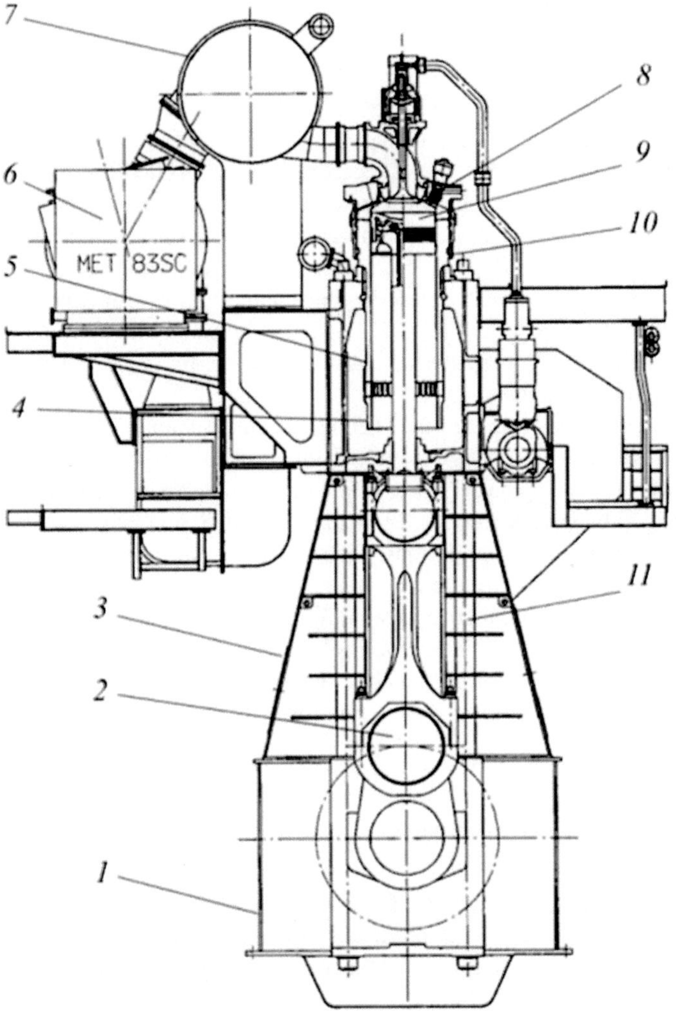

Fig. 3.36 Cross section of the UEC85L5 engine

The transfer of MSE to heavy fuel is a relevant problem because of its lower cost in comparison with diesel fuel and the advantages in using single fuel for the whole MPP. While the main task during the transfer of LSE to heavy fuels was the organization of appropriate fuel preparation, MSE necessitates radical measures for organizing the working process of the engine.

Reducing the duration of fuel supply while increasing the fineness of its spraying (due to an increase in the injection pressure up to 120–150 MPa) allowed the projectors to keep its specific consumption at the previous level and to solve the main problem—ensuring efficient operation of the engines on heavy fuels and reducing

NO_x formation. An increase in the *S/D* ratio up to 1.5 helped to increase the height of the combustion chamber, which provided better conditions for the mixture formation, fuel flame development and combustion. At the same time, the rotation speed was reduced, which in turn reduced the average piston speed, and, correspondingly, the wear rate of the cylinder-piston group. The technique of the combustion chamber cooling was improved to reduce the temperature differences and the increased thermal stress formed due to forcing of the engines and application of heavy fuels.

The constant reduction in the number of the engine crew on the ship with the purpose of decreasing the operating costs necessitates the reduction in the amount of maintenance work on the ship and the transfer of a larger share of these works to coastal enterprises.

A new generation of engines is characterized by a radical change in their layout [2]. The concept of this program is the use of modular solutions involving integration of structural elements into separate modules—monoblocks, elimination of a number of external pipelines and their placement in the monoblocks.

For example, engine cooling and lubrication pumps, coolers, thermostats, a self-cleaning filter and pipelines are combined into one box structure mounted on the front part of the engine framework. The entire cylinder-piston group, cylinder liners and cover can be dismantled as one unit. Then the ship crew substitutes these combined elements with new parts that are supplied by the manufacturer or repaired in advance [2].

B&W Diesel Group medium-speed engines. The most widespread is the model range of L28/32, L/V32/40, L40/54, L/V48/60, L/V51/60, L58/64 engines, which covers capacities of 3000–21000 kW.

These engines have received almost identical design. Let us consider the special features of these engine models through the example of L48/60 and V48/60 MSEs (Figs. 3.37 and 3.38).

The engine framework is a single unit consisting of the foundation frame and the cylinder block. It is protected from tensile loads and deformations by means of anchor ties and long studs for securing the cylinder covers.

Separate cylinder bodies cause the lowest rate of deformation and wear of the liners; there is no deforming effect on adjacent cylinders. The cylinder liners are mounted on the high jackets located on the top of the block. Cooling of the liners occurs only in the jacket zone; their lower part is not cooled. Thick-walled liners and individual jackets for each cylinder provide minimal deformation of the cylinder mirror, which allows reducing the gaps between the piston and the liner.

Lubrication of the cylinders is ensured by the lubricators through the openings in the liner that are extended to the surface of the cylinder into the zone between the first and second compression rings when the piston is in BDP.

The cylinder cover has a double bottom: a thin bottom to reduce thermal stresses and an upright thicker one to reduce mechanical loads exerted by high-pressure gases [2]. The exhaust and inlet valves of the L58/64 engine (Fig. 3.39) are housed in refrigerated bodies, while in the L/V32/40, L40/54 and L/V48/60 engines only the upper valve bodies are cooled. The exhaust valves are equipped with impellers so that exhaust gases could rotate them.

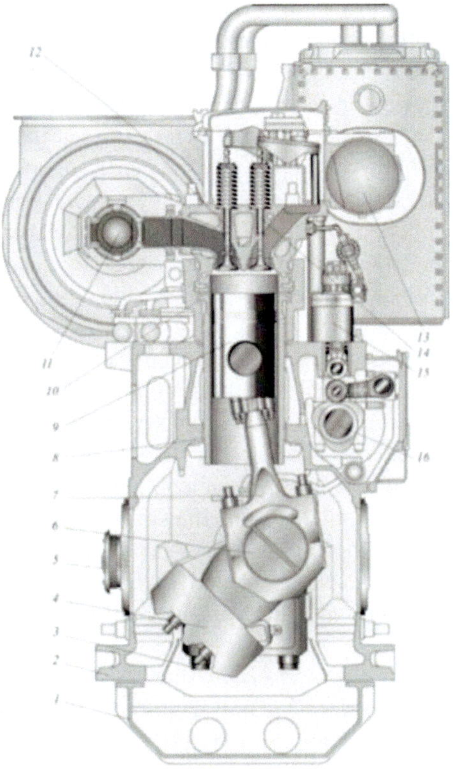

Fig. 3.37 L48/60 medium-speed engine: *1*—pan; *2*—foundation frame; *3*—studs securing frame bearings; *4*—counterweight; *5, 12, 14*—safety, exhaust and inlet valves, respectively; *6, 16*—crankshaft and camshaft, respectively; *7*—connecting rod; *8*—cylinder liner; *9*—piston; *10*—cylinder block; *11, 13*—inlet and exhaust drums, respectively; *15*—fuel pump

The connecting rod is made with a split at the top, which allows the piston to be removed without disassembling the lower head of the connecting rod. Reduction of the length of the V-type diesel is possible with the help of trailing connecting rods.

The piston is composite; its head is forged and made of steel, and the skirt is forged and made of aluminum. Cooling of the piston is carried out by oil, which is fed to the upper head along the connecting rod and then to the cooled cavity of the piston.

To reduce the intensity of soot formation on the exhaust, the fuel supply was intensified and the maximum injection pressure was raised up to 160 MPa. Due to the use of complex technical solutions at the maximum pressure Pz which has grown to 20 MPa, the specific fuel consumption has decreased in comparison to the previous designs.

Figures 3.40, 3.41 demonstrate the cross sections of the Wärtsilä 46 MSE of the line and V-type design.

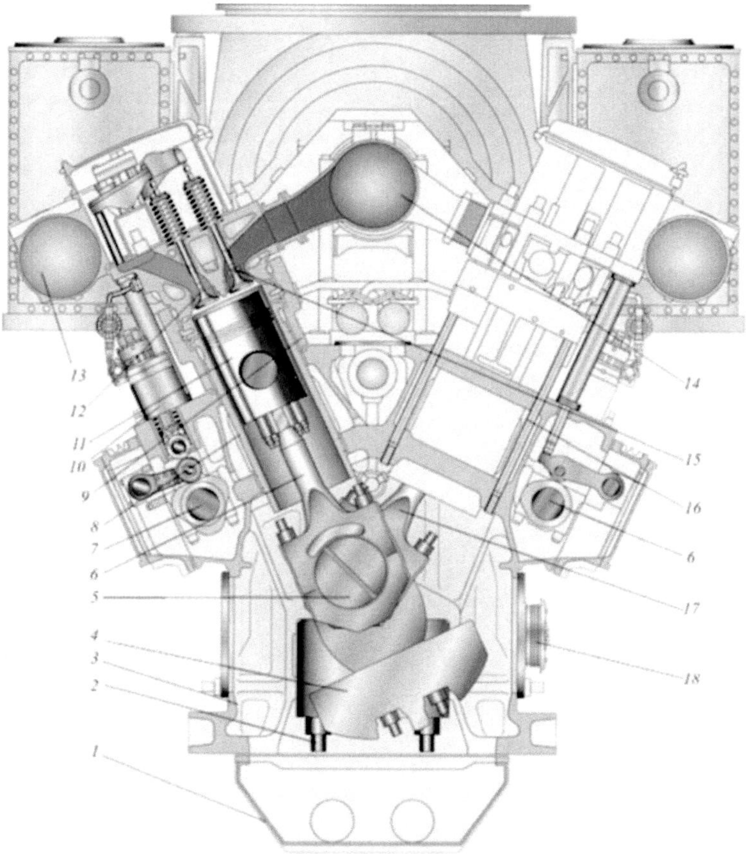

Fig. 3.38 V48/60 medium-speed engine: *1*—pan; *2, 16*—studs securing the bearings and cylinder block; *3*—foundation frame; *4*—counterweight; *5, 6*—crankshaft and camshaft, respectively; *7, 17*—main and trailing connecting rods, respectively; *8*—cylinder liner; *9*—cylinder block; *10*—fuel pump; *11*—piston; *12, 15, 18*—inlet, exhaust and safety valves, respectively; *13, 14*—inlet and exhaust drums, respectively

3.6 Reduction of Harmful Atmospheric Emissions of Marine Diesel Engines

3.6.1 Normalization of Harmful Atmospheric Emissions of Marine Power Plants

Severization of the requirements for economic purity of the MPP has contributed to the implementation of a whole range of studies aimed at creating "ecologically clean engines". These studies were necessitated by the coming into force of the international and regional regulations limiting the emissions from ICE.

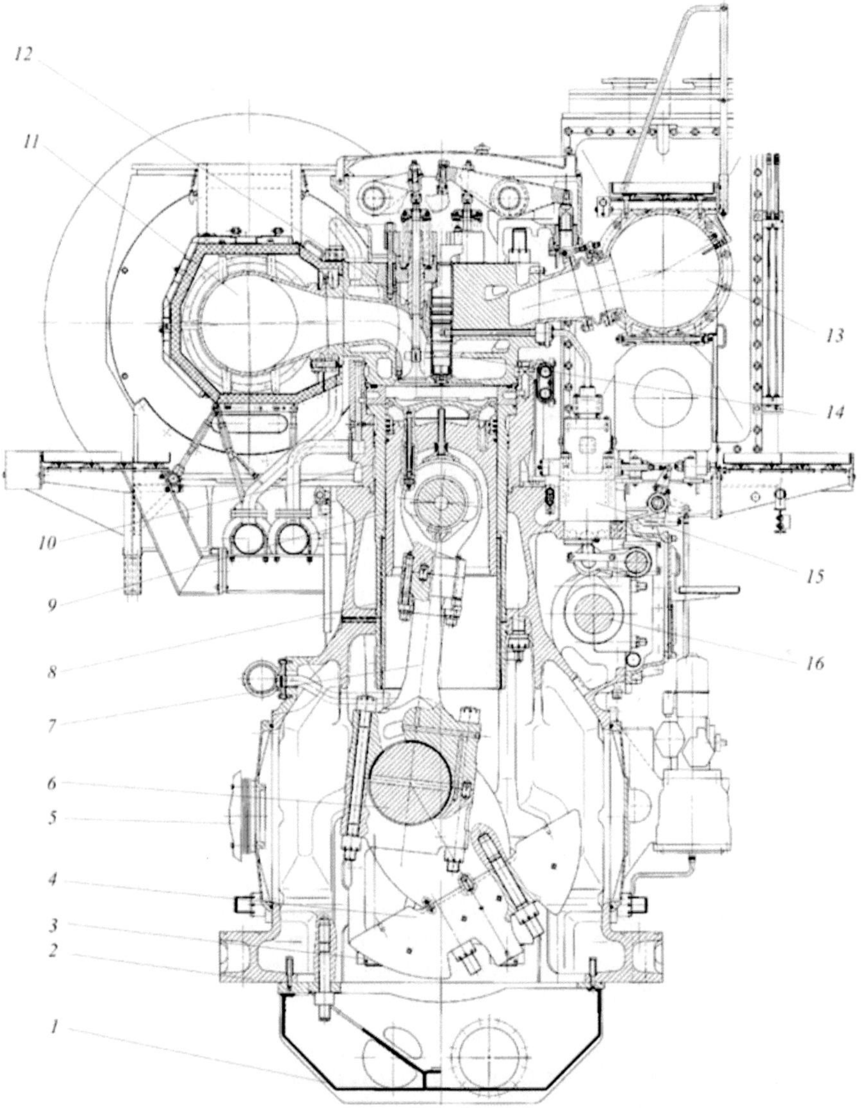

Fig. 3.39 L58/64 medium-speed engine: *1*—pan; *2*—foundation frame; *3*—studs securing the frame bearings; *4*—counterweight; *5, 12*—safety and exhaust valves, respectively; *6, 16*—crankshaft and camshaft, respectively; *7*—connecting rod; *8*—cylinder liner; *9*—piston; *10*—cylinder block; *11, 13*—exhaust and inlet drums, respectively; *14*—nozzle; *15*—fuel pump

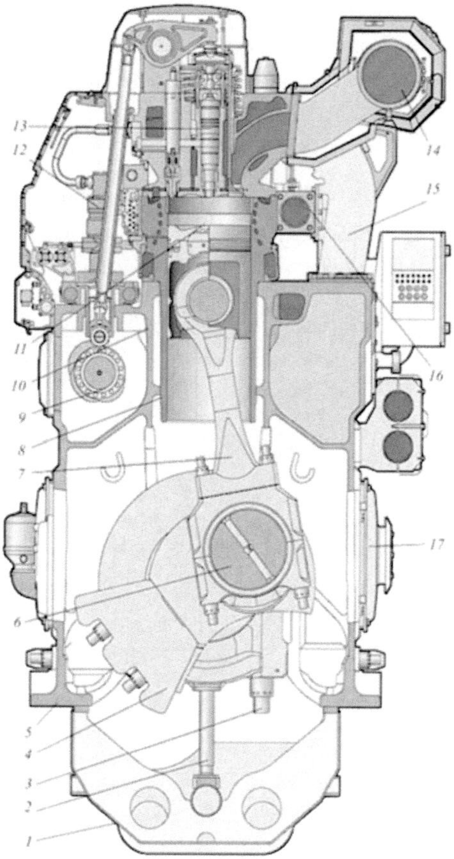

Fig. 3.40 Wärtsilä L46 medium-speed engine: *1*—pan; *2*—jack-screw; *3*—studs securing the frame bearings; *4*—counterweight; *5*—foundation frame; *6, 9*—crankshaft and camshaft, respectively; *7*—connecting rod; *8, 10, 16*—liner, block and cylinders cover, respectively; *11*—piston; *12*—HPFP; *13*—fuel nozzle; *14, 15*—exhaust and inlet drums, respectively; *17*—safety valve

Various indicators are used to estimate the content of harmful emissions. Thus, during gas analysis, the volumetric concentrations of all gaseous components (C_i) are determined; they are measured in % for the components whose concentration is greater than 1%. At lower concentrations, they are determined in *ppm* (parts per million). The ratio between the units of measurement is as follows: 1% = 10^4 ppm; 1 ppm = 10^{-4}%; 1 ppm = $1/10^{-6}$. The concentration of solid parts in the exhaust gases is measured in milligrams per cubic meter of gas under normal conditions (mg/nm^3).

The characteristics of harmful emissions, which quantify their amount, are as follows:

- emission rate E_i, kg/h;
- specific emission e_i, g/(kW·h);
- emission of a harmful component per 1 kg of fuel ε_i, kg/kg of fuel.

These characteristics are interrelated:

$$E_i = \varepsilon_i \cdot G_T;\ e_i = E_i/Ne;\ e_i = \varepsilon_i \cdot g_e,$$

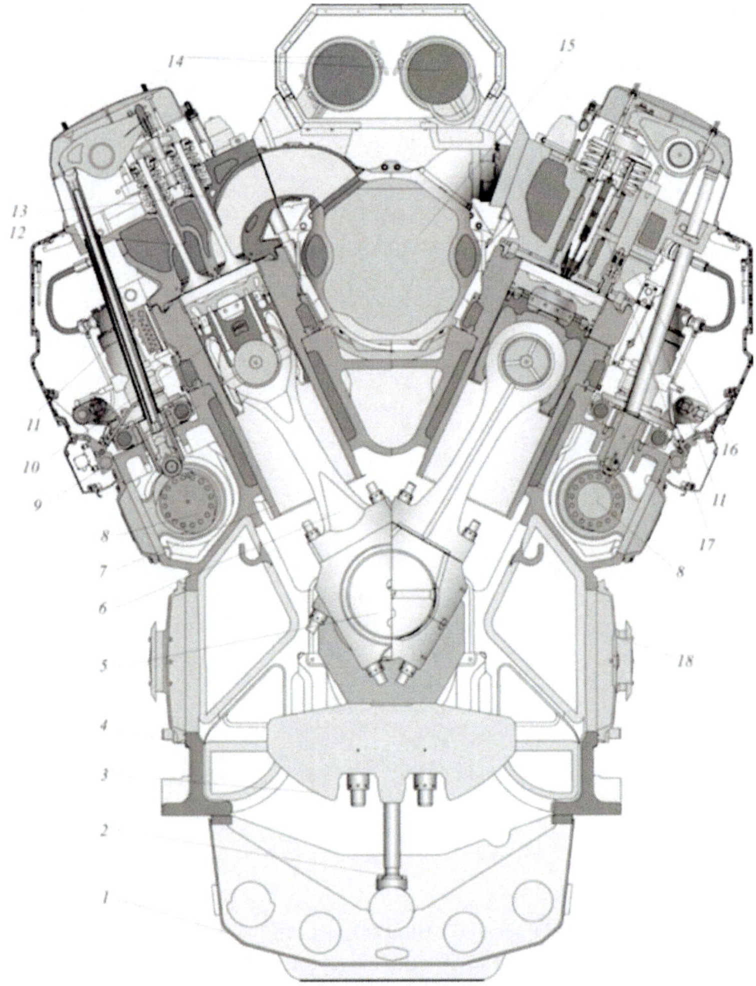

Fig. 3.41 Wärtsilä V46 medium-speed engine: *1*—pan; *2*—jack-screw; *3*—counterweight; *4*—foundation frame; *5, 8*—crankshaft and camshaft, respectively; *6*—connecting rod; *7, 9, 16*—liner, block and cylinder cover, respectively; *10*—piston; *11*—HPFP; *12, 13, 18*—exhaust, inlet and safety valves, respectively; *14, 15*—exhaust and inlet drums, respectively; *17*—fuel nozzle

where N_e is the effective power of the engine, kW; $g_e$is the specific fuel consumption, kg/kW·h; $G_T$is the fuel consumption, kg/h.

The most objective indicator of quantitative estimation of the emission of harmful components from the exhaust gases of ICEs is the specific emission. It is adopted as normative in international and national standards on emission control.

The ratio of different units of measurement of emissions of marine diesel engines is shown in Table 3.1 [2].

Table 3.1 Emissions with exhaust gases of the 10K60MS ship low-speed engine (18900 kW)

Components of exhaustgases	C_i, ppm,%	C_i, g/m^3	E_i, g/h	e_i, g/(kW·h)
NO_x	1570	3.41	352.10	18.63
CO	57	0.08	7.1	0.41
HC	284	0.22	22.2	1.17
SO_x	516	1.56	161.0	8.52
O_2	13.0	196	20225	1070
CO_2	5.2	108	11188	590
H_2O	5.4	0.00	4747	250
PM	–	0.12	12.10	0.66

Given that the volumetric concentration of harmful components in the exhaust gases depends on the rate of their dilution with air, in order to establish the level of emission for different diesel engines, the volumetric concentration is recalculated to the concentration of residual oxygen in the dry exhaust gases (after condensation of H_2O vapor), which is 15% [4].

At the 1997 conference in London, the International Maritime Organization (IMO) initiated the adoption of Annex VI in addition to MARPOL 73/78. It was intended to limit the emission of such harmful components as NO_xandSO_xin the exhaust gases of marine diesel engines. A part of the Annex was the "Technical Code for Controlling the Emission of Nitrogen Oxides from Marine diesel Engines", which is an international standard that establishes the procedure and rules for the economic certification of marine diesel engines at the production plant and during the operation of engines under shipboard conditions. The technical code provided for the limitation of emissions of nitrogen oxides only.

In accordance with Regulation 13 of Annex VI, which is applied to control the emissions of nitrogen oxides by the engines with a power of more than 130 kW installed on ships built after January 2000, except for the engines of ship emergency diesel generators, the engine operation is prohibited if emissions exceed the following limits [3]:

17 g/(kW·h) for $n < 130$ rpm;
$45n^{-0.2}$ g/(kW·h) for $130 < n \ll 2000$ rpm;
9.8 g/(kW·h) for $n > 2000$ rpm,

where n is the nominal engine speed, rpm.

Table 3.2 shows the NO_x emission limits set in the MARPOL 73/78 Convention.

The graphic representation of these limitations is shown in Fig. 3.42.

Along with the restrictions on the emission of nitrogen oxides, which are developed by IMO and are of global international application, the regional regulations have been distributed (the California Rules, Swedish legislative provisions, etc.). The requirements of such rules are quite strict; they apply to the ship ICEs that operate in regional waters.

Table 3.2 Level of NO_x emissions of marine diesels in accordance with MARPOL

Level	Year	NO_x emission limit, g/(kW·h), at the rotation speed		
		$n < 130$	$130 \leq n < 2000$	$n \geq 2000$
Level I	2000	17.0	$45n^{-0.20}$	9.80
Level II	2011	14.4	$44n^{-0.23}$	7.70
Level III[a]	2016	3.4	$9n^{-0.20}$	1.96

[a]In the NOx control zone (level I, II—outside the ECA)

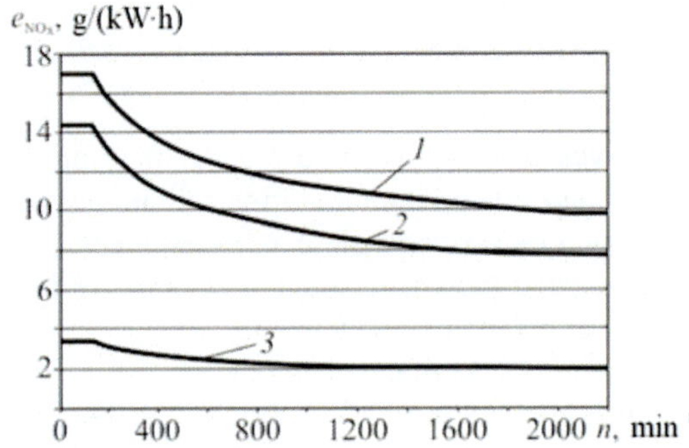

Fig. 3.42 Dependence of specific NO_x emissions on the crankshaft rotation speed of the internal combustion engine: *1*—level I; *2*—level II (global); *3*—level III (in the areas where NO_xemissions are controlled)

The limitation of the emissions of sulfur oxides is set by Regulation 14 of Annex VI of the MARPOL Convention [3] and is as follows.

1. Sulfur content in any liquid fuel used on a ship should not exceed 4.5% (since 2012–3.5%).
2. When ships are located within the SECA (Sulfur Emission Control Area), one of the following conditions must be fulfilled:

   a) sulfur content in the fuel should not exceed 1.5% (before 2010), 1% (before 2015), 0.1% (since 2015);
   b) waste gas cleaning system is applied on the ship, which provides the level of total SO_x emissions by all elements of the MPP below 6.0 g/(kW·h).

These restrictions came into force in May 2006 in the Baltic; in 2007, they were applied to the North Sea and the English Channel.

Figure 3.43 illustrates the nature of the change in the permissible sulfur content in fuel by year.

3.6.2 *Ways to Reduce Harmful Emissions*

Existing methods for reducing NO_x emissions are divided into *primary* methods, aimed to reduce the amount of nitrogen oxides that are formed in the diesel cylinder,

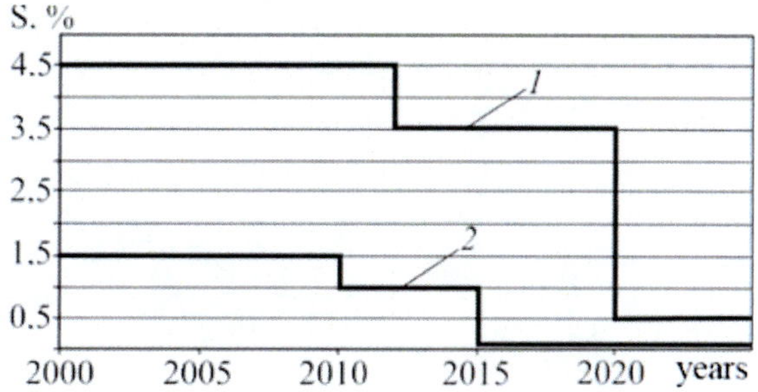

Fig. 3.43 Limitation of sulfur content in fuel: *1*—for all water areas; *2*—for SO_x emission control areas

and *secondary* methods, which involve chemical neutralization of NO_x before exhaust gases are released into the atmosphere.

Primary methods in turn can be divided into two groups:

- methods that involve changing the design of the engine or its separate elements (as a rule, they can be realized only when developing new designs): improvement of the systems for fuel injection, mixture formation and gas exchange; water injection directly into the engine; organization of the vortex motion of the charge in the combustion chamber;
- methods, the implementation of which does not require a significant change in the diesel design; it requires only a minor engine upgrade (the use of alternative fuels, the transfer of the diesel to operation on a fuel-oil emulsion (FOE), humidification of air at the engine inlet, and recycling of a part of the exhaust gases).

Physically, the effect of these methods on NO_x emissions is that they all provide a reduction in the temperature of fuel combustion, which determines the rate of NO_x formation in the cylinder.

Use of engines with electronic control. It is characteristic for such engines that the fuel supply and gas exchange are controlled by a processor, which allows changing the fuel supply pattern and gas distribution phases depending on the loading and speed modes. Electronic control of diesel systems allows for adjustment to the economic mode of operation with an increased NO_x level or to the "low emission mode", at which it can be reduced by 25–40% with slightly increased specific fuel consumption.

The use of *alternative fuels* radically affects the levels of harmful emissions [7]. Liquefied natural gases, biodiesel fuels, synthetic fuels belong to such fuels and can actually be applied in ship engines. Figure 3.44 shows the general view of the power plant of the first ship where LNG was used as fuel. This was the multifunctional dry cargo ship with the displacement of 5600 tons with the Bergen B35: 40CV12PG main four-stroke engine. Compared to ships using heavy fuels, this ship has by 90% lower NO_x emissions, by 20% less CO_2 and zero SO_x level [8].

The use of *water-fuel emulsions* (WFE) for the operation of marine diesel engines has left the experimental stage and has serious prospects for use on ships [7–9]. The mechanism of NO_x reduction with the use of WFE and its injection into the combustion chamber is that while evaporating, water reduces the peak of the maximum

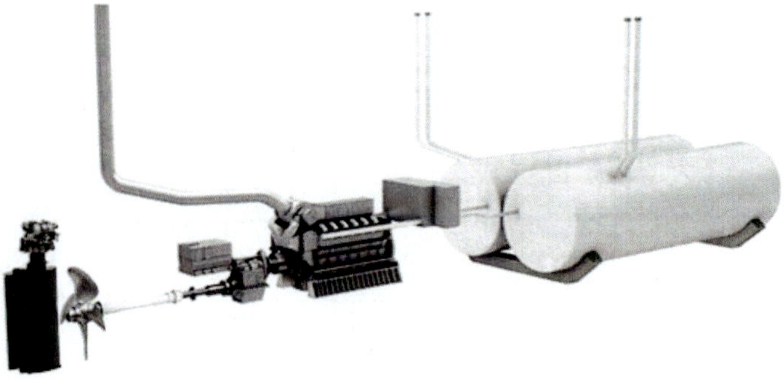

Fig. 3.44 General view of a propulsion complex with a medium-speed engine using LNG as fuel

temperature during combustion and thereby reduces the NO_x emission. During operation of the diesel engine using WFE, the specific fuel consumption slightly increases due to the lowering of the temperature and the combustion pressure. This method does not require changes in the engine design and does not reduce its reliability; with the water content of less than 30% of the fuel, the specific fuel consumption rises slightly, and the smokiness of the exhaust gases decreases. Presently, there is a number of examples of the use of WFE for the operation of diesel engines. Thus, at the beginning of 2009, the WFE preparation system was used as a part of the MPP for the 11K90MS-C engine of a container ship [3]. Reduction of NO_x emissions, as demonstrated by the operation of the MAN B&W two-stroke engines, is about 1% for every 1% of water in fuel [4]. Figure 3.45 shows the principal scheme of WFE preparation [7].

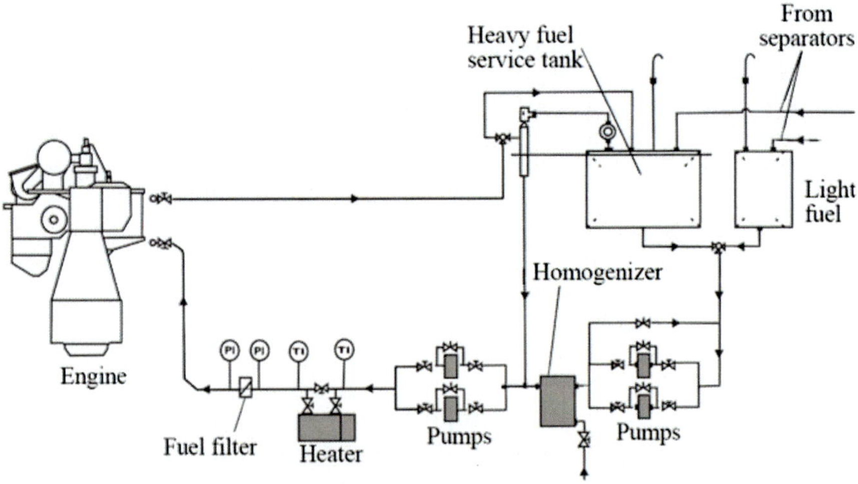

Fig. 3.45 Principal scheme of preparation of the water-fuel emulsion for a marine diesel engine

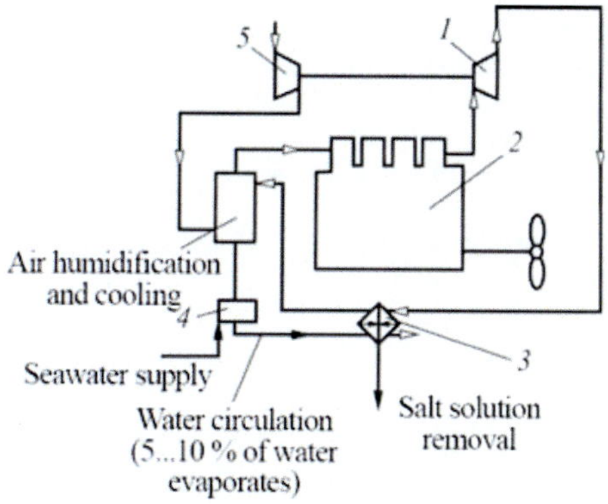

Fig. 3.46 Scheme of the charge air humidification system: *1*—turbine; *2*—ME; *3*—heat exchanger; *4*—recirculating tank; *5*—compressor

Water injection directly into the engine cylinder. As a way to reduce NO_xemissions, it is suggested to inject distilled water directly into the cylinder with the help of a special high-pressure fuel pump (30–50 MPa) through a nozzle with two separate nosepieces. The maximum NO_xreduction is achieved when water is injected at the end of the compression stroke before feeding the fuel [2]. Application of this method makes it possible to reduce the NO_xemission by 60–70% from the original with the water content of 40–60%.

The disadvantages include significant costs of the engine modernization; increase in the specific fuel consumption (by 5–7% with a decrease in NO_xemissions by 60–70%); considerable consumption of distilled water. Despite this, such systems find individual application on ships.

Humidification of the charge air. The abbreviations HAM (Humid Air Motor) and SAM (Scavenge Air Moisturizing) are used when describing such a way of reducing NO_xemissions.

The humid air supply system (Fig. 3.46) makes use of preheated charge air saturated with water vapor, which is obtained by evaporating seawater with the recycled heat of the engine exhaust gases or the fresh water heat from the engine cooling system. The amount of water supplied to the engine is three times larger by mass than the amount of fuel burned. Such a water-fuel ratio leads to a reduction in the emissions of nitrogen oxides by 70–80%.

Reduction of the amount of nitrogen oxides formed in the combustion chamber of the diesel engine during humidification of the working fluid is due to the fact that water vapor takes a significant amount of heat for steam generation because of its high heat capacity, which leads to a decrease in the temperature in the combustion chamber.

Despite a slight increase in specific fuel consumption, these systems have a prospect due to the significant potential for reducing nitrogen oxide emissions.

EGR (Exhaust Gas Recirculation) is a process when a part of the exhaust gases from the exhaust drum is supplied to the scavenging receiver.

Decrease in the amount of NO_x in the recirculation method is caused by the presence of carbon dioxide with a high heat capacity in the exhaust gases, which

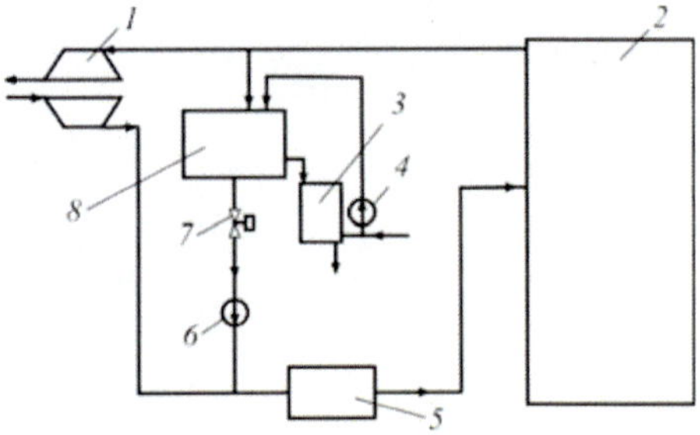

Fig. 3.47 Exhaust gas recirculation system: *1*—turbocharger; *2*—ME; *3*—water purification system; *4*—seawater pump; *5*—charge air cooler; *6*—electrocompressor; *7*—recirculation valve; *8*—scrubber

reduces the temperature in the combustion chamber. Along with this, the oxygen concentration in the combustion zone decreases and less NO_x is formed because of the partial replacement of air by the exhaust gases. To prevent incomplete combustion due to the decrease in the oxygen content of the charge, the fraction of gases should not exceed 15% [3, 8].

The principal scheme of the exhaust gas recirculation system is shown in Fig. 3.47.

The exhaust gases are preliminarily passed through sprayed water in a special device (scrubber), where they are cooled, cleaned of soot, solid particles and sulfur oxides. After cleaning, the exhaust gases are cooled and supplied to the scavenging receiver via the electrocompressor.

When recycling 15% of the exhaust gases, the NO_x reduction makes up 40–50% [8].

At the same time, it should be noted that the gas bypass leads to a decrease in the air excess ratio during combustion and an increase in the carbon oxide content due to incomplete combustion of fuel.

Application of the considered primary methods can actually ensure the reduction of NO_x emissions by 30–50%.

To ensure that the power plants of ships that are currently being built or have just been designed meet the stringent requirements set by Annex VI to MARPOL (recap: in 2016, the level of Tier III already provided for limiting the NO_x content to 3.4 g/(kW·h)), it is necessary to use *secondary methods* or to combine them with the primary ones. The following technologies belong to such methods.

Selective Catalytic Reduction (SCR) refers to the most effective means of reducing NO_x emissions (Fig. 3.48). This purification system suggests passing the exhaust gases through the reactor that contains a catalyst. Gaseous ammonia NH_3 is fed to the reactor. In the SCR reactor, the exhaust gases are mixed with ammonia, which results in the formation of harmless products from nitrogen oxides—nitrogen and water vapor.

The interaction of the components that are fed to the SCR reactor occurs according to the following reactions [6]:

$$4NO + 4NH_3 + O_2 \quad 4N_2 + 6H_2O;$$

$$2NO_2 + 4NH_3 + O_2 \quad 3N_2 + 6H_2O.$$

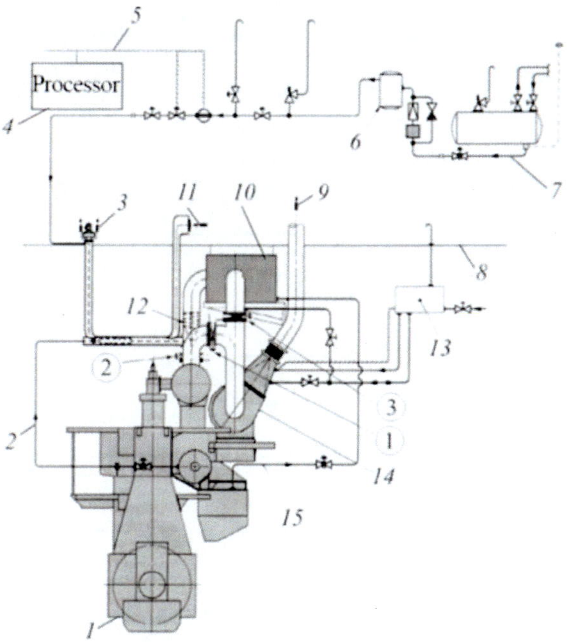

Fig. 3.48 Scheme of the SCR system: *1*—engine; *2*—direction of air movement; *3, 9*—air and exhaust gas outlets, respectively; *4*—processor; *5, 11*—air supply and outlet, respectively; *6*—evaporator; *7*—tank with urea; *8*—deck; *10*—SCR reactor; *12*—static mixer; *13*—NOx and O_2analyzers; *14*—turbocharger; *15*—heating and sealing air; ①, ②, ③– flaps

Both reactions take place independently; in addition, nitrogen oxides and ammonia in the presence of a catalyst interact with each other:

$$NO + NO_2 + 2NH_3 \quad 2N_2 + 3H_2O.$$

The use of gaseous ammonia s quite problematic due to its toxicity. This issue is partially resolved by using the aqueous urea solution instead, which does not require any special storage conditions. In the reaction zone, the aqueous urea solution decomposes to form ammonia:

$$CO(NH_2)_2 + H_2O \quad 2NH_3 + CO_2.$$

In the SCR reactor, the exhaust gases are mixed with urea before passing through the layer of a special titanium vanadium catalyst (Fig. 3.49).

The effective reduction of NO_x emissions (up to 98%) in comparison with the basic (noncatalytic) option when using the SCR system can be guaranteed only in a fairly narrow range of temperatures of the exhaust gases (temperature window)—300–400 °C. If the temperature is above the specified limits, NH_3 will burn before reacting with NO or NO_2. At temperatures below 300 °C, the reaction rate will be much lower, therefore, the condensation of ammonium sulphates will destroy the catalyst.

The 90% decrease in NO_x content is ensured by ammonia supply at the level of 15 g/kW [6].

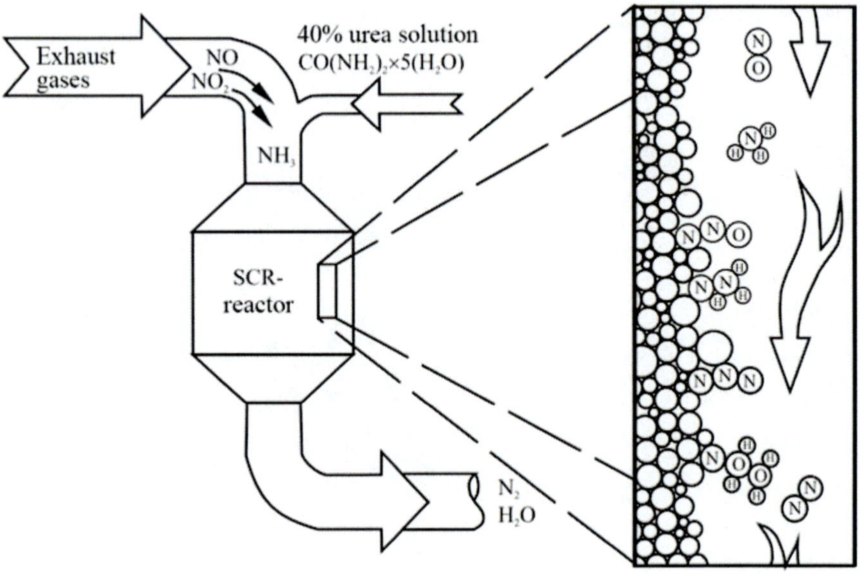

Fig. 3.49 Principal scheme of NO_X neutralization in the SCR reactor

The placement of the SCR reactor depends on the type of the engine. Given the limited "temperature window" within which an effective functioning of the SCR reactor is ensured in the power plants with MSE, the temperature of the exhaust gases is sufficiently high, so the reactor is placed behind the turbine of the turbocharging unit. In LSEs, the SCR reactor is located before the gas turbocharger, which leads to a number of operational problems. A partial solution to these problems is the use of high-performance turbochargers or, if necessary, additional electric-powered superchargers as a part of the LSE with SCR.

Figure 3.50 shows the variants of placement of the SCR reactor on the ICE [4].

An insight into the dimensions of the SCR system is given by the characteristics of the reactor installed on the 11K90MS engine with the power of 57200 kW: diameter—2.4 m, height—4.5 m, length—15 m, weight—42 tons (taking into account the catalyst) [4].

The design and dimensions of the SCR reactor as the largest and heaviest element of the system depend on the consumption of exhaust gases, their temperature, and level of the NO_X emission reduction. Placing pipelines, tanks with ammonia and a mixer in the engine room also requires additional space.

At present, several dozens of ships equipped with SCR systems are in operation, and new ships are being built with the application of the considered system for cleaning exhaust gases. The efficiency of purification of exhaust gases from NO_X in such systems reaches 98%.

For ships operating in ECA, the use of SCR systems is the only way to comply with Tier III rules.

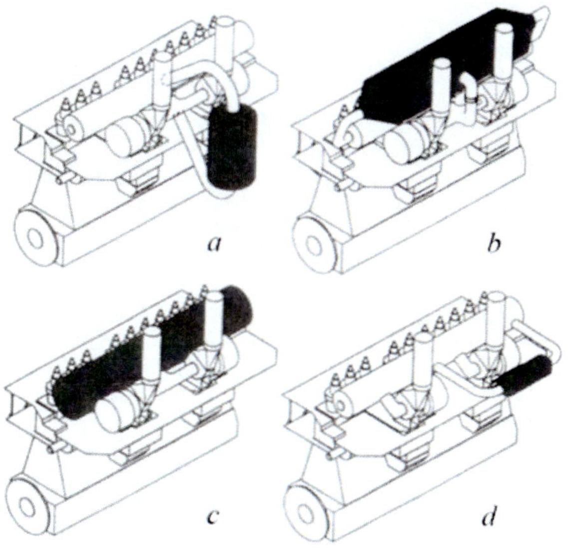

Fig. 3.50 Variants of placement of the SCR reactor on the diesel engine: *a*—vertical; *b*—horizontal; *c*—reactor built into the engine; *d*—reactor for a group of cylinders

Exhaust gases purification from sulfur compounds. SO_x emissions are caused by the maintenance of sulfur in fuel. The simplest way to bring the content of sulfur compounds to existing standards is to use low-sulfur fuels. Along with this, the purification of exhaust gases by means of scrubbing technologies belongs to radical and fairly simple ways of reducing the content of sulfur compounds.

The principal scheme of the purification system is shown in Fig. 3.51. In the scrubber, there is heat and mass exchange between the exhaust gases and seawater sprayed in countercurrent to the gases. Sulfur oxides and suspended particles are absorbed by seawater. After the scrubber, seawater is supplied to the purification and neutralization system, which includes two hydrocyclones and a separator. When mixing oxidized water from a scrubber and fresh seawater, the sulphate is neutralized with calcium carbonate, which is contained in seawater. As a result, calcium sulfate (gypsum) and carbon dioxide are formed. The final completion of the neutralization

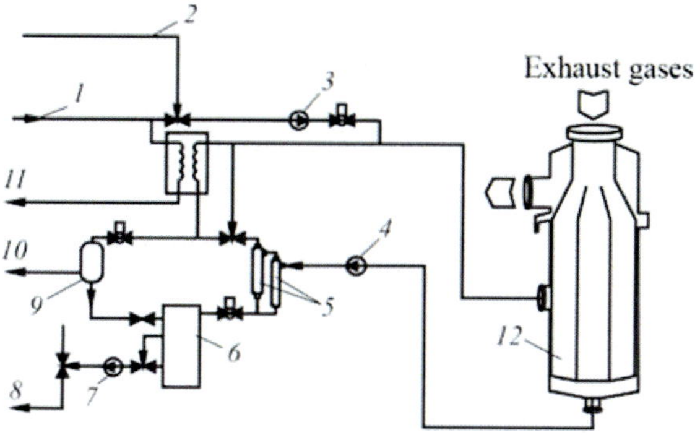

Fig. 3.51 Exhaust gas purification system: *1, 2*—sea and fresh water, respectively; *3, 4, 7*—feed, circulating and sludge pumps, respectively; *5*—hydrocyclone; *6*—tank of sludge; *8*—overboard discharge of sludge; *9*—slurry separator; *10, 11*—discharge of purified and cooling seawater; *12*—scrubber

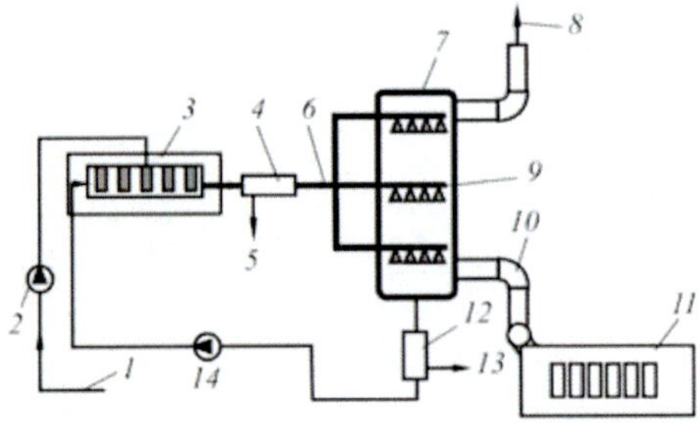

Fig. 3.52 Principal scheme of the $CSNO_x$ system: *1*—seawater from the Kingston box; *2, 14*—seawater pumps; *3*—water preparation generator with enhanced absorption properties; *4, 12*—filtration plant; *5, 13*—waste discharge from the filtration plant; *6, 10*—water and exhaust gas supply to the scrubber, respectively; *7*—scrubber; *8*—yield of purified gases of the ME; *9*—spraying of water in the scrubber; *11*—ME

process takes place overboard at a distance of several dozen meters from the point of discharge [2, 8].

An alternative to the above method of purification of the exhaust gases is the $CSNO_x$ system developed by Ecospec Global Technology [8]. It provides a reduction in the content of CO_2, SO_2, and NO_x (Fig. 3.52).

A special feature of the $CSNO_x$ system is the use of pretreated seawater in order to increase alkalinity and pH.

Increased pH and alkalinity sharply reduce the oxidative action of CO_2, SO_2 and NO_x during scrubbing. The treatment, which significantly increases the absorption properties of water, is carried out with the help of electrical devices without chemical reagents. When exhaust gas reacts with sprayed pretreated water with increased absorption properties, the following occurs:

- CO_2 is converted into bicarbonates, which are natural for seawater;
- SO_2 and NO_x are converted to sulfates and nitrates, respectively, which are also natural for the water environment;
- solid particles (SP) after treatment of the exhaust gases in the scrubber are removed and separated with liquid waste.

The test of the $CSNO_x$ system installed on a 105660-ton deadweight tanker built under the supervision of the ABC Classification Society (USA) showed that emissions of the main pollutants decreased by 92.9%, 82.2% and 74.4% for SO_2, NO_x and CO_2, respectively. The content of the listed substances at the engine outlet was 669.3 ppm for SO_2, 158.5 ppm for $NO_x$and 5.16 ppm for CO_2, and at the scrubber outlet—47.4 ppm of SO_2, 28.2 ppm of NO_x and 1.32 ppm of CO_2 [8].

The peculiarity of this method is that it practically excludes secondary pollution of the marine environment and does not deoxidize water, thus ensuring environmental protection. The estimated cost of such asystem is 100 $/kW of power of the main engine [8].

3.7 Placement of Diesel Power Plant Equipment in the Engine Room of the Ship

3.7.1 General Provisions

The first rule which defines the basic principles of placement of the MPP mechanisms is that they should be located at the spots where they will most effectively perform their functions.

The second rule: since the purpose of a commercial or passenger ship is transportation of goods or passengers, and this is what generates the profit, the volume of the hull that is allocated for the placement of machines, mechanisms, apparatus and devices should be minimal. The location of the power equipment is directly related to the location of the engine room (ER) along the length of the ship, which is determined by the purpose of the ship and its features.

In the world practice, four variants of the arrangement of the power plant premises along the ship length have been applied: stern, medium, intermediate (shifted to the aft from the midship frame) and bow. Each of the options has its advantages and disadvantages.

The stern arrangement of the power plant prevails on modern ships of different types and purposes. With this arrangement of ER, the conditions for cargo placement and handling are improved, the cargo capacity is increased, and the mass of the power plant is reduced because of the shortening of the shaft line and elimination of such a structural element as the tunnel of the propeller shaft. The upper deck does not have any superstructures along the entire length of the cargo holds.

The stern disposition of the ER has considerable disadvantages. Displacing the ER in the stern and cargo holds in the bow creates a problem of ship differentiation (equilibration in the longitudinal direction) both at full load and at ballast run. The location of the navigation bridge on the yut superstructure obstructs the visibility of the space in front of the ship and its controllability. In the living areas of the stern superstructure, noise and vibration from the operation of the propellers are more pronounced.

Recently, a certain number of ships with the bow arrangement of the ER have appeared; they are ships serving oil and gas production platforms, ships for transporting heavy cargo, ferries. The intermediate and medium ER positions are characteristic for special-purpose ships, icebreakers, lifeguards, modern container ships of a high cargo capacity, passenger ships, tug-lifters.

A special feature of the medium placement of the ER is a significant length of shaft lines which pass through cargo holds in watertight tunnels. The tunnels not only reduce the useful volume of cargo holds, but also complicate the mechanized intra-hold works.

The use of the intermediate placement of the ER is caused by the need to free the stern deck from the superstructure.

The number of compartments for the installation of the power plant should be minimal with their length being the shortest. On transport ships, the power plants are usually located in one compartment if its length does not exceed the permissible length at which the emergency unsinkability of the ship is ensured.

The power plant is divided into autonomous groups and placed in several compartments on passenger ships, icebreakers, nuclear ships and other ships for which there are increased requirements for the unsinkability and survivability of the MPP.

Placement of the power plant and its equipment on the ship must meet the requirements of the Rules of the Classification Society which are followed during design and construction of this ship.

The power plant should operate reliably in all possible operating modes, with a continuous roll of the ship up to 15° and a differential of up to 5°, emergency power sources and mechanisms—with a continuous roll up to 22.5° and a trim of up to 10° [3].

The machines, mechanisms, apparatuses and devices of the power plant are arranged in such a way that they can be approached freely and conveniently for maintenance and emergency repairs. There should be passages between the equipment with a width of not less than 0.6 m [10]. There should be vacant space for dismantled parts in the area of large mechanisms. At the same time, it is necessary to provide vacant space for lifting stators and rotors of turbines and turbochargers, covers and pistons of the engines, for replacing tubes in the heat exchangers.

To improve the survivability of the MPP, it is assumed to reserve individual mechanisms and apparatuses, the damage of which may lead to an accident or a failure of the plant. The list of mechanisms and apparatuses that are subject to reservation is regulated by the Rules of the Classification Society. Typically, such mechanisms include oil, fuel, feed pumps, cooling system pumps, oil coolers, fans, start-up air compressors, marine power plant generators.

Mechanisms and apparatuses are installed and secured on solid and rigid foundations. The foundations should be light, so one should place the equipment above the floras, stringers and around the bends. Decks and platforms are equipped with special reinforcement of the set if the foundation is not installed directly on the rigid connections of the floor covering.

It is allowed to install mechanisms and other equipment on the outer shell of the hull, watertight bulkheads, the walls of the shaftline tunnel or the fuel and oil tanks, provided they are fastened to the stiffeners or on the brackets that are welded to the ship plating near the stiffeners.

In addition to the ER premises, placement of the equipment of the power plant makes use of additional volumes of the shaftline corridor, engine shafts and chimney casings.

3.7.2 Arrangement of Machines, Mechanisms, Apparatuses, Devices and Systems

The placement of power plant equipment can be single- and multi-tiered. The latter provides that the power plant equipment is located on several levels: in the hold, on the platforms, on the intermediate decks and in the chimney hood. This increases the useful height of the ER and reduces its length. The single-tiered arrangement is widely applied on ferries, passenger ships and in other cases when not the length but the length of compartments is limited, and more stringent requirements are imposed on the survivability of the plant.

The main engines and gears are located in the lower part of the ER, in the hold, on the foundations connected with the bottom framing of the ship. Their placement should be such as to ensure reliability, maneuverability, ease of maintenance and operation of the ship. A sufficient area must be reserved around the ME for the placement of large parts extracted from the engine during repair. Above the engine, it is necessary to provide vacant space for removing pistons from the cylinders with a stem or connecting rod using special crane devices.

The auxiliary equipment must be located closer to the ME that it serves, close to the units connected by the common working fluid that serve the same ME. There should be space between the mechanisms for the possibility of their servicing. The use of diesel-reducer units fixes the position of MSE in relation to the gear. With the use of MSE and HSE with electric gear, the internal combustion engine with generators and MPE can be located in the same room or in different rooms. In the first case, the engines and generators are mounted on the platform above the MPE. The main propulsion engine is always (with the exception of the use of the propulsion-steering complex "Azipod"), even at the middle position of the ER, located in the last stern compartment, which is not suitable for cargo transportation.

Marine power station. Diesel generators, which are an important source of noise and vibration, are advisable to install on the dampers in the hold or on a platform parallel to the DP. When installing the DG on the platform, they must be reinforced. In order to reduce the noise of the DG, it is advisable to enclose it with screens or to place it in specially enclosed rooms.

The marine power station of special ships, the power of which is comparable to the power of the ME, can be completely taken to a separate noise-proof room—an auxiliary ER.

Placement of the shaft generator is associated with the location of power take-off: directly from the ME (in the bow or stern), the gearbox, or the propeller shaft line. Most often, the shaft generator is installed above the ER deck in the aft.

In case of use of a steam turbine generator in the SPP, it is placed on a platform, near the AB and RB.

An emergency DG is located in the room together with the switchboard, fuel tank and equipment that provides the DG operation; it is above the bulkhead deck and has an exit to the open deck.

The main switchboard is placed as close as possible to the current sources, as a rule, perpendicular to the diametric surface in places with the lowest concentration of gases, steam, dust and moisture. It can be located both in the hold and on the platform; if there is a central control station for the power plant, the MSB is placed there. The passage before the MSB must be at least 0.8–1.0 m (depending on the length of the board) and not less than 0.6–0.8 m from the back side [10].

Auxiliary boiler plant. The auxiliary boilers are most often installed on the platforms near the front bulkhead of the ER, closer to the places of steam consumption, such as tanks, living and service rooms, and cargo pump compartments with the steam drive. The boilers are enclosed by an oil-proof coaming with the height of at least 0.2 m and by a metal fence in the furnace area; the latter protects the equipment from the impact of flame if there will be any. The recovery boiler is placed in the ER shaft on the path of the gas outlet pipeline.

Mechanisms, apparatuses, devices, capacities of MPP systems. Fuel tanks cannot be located in the ER. Storage and sludge collecting tanks are adjacent to the ER; in this case, they have only one common wall. On the height of the ER, they are located at the platform level.

It is forbidden to place fuel tanks above the ramps, ME and machinery, steam boilers, gas pipes, electrical equipment and control posts. If this is necessary, there should be trays under the entire bottom area of the tank.

Fuel in an amount not less than the daily stock should be kept outside the double bottom.

Pumps, separators, filters and other equipment of the fuel system should be placed in the hold of the ER in a compact, preferably aggregate area, near the cleaned fuel storage tanks. Fuel pumping must be provided at least by two pumps, one of which is a reserve. If the power plant operates on heavy fuel, at least two separators are installed.

The main engine must have at least two fuel pumps: one is a reserve pump, and the other is with an independent drive. At that, the feed of the reserve pump should not be less than the feed of the main pump. With two or more MEs are located in the same ER, it is sufficient to have a single reserve pump with an independent drive. Screw and gear pumps are used for pumping fuel.

Oil reserves are stored in one or more compartments located in or near the ER. It is not recommended to store oil reserves in double-bottom tanks. Under the ME, there are waste oil tanks separated by cofferdams from the outer bottom and other compartments of the double-bottom space.

Each ME must have at least two circulating lubrication pumps with the feed of both of them that meets the maximum power mode. One of these pumps with an independent drive is the reserve pump. With two MEs in the same ER, it is sufficient to have one pump for each engine and one common reserve pump with an independent drive.

Pumps of the circulating cooling system are placed in the lowest part of the circuit. It is better to position the pump so that it receives liquid from the cooler and pumps it into the cooling cavity.

Seawater enters the cooling system pumps through the receiving fittings (kingston, clinket latch, swing gate) located on the kingston boxes or in the kingston-distributive channels. Location of the sea inlets for receiving cooling water protected by gratings should be such as to ensure the normal flow of water under different conditions of navigation. The seawater intake is assumed to take place mainly in the bow of the ER, and the discharge—in the stern of the port side.

Cooling systems are equipped with two pumps of the same power: the main and reserve one, located below the waterline.

On icebreakers and ice navigation ships, the seawater intake is carried out through ice boxes.

Mechanisms and apparatuses of ship systems (hold system pumps, hold water separators, fire pumps, pumps and apparatuses of water supply systems, incinerators) are placed in the hold of the ER, grouped according to functional characteristics.

Traps, decking, exits. Special platforms are built in several tiers for the convenience of servicing of the mechanisms and apparatuses at a significant height of the power plant premises. Decking is made of removable metal corrugated sheets on a metal grating.

Traps for moving in the vertical direction and exit from the premises of the power plant must be made of metal, with the width of at least 600 mm and the angle of inclination to the horizontal surface of 60°. The length of the trap march is up to 6 m.

Exits from the premises of the power plant should ensure the possibility of fast evacuation of the operating personnel in emergency situations. Each premise of the power plant must have at least two exits located at the opposite ends.

3.7.3 An Example of the Placement of Equipment for a Power Plant with a Low-Speed Engine in the Engine Room of a Container Ship

Let us consider the structure of the diesel MPP, the characteristics of its elements and the placement of the power equipment in the ER on the example of a real container ship "Jens Maersk" (Table 3.3) [3].

The characteristics of the MPP elements are given in accordance with the dimensions used by the firms producing power equipment.

The following materials are obligatory components of the ship technical documentation related to the MPP.

General characteristics of the ship. The "Jens Maersk" container ship is intended for transportation of 3000 TEU. The total displacement is 30166 tons, the length is 213 meters, the width is 31.7 meters, the draft is 10.1 meters, the speed is 14.2 knots.

Figures 3.53,3.54, 3.55, 3.56 and 3.57 show the arrangement of machinery and equipment in the engine room of this ship.

Table 3.3 Principal machinery

Engines, apparatus, boilers			
Main Engine—ME		Main Generator Engines	
Maker	NSD–MAN B &W	Maker	MAN B &W
Model	7K90MC-C Mk6	Type	Hole by 8L27/38
Type	Two stroke, single acting direct reversible, crosshead diesel engine with two constant pressure turbochargers and air coolers	Number of cylinders	8
		Bore	270 mm
		Stroke	380 mm
		Output MCR	2400 kW
Number of cylinders	7	Rating	2290 kW, 450 V AC, 60 Hz, at 720 rpm
Bore	900 mm	Turbocharger	MAN-B &W NR24/S
Stroke	2300 mm	Maker	MAN B &W
Output MCR	31990 kWat 104 rpm	Type	Holeby 6 L 27/38
Directionofrotation	Clockwiselookingfromaft	Number of cylinders	6
Specific fuel oil consumption	127.9 g/bhp per hour at 85% MCR	Bore	270 mm
Daily fuel oil consumption	111.6 t/day at MCR	Stroke	380 mm
Mean effective pressure	18.0 kg/cm^2	Output (MCR)	1800 kW
Maximum cylinder pressure at MCR	140 kg/cm^2	Rating	1720 kW, 450 V AC, 60 Hz, at 720 rpm
Turbochargers (2)	ABB-TPL85-B11	Turbocharger (1)	MAN-B&W NR20/S
Emergency Generator Engine		**Oil Fired Boiler**	
Maker	Demp A/S	Maker	AalborgIndustries
Type	MAN/D 2886 TO	Type	AQ-10/12 W

(continued)

Table 3.3 (continued)

Engines, apparatus, boilers			
Main Engine—ME		Main Generator Engines	
Number of cylinders	6	Maximumpower	3000 kg/h at 174 °C and 7 kg/cm^2
Bore	121 mm	**Exhaust Gas Boiler**	
Stroke	152 mm	Maker	AalborgIndustrie
Rating	218 kW, 450 V AC, 60 Hz, at 1800 rpm	Type	AQ-2
		Powerat 85% MCR	3000 kg/h at 174 °C and 7 kg/cm^2
Mechanisms, apparatuses and devices			
Machinery Name	Maker	Type	Power
Pumps			
ME Lubricating Oil Pumps	Mashinfabriken IRON A/S	CVLS 2-300/315 submerged vertical single stage centrifugal	698 m^3/h at4.5 kg/cm^2
ME/Generator Engines Fuel Oil Supply Pumps	IMO	ACG 052 N6 NTBO	11 m^3/h at4 kg/cm^2
Main Engine/Generator Engines Fuel Oil Circulating Pumps	IMO	ACG 070 N6 IVBP	21 m^3/h at4 kg/cm^2
ME Turbocharger Lubricating Oil Pump	IMO	ACG 070 N6 IVBO	30 m^3/h at 4 kg/cm^2
ME Cylinder Lubricating Oil Transfer Pump	IMO	ACE 032 L3 NVBP	1.9 m^3/h at 3 kg/cm^2
ME Lubricating Oil Transfer Pump	IMO	ACG 052 K6 IVBO5	10 m^3/h at 3 kg/cm^2
ME Camshaft Lubricating Oil Pump	IMO	ACG 052 N6 IVBO	12 m^3/h at4 kg/cm^2
ME Air Cooler Cleaning Unit	C.C. Jensen	Grundfos CRN2-40 APG BUBV	3 m^3/h at3 kg/cm^2

(continued)

Table 3.3 (continued)

Engines, apparatus, boilers			
Main Engine—ME		Main Generator Engines	
ME Camshaft Lubricating Oil Bypass Cleaning Unit	Karberg and Hennemann	CJC	0.3 m^3/h at 0.8–2.8 kg/cm^2
Stern Tube Bearing	Consilium	Railko WA80 HS	–
Stern Tube Bearing Lubricating Oil Circulating Pump	IMO	ACE 038 N3 NVBP	4.5 m^3/h at 2.0 kg/cm^2
Stern Tube Dosing Unit	B & V Industrietechnik	VA-180	3.1 L/h at 46.0 kg/cm^2
Stern Tube Lubricating Oil Transfer Pumps	IMO	ACE 038 N3 NVBP	4.5 m^3/h at 2.0 kg/cm^2
Main Generator Engines Diesel Oil Supply Pump	IMO	ACG 052 N6 IVBP	9.3 m^3/h at 8.0 kg/cm^2
Main Generator Engines Emergency Diesel Oil Service Pump	IMO	ACG 052 N6 IVBP Air operated	8.1 m^3/h at 8.0 kg/cm^2
Main Generator Engines Lubricating Oil Transfer Pump	IMO	ACE 032 L3 NVBP	1.9 m^3/h at 3.0 kg/cm^2
Boiler Feed Water Pumps	Aalborg Industries	CR 4-120 centrifugal	12 kg/cm^2 at 4 m^3/h
Boiler Fuel Oil Service Pump Unit	Aalborg Industries	ZALV/ZASV 850	18 L/min at 5.0 kg/cm^2
Boiler Waste Oil Transfer Pump	Aalborg Industries	ANBP 6.2 HV01 rotary positive displacement	0.95 m^3/hat 5 kg/cm^2
Diesel Oil Ignition Pump	Aalborg Industries	2	61 L/h at 7 kg/cm^2
Ballast Pump	Behrens	VRW 7/350S G vertical single stage centrifugal	500 m^3/h at 2.3 kg/cm^2

(continued)

Table 3.3 (continued)

Engines, apparatus, boilers			
Main Engine—ME		Main Generator Engines	
Ballast/Bilge Pump	Behrens	VRW 7/350S G vertical single stage centrifugal	500/250 m^3/h at 2.3/3.4 kg/cm^2
Bilge/Fire Pump	Behrens	VRW 5/420 vertical single stage centrifugal	220/140 m^3/h at 5.8/7.0 kg/cm^2
Ballast/Bilge Ejector	Korting Hannover AG	13.37. So	100 m^3/h at 1.0 kg/cm^2
Engine Room Bilge Pump	Behrens	V 136-70 Q M HD	25 m^3/h at 5.0 kg/cm^2
Bilge Water Separator	B + V Industrietechnik Gmb	Turbolo—Oily Water Separator TCS 5 HD	5.0 m^3/h
Cargo Hold Bilge Pump	Behrens	V 136-210 Q M HD vertical single stage centrifugal	90 m^3/h at 3.0 kg/cm^2
Bow Thruster Room Emergency Fire Pump	Behrens	VRW 5/350 G vertical single stage centrifugal	75 m^3/h at 7.0 kg/cm^2
Engine Room Fire Pump	Behrens	VRW 5/350 G vertical single stage centrifugal	140 m^3/h at 7.0 kg/cm^2
Heeling Pump	Iron	QT2—300 horizontal reversible two stage propellor	550 m^3/h at 14 mth
Seawater Pump- Fresh Water Evaporator	Behrens	VRW 3/350 G vertical single stage centrifugal	70 m^3/h at 5.0 kg/cm^2
Main Sea Water Pumps Central Fresh Water Coolers	Behrens	VRW 9/350G	725 m^3/h at 2.5 kg/cm^2
Fresh Water Cooling Pumps Low Temperature	Behrens	VRF 9/350 G	677 m^3/h at 3.0 kg/cm^2

(continued)

Table 3.3 (continued)

Engines, apparatus, boilers			
Main Engine—ME		Main Generator Engines	
Fresh Water Cooling Pumps High Temperature	Behrens	VRF 5/350S G	230 m^3/h at 3.0 kg/cm^2
Diesel Oil Transfer Pump	IMO	ACG 070 K6 IVBO	22 m^3/h at 3.0 kg/cm^2
Heavy Fuel Oil Transfer Pump	IMO	ACF 110 L4 IVBO	110 m^3/h at 4.0 kg/cm^2
SludgeTransferPump	IMO	AEB1E750	40 m^3/h at 5 kg/cm^2
Calorifier	Osco	17RRDE	2.000 L/h at 40 k steam, 20 k electrical
Hot Water Circulating Pump	Allweiler	UP 25-72	1–3 m^3/h at 0.3–0.5 kg/cm^2
FreshWaterHydrophorePumps	Allweiler	SOB 223 W G2V	5 m^3/h at 6.0 kg/cm^2
Fresh Water Pump for Water Fog Firefighting Plant	Danserv A/S	S 70-50-220	–
FreshWaterEvaporator	Sondex A/S	SFD 23/30	30 $m^{3/24}$ h
Reefer Fresh Water Cooling Pumps	Berhens	VRF 7/350 vertical single stage centrifugal	400 m^3/h at 5.0 kg/cm^2
Reefer Sea Water Cooling Pumps	Berhens	VRF 5/350S G vertical single stage centrifugal	240 m^3/h at 3.0 kg/cm^2
Purifiers			
HeavyFuelOil	Westfalia	OSC 50-0136-066—self-cleaning	5400 L/h
DieselOil	Westfalia	OSC 50-0136-066–self-cleaning	5400 L/h
LubricatingOil ME	Westfalia	OSC 30-0196-066/2–self-cleaning	5500 L/h

(continued)

Table 3.3 (continued)

Engines, apparatus, boilers			
Main Engine—ME		Main Generator Engines	
Lubricating Oil Diesel Generator Engines	Westfalia	OSC 15-0196-067/7–self-cleaning	1420 L/h
Aircompressors			
StartingAirCompressors	Tanabe	H-274 piston	30 kg/cm^2 at 400 m^3/h
Working Air Compressors	Tanabe	TASK-1018G	8.0 kg/cm^2 at 150 m^3/h
Provisions Refrigerating Compressors		**Workshop Air Conditioning Plant**	
Maker	York	Maker	York
Type	4 V-62Y	Type	KGS 120 F 60
Power	20 k	Power	12 k—cooling, 15 k—heating
Refrigerant	R134A	Refrigerant	R134A
Accommodation Air Conditioning Plant Compressors		**sSteering Gear**	
Maker	York	Maker	Hatlapa
Type	SMC-108S	Type	Teleram R4ST 550 h
SerialNos	115639 and 115640	Pumptype	A4VG 56
Refrigerant	R134A	**Stern Tube Bearing**	
Control Room Air Conditioning Plant		Maker	Consilium
Maker	York	Type	Railko WA80 HS
Type	KGS 180 F 60	**Thrusters Bow and Stern**	
Power	18 k cooling, 9 k heating	Maker	Kawasaki
Refrigerant	R134A	Type	Electrically driven KT-130B3

(continued)

Table 3.3 (continued)

Engines, apparatus, boilers		
Main Engine—ME	Main Generator Engines	
	Thrust	159 kN
	Primemover	1100 kW
	Propellerspeed	195 rpm

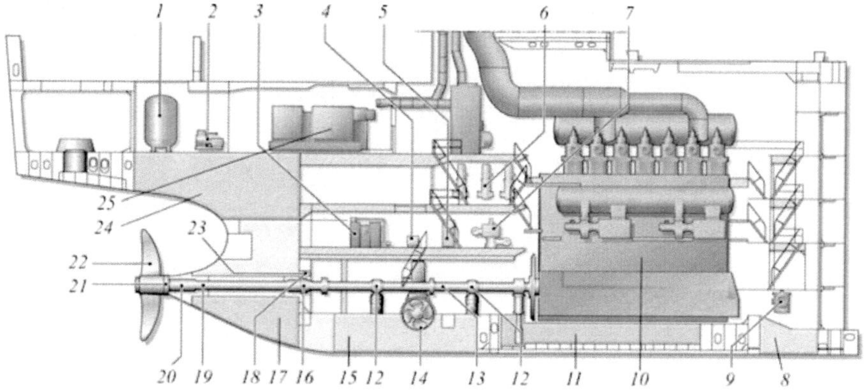

Fig. 3.53 Location plan of engine room—main engine room elevation: *1*—start air vessel for main and auxiliary engines; *2*—start air compressor; *3*—sewage collecting unit; *4*—hydraulic power pack for stern thruster; *5*—pmcs4 control cabinet; *6*—fresh water cooling pumps (low temperature); *7*—LO automatic filter with bypass filter; *8*—waste oil tank; *9*—main seawater filter; *10*—me; *11*—lubricating oil tank; *12*—intermediate shaft bearing; *13*—intermediate shaft (1); *14*—stern thruster; *15*—bilge water tank; *16*—stern tube seal (forward); *17*—fresh water cooling pump; *18*—forward seal gravity tank; *19*—propeller shaft; 20—stern tube bearing; *21*—stern tube seal (aft); *22*—fixed propeller; *23*—forward gravity tank; *24*—water ballast tank; *25*—auxiliary engine (4)

3.8 The World's Leading Manufacturers of Marine Diesels

The world leader in the production of low-speed engines is MAN Diesel & Turbo, which was formed in 1980 by merging of two diesel-building companies—MAN (Germany) and Burmeister & Wain (Denmark). These were two famous, large firms that had had considerable experience in the production of diesel engines.

Burmeister & Wain was established in 1843. Manufacture of diesel engines was started in 1904, and the first ship four-stroke eight-cylinder crosshead diesel engine with the capacity of 1250 hp at the speed of 140 rpm was produced in 1912. The first two-stroke diesel with the power of 1000 hp with oppositely moving pistons was put into operation in 1930. The production of two-stroke diesel engines with the direct-valve scavenging was mastered in 1939; the first of them was a diesel with a cylinder power of 610 hp at the rotation speed of 122 rpm [11].

Introduction of the gas turbine charging of two-stroke diesel engines in 1952 allowed increasing the cylinder power by 35% and reaching the maximum aggregate power of 15000 hp. In 1959, there was launched production of the engines with the cylinder diameter of 840 mm and the maximum aggregate power of 25200 hp. In the late 1960s, a diesel engine with the cylinder diameter of 980 mm was created, which allowed offering engines with an aggregate power of up to 70000 hp. By that time, the company included 11 production and research departments, some of which used to be separate enterprises. For instance, the plant Alfa Diesel established in 1883 became a subsidiary of the concern in 1938; it is presently engaged in the production of four-stroke diesel engines in conjunction with a propulsion plant.

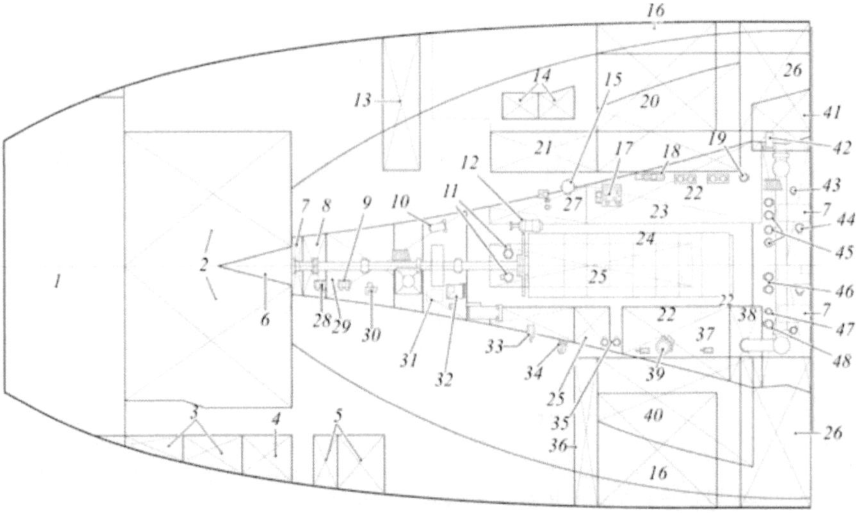

Fig. 3.54 Location plan of engine room—tank top level: *1*—void; *2*—water ballast; *3*—cylinder oil storage tank; *4*—LO cleaning tank; *5*—LO storage tank—generator engines and main engine; *6*—LO purifiers sludge tank; *7*—bilge well; *8*—stern tube lubricating oil drain tank; *9*—stern tube LO dosing pump; *10*—bilge pump; *11*—main engine lubricating oil pump; *12*—main engine turning gear; *13*—distilled water tank; *14*—HFO service tank; *15*—scavenge air drain tank; *16*—marine diesel oil storage tank; *17*—camshaft lubricating oil pumps and cooler; *18*—sludge transfer pump; *19*—fire pump; *20*—cargo hold bilge tank; *21*—sludge tank for fuel oil purifiers; *22*—HFO and do purifiers; *23*—HFO and MDO overflow tank; *24*—main engine; *25*—main engine LO sump tank; *26*—fresh water tank; *27*—fuel oil drain tank; *28*—stern tube lubricating oil circulating pump; *29*—bilge water tank (clean); *30*—hydraulic power pack for stern thruster; *31*—turbocharger air cooler cleaning water drain tank; *32*—main engine turbocharger LO unit; *33*—main engine LO purifier supply pump; *34*—LO transfer pump; *35*—turbocharger LO pumps; *36*—marine diesel oil service tank; *37*—MDO transfer pump; *38*—deep sea chest; *39*—bilge water separator; *40*—water ballast side tank; *41*—high sea chest; *42*—cargo hold bilge pump; *43*—bilge/fire pump; *44*—ballast/bilge pump; *45*—reefer SW circulating pumps; *46*, *48*—main SW supply pumps № 1, 2, 3; *47*—evaporator SW supply pump

In the 1970s, B&W was the second with its annual production of 2.5–3.0 million hp (Sulzer was the first, and MAN was the third). At that time, the French company Pielstik was in the lead of the MSE production, with MAN and Sulzer making up the top three. After MAN had acquired the controlling stake of the Danish firm that was on the border of bankruptcy and the two powerful firms merged in the late 1970s, the newly established company took the leading position in the world market of the LSE production.

MAN (Mashinenfabric Augsburg Nurnberg AG) is the oldest and one of the world's largest designers and manufacturers of diesel engines. From 1897 (when the first Rudolf Diesel engine with the power of 20 hp was built in Augsburg) to 1979, the total power of the diesel engines produced by MAN was 60.5 million hp [11].

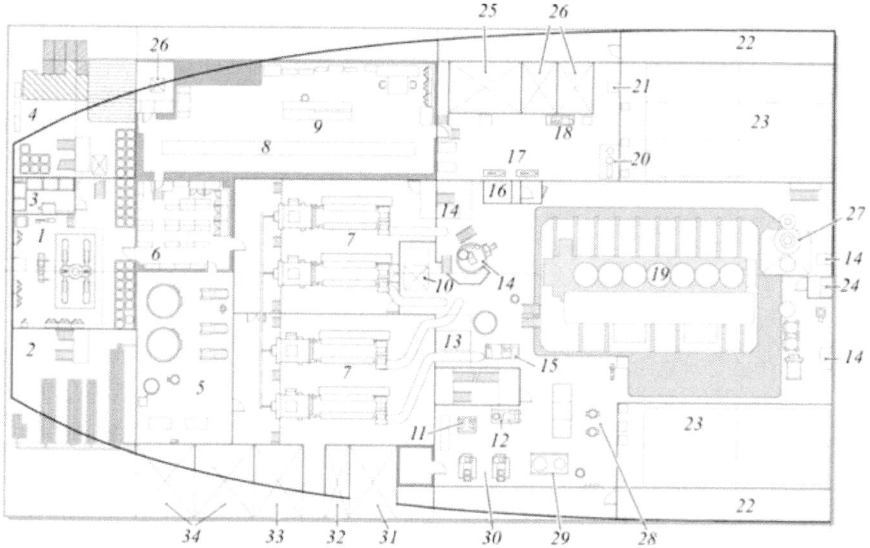

Fig. 3.55 Location plan of engine room—3rd deck level: *1*—steering gear room; *2*—CO_2 room; *3*—paint store; *4*—rope store; *5*—compressor room; *6*—engine room store; *7*—diesel generators № 1, 2, 3, 4; *8*—main switchboard; *9*—engine control room; *10*—hatch; *11*—provisions refrigeration plant; *12*—accommodation hot water heating unit; *13*—vent; *14*—oil fired boiler; *15*—condensate collection tank; *16*—elevator shaft; *17*—boiler HFO and waste oil heaters; *18*—hydraulic valve power pack; *19*—ME; *20*—waste oil tank and pump; *21*—cylinder oil service tank; *22*—passageway; *23*—cargo hold port/starboard № 6; *24*—emergency exit; *25*—HFO settling tank; *26*—HFO daily service tank; *27*—spare main engine piston, cylinder cover and liner; *28*—fresh water cooling pumps (high temperature); *29*—calorifier and hydrophore tank; *30*—accommodation air conditioning plant; *31*—LO storage tank—main engine; *32*—LO storage tank—generator engines; *33*—LO cleaning tank; *34*—cylinder oil storage tank

Since 1934, the company has been developing and producing turbochargers of its own design for charging diesel engines. To date, almost all the engines of the merged company (both LSEs and MSEs) are equipped with proprietary turbochargers.

The diesels of MAN Burmeister & Wain (known as MAN Diesel & Turbo since 2010) are produced by 34 licensees in 19 countries. To service the engines in operation, there have been created 30 service agencies in 18 countries and 10 foreign branches for selling products.

The main scientific and technical centers of the company are located in the cities of Augsburg (Germany) and Copenhagen (Denmark). At present, the company has the largest power range of LSE and MSE, which meets all the requirements of real and potential customers. The power of the LSEs produced annually by the company over the past 10 years is 58–65% of the world's engine power production; the same indicator for MSE is about 20%.

The Swiss company Sulzer started production of ICEs in 1898. The first ship non-reversible diesel engine with the power of 40 hp was built in 1904, and in 1909 the first reversible four-cylinder two-stroke diesel engine was produced. It has the

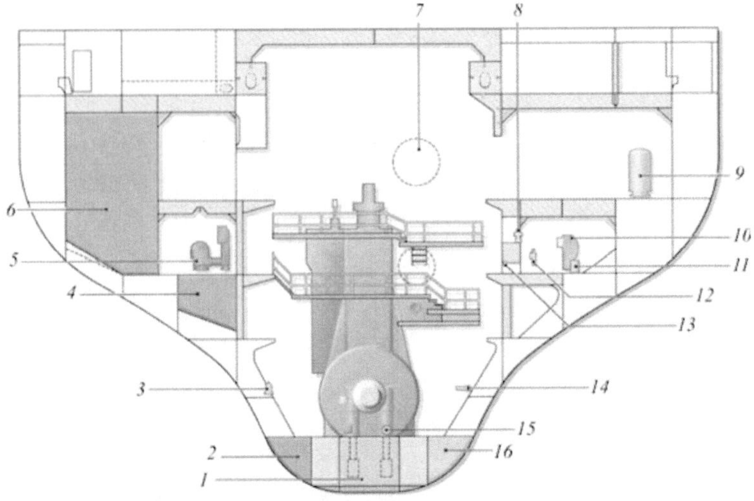

Fig. 3.56 Location plan cross section—main engine room frame 40: *1*—lubricating oil circulating tank; *2*—fuel oil drain tank; *3*—camshaft lubricating oil bypass cleaning unit; *4*—separators fuel oil sludge tank; *5*—fuel oil purifier; *6*—fuel oil daily service tank; *7*—ME exhaust drum; *8*—auxiliary diesel oil supply pump; *9*—hydrophore tank; *10*—evaporator; *11*—evaporator dosing unit; *12*—suction filter; *13*—boiler diesel oil ignition pump; *14*—ME purifier supply pump; *15*—ME lubricating oil pump; *16*—cooling water drain tank

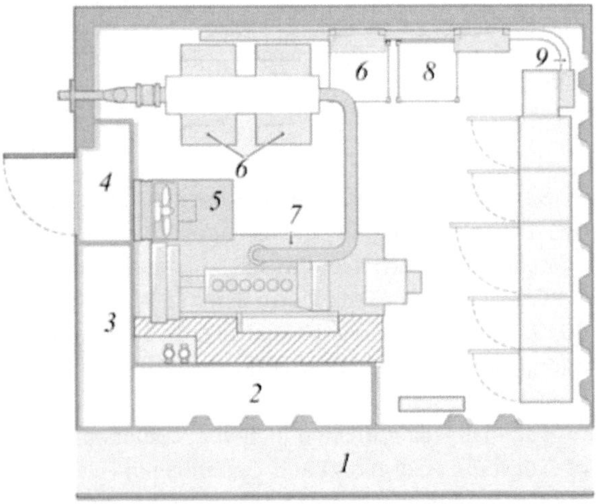

Fig. 3.57 Location of the engine room—emergency generator room: *1*—cofferdam; *2*—diesel oil daily use tank for emergency engine 3.6 m^3; *3*—exhaust air; *4*—supply air above; *5*—vent fan above; *6*—emergency transformers 440/230 V; *7*—emergency generator engine 180 kW/1800 rpm; *8*—starting battery; *9*—cables

power of 90 hp, the cylinder diameter of 175 mm, the stroke of the piston of 250 mm and the rotation speed of 375 rpm.

The diesel had a direct scavenging with exhaust valves in the cylinder cover and inlet ports in the cylinder liner.

Loop scavenging was used in the further designs of two-stroke Sulzer engines. In 1950, the company replaced the massive cast-iron structures of the engine housing with welded structures, which led to the reduction in the engine mass by almost 15%.

Sulzer has always paid a lot of attention to charging in diesel engines. The first research sample of the charged four-cylinder diesel engine was tested in 1924. The engine power was increased by 25%. In 1946, a two-stroke trunk six-cylinder diesel with the cylinder diameter of 480 mm was produced, and its power was increased by 20% due to turbocharging.

For a long time from 1950 to 1980, the company (Sulzer Brothers Ltd by that time) was ranked the first in the world for the production of low-speed diesel engines.

At present, New Sulzer Diesel Ltd is a part of the concern Wärtsilä. It is engaged in various activities, and the main one is the development and production of LSE, MSE and HSE.

In the period of its establishment, Wärtsilä Diesel (founded in 1954) produced two-stroke diesel engines of the K50E series under the license of the Swedish company AB BoforsNohab. It launched production of two-stroke M51 engines in 1957, and Z40/48 engines in 1964 (under license of Sulzer).

In 1960, the company started producing a proprietary engine of the type 14; its cylinder volume was 14 L. Around 770 pieces of such engines were made; they were made without charging first, and then the feature was added to the engines of the 14T and 14TK series. Based on the advanced 14TK engine, a Vasa diesel engine of the 24 series was created; its cylinder diameter was 240 mm. These engines were the forerunners of modern engines, being produced in 1964; they were supplied in the form of a series model with four, five, six, or eight cylinders.

In January 1979, the diesel-building sector of Wärtsilä bought Nohab Diesel Division of the Swedish company AB BoforsNohab in the city of Trollhättien and established the subsidiary BoforsNohab AB on its basis for the creation of a full range of ship MSEs. This allowed Wärtsilä to launch the engines with the power of 700–5800 kW into the world market and accelerate their further improvement.

About 14500 workers were employed at 18 different Wärtsilä enterprises. By the annual production of four-stroke marine diesel engines, Wärtsilä was ranked as the world's 6th and 7th in 1981 and 1982 (compare to 18th and 25th in 1980 and 1979, respectively). In the late 1980s, the company opened Stork-Wärtsilä Diesel in the Netherland, WärtsiläSacm Diesel in France, branches in Sweden and Norway.

As a result of such a rigorous activity, in 1991 the company was ranked the 4th in the world in terms of the total power of the produced MSEs and HSEs of its own design, inferior only to such companies as MAN, Pielstik and Sulzer. After the acquisition of the controlling stake of New Sulzer Diesel in 1997, Wärtsilä NSD became the world's leader in the MSE production.

Currently, in addition to its 10 enterprises in 7 countries, Wärtsilä Corporation has licensees in Brazil and South Korea and representative offices in 50 countries. The company offers the power range of 5000–80000 kW for LSEs and 700–20000 kW for MSEs, which are in demand on the world market due to their economic efficiency, reliability and ability to work on heavy fuel.

The Japanese company Mitsubishi Heavy Industries Ltd was founded in 1857. At first, it was a small metal-processing plant in the city of Nagasaki, which was specialized in shipbuilding. Then in 1904, a scientific research institute was founded there and launched the creation of an internal combustion engine in 1925. The development of marine diesel engines was started in 1930. From 1932 to 1955, Mitsubishi was producing two-stroke ship engines with transverse scavenging. In 1948, the development of UE diesel engines with direct-flow scavenging and gas turbine charging was started in Nagasaki; their production began in 1955. In the same year, the company began to produce trunk MSEs of its own design and made extensive steps to create a single standard size range for two- and four-stroke diesel engines [11].

The research conducted by the company became the basis for the creation of a basic model of the LSE of the UEC series, a basic model of the MSE of the UET series, and a standard size range of HSEs of proprietary design. Nevertheless, the main products of the company were the LSEs.

Development of the turbochargers of the MET series allowed the company to return to the single-stage charging instead of the two-stage system, which had been used in diesel engines since 1975.

Over the past 20 years, the company has developed seven modifications of its diesel engines, raising the cylinder power capacity almost by twice.

The aspiration to increase the efficiency of the engines prompted the company to create long-running models with the reduced rotation speed.

The three major Mitsubishi diesel-engine plants are located in the cities of Nagasaki, Kobe and Yokohama. In addition to its own diesel engines, the company produces marine diesels under the license of MAN B&W. A substantial part of diesel engines is produced by such licensees as the Japanese firms Kobe Diesel, Akasaka Diesel and Ube Industries. The company currently produces 19 types of LSEs with the power of 1120–46800 kW (Table 3.4). Along with diesel engines, the company produces other ship equipment: turbochargers, steam turbines and boilers, turbine generators, pumps, cranes, as well as propellers under the license of the CaMeWa.

Within the last 5-7 years, Caterpillar Marine Power Systems has taken the leading position in the world for the production of HSE and MSE for use on the ships of the river and marine fleet both as main and auxiliary engines. The power range of this company's engines is 90–16000 kW. In 1997, Caterpillar bought the shares of MaK-Motoren.

The German machine-building company MaK-Mashinenbau Kiel, which was a subsidiary of F. Krupp, was founded in 1997 on the basis of the Deutsche Worke plants. Production of diesel engines began in 1930; and since 1964, the company belongs to Freed Krupp AG. Before joining Caterpillar, it was called Krupp MaKMashinenbau GmbH; its main enterprise is in Kiel.

While the power of the engines produced before 1948 by the Deutsche Worke plants did not exceed 440 kW, MaK produced the engines with the power of 1000–16000 kW. For the total power of MSEs that were used on ships as the main ones MaK was constantly in the top five.

Table 3.4 shows the manufacturers of ship ICEs. The characteristics of modern ICEs are presented in Appendix 1.

Table 3.4 Manufacturers of ship internal combustion engines [3]

Company	Country	Engine type, power rate, kW
ABC Anglo Belgian Corporation	Belgium	MSE, 330–1060 HSE, 1100–3530
Calesen Diesel	Denmark	MSE, 130–610
Caterpillar Marine-Power Systems	Germany	MSE, 1800–16000 HSE, 93–5400
CRM Spa Motori Marins	Italy	HSE, 687–1545
Cummins Engine Company Ltd	Great Britain	HSE, 373–1864
Daihatsu Diesel MFG Co Ltd	Japan	HSE, 56–3844
Fornaut (SA)	France	HSE, 250–400
GE Marine	The USA	HSE, 1308–4661
Grenaa Motorfabrik AS	Denmark	MSE, 442–758 HSE, 780–1200
Guascor SA	Spain	HSE, 150–960
The Hanshin Diesel Works Ltd	Japan	MSE, 478–1471
Isotta Fraschini Motori SpA (Fincantieri group)	Italy	HSE, 220–2600
Kelvin Diesels	Great Britain	HSE, 168–560
Makita Corporation	Japan	MSE, 1600–12640 HSE, 1105–1470
MAN Diesel & Turbo	Denmark	LSE, 1350–87220 MSE, 7860–21600 HSE, 960–9000
Matsui Iron Works Co Ltd	Japan	MSE, 147–1690
Mitsubishi Heavy Industries Ltd	Japan	LSE, 1120–46800
MTU Friedrichshafen GmbH	Germany	HSE, 125–9000
Niigata Power Systems Co, Ltd	Japan	MSE, 882–2794 HSE, 404–4000
Rolls-Royce Marine AS, Engine–Bergen	Norway	MSE, 1440–8750
Scania	Sweden	HSE, 155–588
SEMT Pielstik	France	MSE, 3300–28000 HSE, 740–8100
SKL Motoren und Systemtechnik GmbH	Germany	MSE, 736–1100 HSE, 670–2350
Sisu Diesel Inc.	Finland	HSE, 140–300
Volvo Penta AB	Sweden	HSE, 81–1380
Wartsila Corporation	Finland	LSE, 5100–80080 MSE, 2010–20000 HSE, 720–5200
Yanmar Diesel Engine Co Ltd	Japan	MSE, 1470–3310 HSE, 368–1324

References

1. *Morskoyentsiklopedicheskiyslovar: v 3 t.* [Marine Encyclopedic Dictionary: in 3 volumes]. Leningrad, Sudostroenie Publ., 1991, Vol. 3, 488 p.
2. Voznitskiy I. V. *Sudovyedvigatelivnutrennegosgoraniya: uchebnik: v 2 t. T. 1* [Ship internal combustion engines: textbook. Ch. 1]. Moscow, Morkniga Publ., 2008. 283 p.
3. Horbaov V. M. *Entsyklopediiasudnovoienerhetyky: pidruchnyk* [Encyclopedia of marine power engineering: textbook]. Mykolaiv, NUK Publ., 2010. 624 p.
4. Pakhomov Yu. A. *Sudovyeenergeticheskieustanovki s dvigatelyamivnut-rennegosgoraniya: uchebnik* [Marine power plants with internal combustion engines: textbook]. Moscow, TransLit Publ., 2007. 528 p.
5. *Morskoyentsiklopedicheskiyslovar: v 3 t.* [Marine Encyclopedic Dictionary: in 3 volumes]. Leningrad, Sudostroenie Publ., 1991, Vol. 1, 504 p.
6. Horbov V. M., Ratushniak I. O., Trushliakov Ye. I., Cherednichenko O. K. *Sudnovaenerhetyka ta Svitovyiokean:pidruchnyk* [Marine power engineering and the World Ocean: textbook]. Mykolaiv, NUK Publ., 2007. 596 p.
7. Gorbov V. M., Mitenkova V. S. *Alternativnyetopliva v sudovykhenergeticheskikhustanovkakh* [Alternative fuels in marine power plants]. *Sudokhodstvo* [ShipNavigation], 2007, no. 1–2 (127), pp. 64–65.
8. Fournier A. Controlling air emissions from marine vessels: Problem and opportunities. University of California, 2006. 85 p.
9. Gorbov V. M. *Primenenievodotoplivnykhemulsiy v sudovoyenergetike: ucheb. posobie* [Application of water fuel emulsions in marine power engineering: textbook]. Nikolaev, NKI Publ., 1991. 54 p.
10. *RehistrsudnoplavstvaUkrainy. Pravylaklasyfikatsii ta pobudovymorskykhsuden* [The Register of Shipping of Ukraine. Rules for classification and construction of sea vessels]. Kyiv, RehistrsudnoplavstvaUkrainy Publ., 2002, ch. 1, 394 p.
11. Horbov V. M., Shapovalov Yu. O., Ratushniak I. O. *Holovnidvyhunysuchasnykhtransport-nykhsuden:navch. posib.* [The main engines of modern transport ships: textbook]. Mykolaiv, UDMTU Publ., 1998. 72 p.

Chapter 4
Matching Characteristic of Hull, Enging and Propeller

To get higher efficiency of propeller and lower fuel consumption of main engine, it is necessary to study the match characteristic of hull, engine and propeller and find suitable propeller and main engine.

4.1 Operation Characteristics of Ship, Propeller and Main Engine

To know the match characteristic of hull, engine and propeller, the characteristic of each part should be understood.

4.1.1 The Speed Characteristics of Diesel Engine

For the diesel engine, there are several characteristics such as speed characteristics (sometimes are called external characteristics), load characteristics, governing characteristics, propulsion characteristics etc. The engine used to drive a propeller will follow a propulsion characteristics, but to find the match between the engine and propeller, the speed characteristics is used. For a diesel engine, it is known:

$$P_e = \frac{p_e \cdot V_h \cdot n \cdot i}{30\,m} = Cp_e \cdot n$$

$$M_e = 9.55\frac{P_e}{n_e} = C'p_e$$

Z. Yang et al., *Marine Power Plant*,
https://doi.org/10.1007/978-981-33-4935-3_4

where:

P_e—effective power output of engine.
M_e—effective torque.
p_e—mean effective pressure in the cylinder.
V_h—working volume of each cylinder.
n—speed of engine.
i—number of cylinders.
C, C′—constant.

The speed characteristics will be got through the real test for different engine, see Fig. 4.1. It is nearly a linear relationship between power and speed of engine. When the position of injection racker is fixed, the mean effective pressure is almost fixed, but in a real test, as a difference of friction at different speed, the mean effective pressure may change a little. The relationship between torque and speed is nearly a horizontal line, it is just the same reason. The different line show different racker position, that is different fuel injection quantities of each working cycle. Line 1 is 1 h characteristics, the engine can not be run for long time, line 2 is 12 h characteristics, overload characteristics, line 3 is rated load characteristics, line 4 is part load characteristics. Different line shows the different position of pump's racker. Normally, the engine should be working under the line 3.

To keep an engine work in order, it is limited by several factors, see Fig. 4.2, the operation range of the engine. MCR is maximum continuous rating of engine. It corresponds with 100% effective power output (P_e) and 100% engine speed (n_e). Line 1 is limitation of mechanical load, it shows the relationship between effective power and engine speed at rated torque moment (M_e). Line 2 is the limitation of

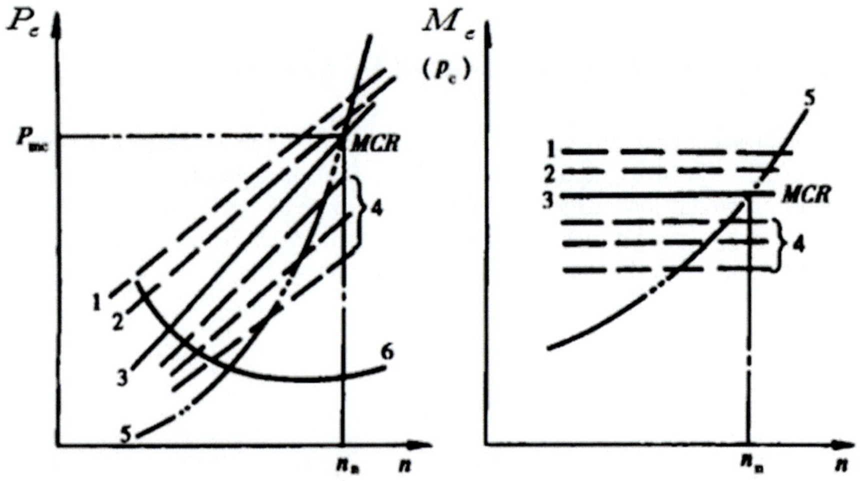

Fig. 4.1 The speed characteristics of diesel engine. *1*-one hour power limit 2–12 h limit *3*-rated power line *4*-part load line *5*-propulsion curve

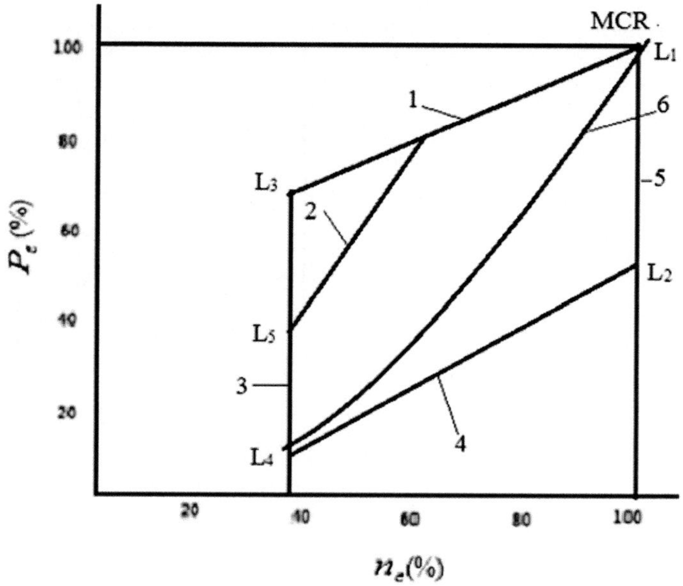

Fig. 4.2 Operation range of diesel engine

thermal load. Line 3 is minimum speed limitation, and line 5 is maximum speed limitation. Line 4 is lowest load limitation. The engine should be operated in area of $L_1L_2L_4L_5$. Line 6 is propulsion characteristics line.

4.1.2 *Propeller Characteristic*

According to the Ship propulsion principle, there are following characteristics:

$$T = K_T \rho n_p^2 D^4$$
$$M_p = K_Q \rho n_p^2 D^5$$
$$K_T, K_Q = f(J)$$
$$J = \frac{h_p}{D} = \frac{v_p}{Dn_p}$$
$$\eta_p = \frac{Tv_p}{2\pi n_p M_p} = \frac{K_T \rho n_p^2 D^4 v_p}{2\pi n_p K_Q \rho n_p^2 D^5} = \frac{K_T}{K_Q} \frac{v_p}{2\pi n_p D} = \frac{K_T}{K_Q} \frac{J}{2\pi} = f(J)$$

where:

T—thrust created by the propeller.
M_p—torque received.
K_T—thrust coefficient.
K_Q—torque coefficient.

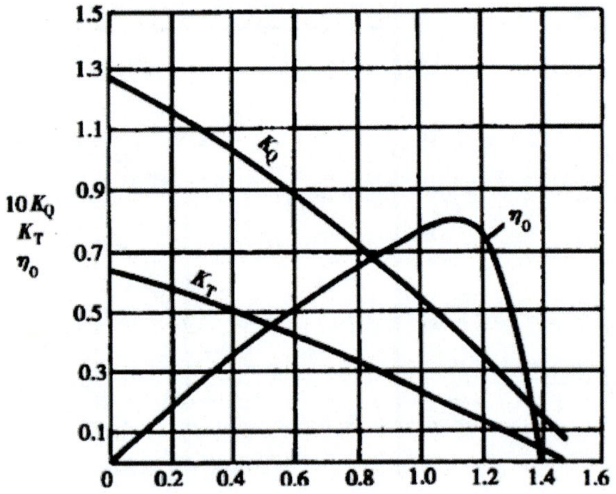

Fig. 4.3 Propeller characteristics

J—velocity coefficient of the propeller.
η_p—efficiency of propeller.
D—diameter of propeller.
ν_p—velocity of propeller at speed of n_p.

K_T, K_Q, η_p are function of J, see Fig. 4.3, K_T, K_Q both decrease with increase of J, but η_p increases first then decreases.

For a fixed propeller, at a stable constant speed, J keep constant, K_T, K_Q, could be treated as constant, so T, M_p, P_p are function of n_p:

$$T = C_1 n_p^2 \quad M_p = C_2 n_p^2$$
$$P_p = C n_p^3$$

When the operation condition changed and at new stable status, J would be at new value, and C_1, C_2, C also at new value, see Fig. 4.4, different J represent different operation condition.

1) $J\downarrow$, R $\uparrow$, $\nu_p\downarrow \rightarrow K_T$,$K_Q$ $\uparrow$thenT,$M_p\uparrow \rightarrow M_e\uparrow$
2) J $\uparrow$, R $\downarrow$, $\nu_p\uparrow \rightarrow K_T$,$K_Q$ $\downarrow$thenT,$M_p\downarrow \rightarrow M_e\downarrow$,
3) $J = 0$, $\nu_p = 0$ (Mooring condition), K_T,K_Q at maximum, T,$M_p$at maximum, to prevent the engine overload of engine normally, $n_e \leq 75\% n_H$, $\eta_p = 0$.

The change of J represents the change of propeller load, i. e. the change of external load, when J is small, the properller is heavy, and vice versa.

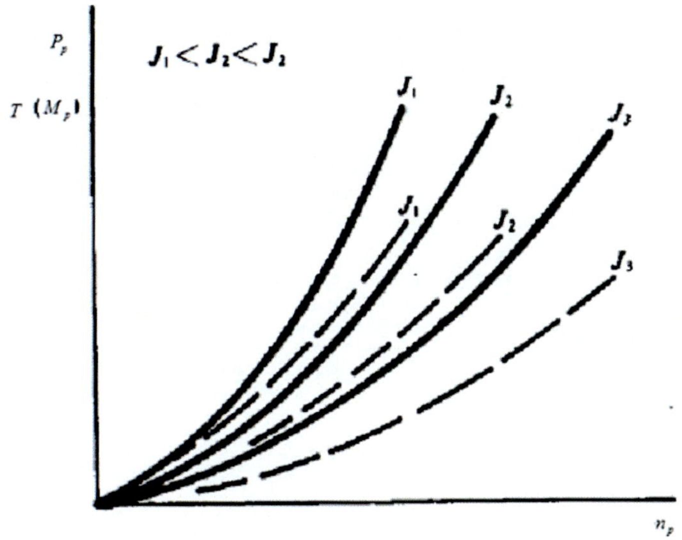

Fig. 4.4 Propulsion characteristic of propeller

4.1.3 *The Resistance Characteristics*

The resistance of a ship is very complex, there are friction resistance R, vortex resistance R_F, wave resistance R_w, it is different with different type of ships, see Fig. 4.5. When the velocity of ship is slower, the friction resistance is the mainly factor, withe the increase of the velocity, the wave resistance will become to the mainly factor. From the ship experiment it can be found:

$$R \propto v_s^{1.86} \quad R_F \propto v_s^2 \quad R_w \propto v_s^4$$

For the displacement ship, the friction resistance is the main resistance, so it is considered:

$$R = \alpha_R v_s^2$$

where:

R—resistance

v_s^2 —velocity of ship

α_R —resistance coefficient, it is constant at a fixed navigation condition, but changes at different navigation condition

The resistance characteristic could be a bunch of curve, see Fig. 4.6, different α_R represents different navigation condition. Since $P_R \propto R v_s \quad P_R \propto \alpha_R v_s^3$, that is the effective power of the hull is proportional to the third power of the ship velocity.

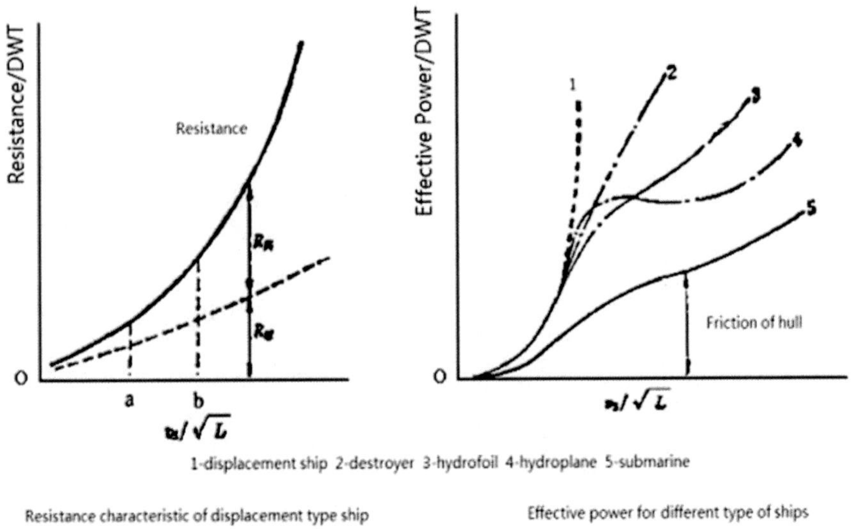

Fig. 4.5 Resistance of various type ship

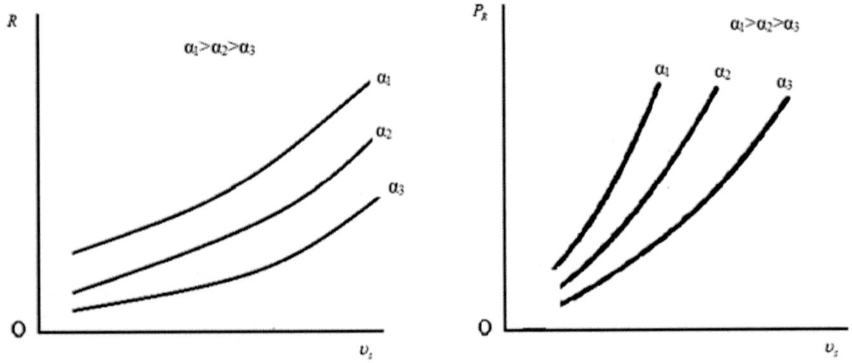

Fig. 4.6 Resistance characteristic

The relationships between the engine and the propeller when the ship navigating at a constant speed are as follows:

$$n_e/i = n_p$$
$$M_e \cdot i - M_T = M_p$$
$$P_e \cdot \eta_c = P_p$$

The relationships between the ship and the propeller:

$$\nu_p = \nu_s - u = \nu_s(1 - \omega)$$
$$T_e = R$$
$$P_p \cdot \eta_p \eta_r \eta_s = P_T \cdot \eta_s = P_p \cdot \eta_t = P_R$$

where:

- i—transmission ratio
- M_T—friction loss moment
- η_c—transmission efficiency
- u—wake velocity
- ω—wake coefficient
- T_e—efficient thrust
- η_r, η_s—relative efficiency and hull efficiency

At the stable status:

$$T_e = C_T n_p^2 = R = a_R v_s^2$$
$$v_s = \left(\frac{C_T}{a_R}\right)^{1/2} n_p = C n_p \quad R \propto n_p^2$$
$$P_R = a_R' v_s^3 \quad P_R \propto n_p^3$$
$$M_e = C' p_e \quad P_e = C p_e n_e$$
$$M_p = C_2 n_p^2 \quad P_p = C n_p^3$$

As long as to adjust the scale, P, n_p, n_e and P_e, P_p, P_R could put in the same coordinate i. e. ν_s, n_p, n_e in X- coordinate, P_e, P_p, P_R in Y- coordinate, see Fig. 4.7. The resistance of the ship and the thrust of the propeller are mutual transformation, so that the match among the ship, the engine and the propeller can be simplified to the match between the engine and the propeller.

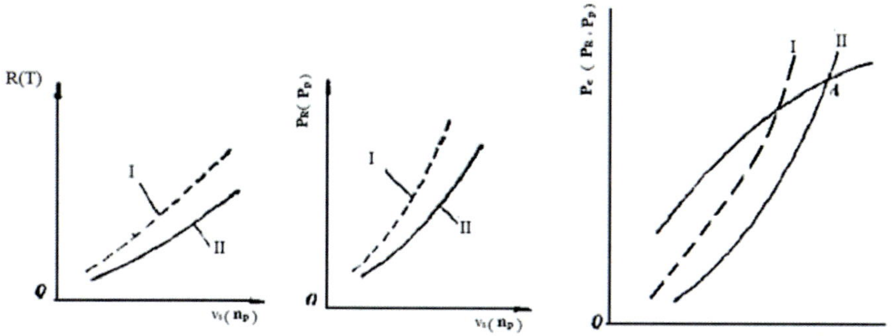

Fig. 4.7 Characteristic curve at the uniform coordinates

In the same coordinate, the cross point is the match point such as A, the ship is at a stable state.

For a propulsion system, it not only one load or one power sometimes, like shaft generator, or two engines driving one propeller, before matching, sum the powers and loads frist, there are some basic rules of matching.

Before matching.
with gearbox:
Torque times transmission ratio and speed divided by transmission ratio:
For axial load, sum the all loads.
For the multi-paralleled: sum all powers.

To choose the match point, the aim is to to make the efficiency of the propeller at the maximum when the engine at the rated point. But sometimes to consider a higher propulsion efficiency, the main engine may not work at the rated point.

After merging the power and load, the relation between speed and power (or torque) can be showed with the match characteristic, see Fig. 4.8.

Matching with dimensionless parameters.

If the reasonable reference parameters are selected, dimensionless parameters could be used to simplify the relationship of the matching parameters. If the rated parameters are used to the reference, other parameters are represented by dimensionless parameters as followings:

$$\textit{rotational speed} \quad \underset{*}{n} = n/n_H$$

$$\textit{speed of ship} \quad \underset{*}{v_s} = v_s/v_{sH}$$

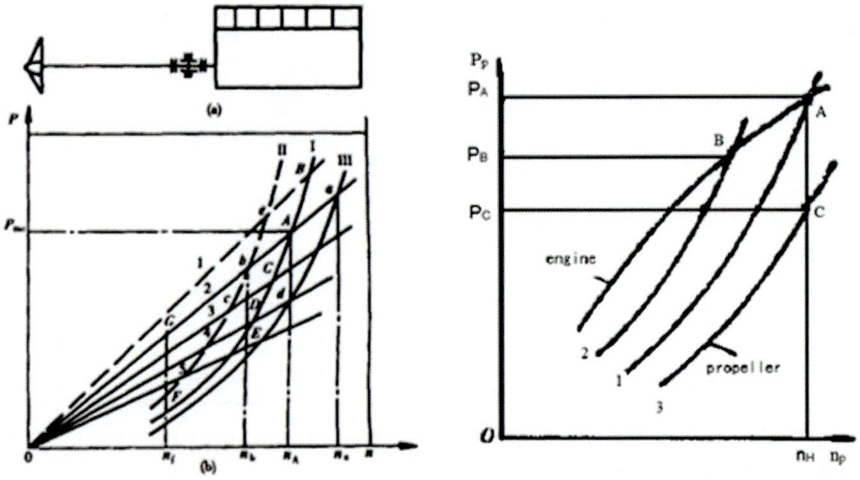

Fig. 4.8 Match point changes with different external load characteristic

$$mean\ effective\ pressure \quad \underset{*}{p_e} = p_e/p_{eH}$$

$$torque \quad \underset{*}{M} = M/M_H = Cn^2/Cn_H^2 = \underset{*}{n}^2$$

$$thrust \quad \underset{*}{T} = T/T_H = \underset{*}{n}_p^2 = \underset{*}{M}_p$$

$$power \quad \underset{*}{P} = P/P_H \quad \underset{*}{P}_p = \underset{*}{n}_p^3$$

$$\underset{*}{P_e} = \underset{*}{P_p} = \underset{*}{P_R} = \underset{*}{n}_p^3 = \underset{*}{v}_s^3 \quad \underset{*}{n_e} = \underset{*}{n_p} = \underset{*}{v_s}$$

It makes more easy to find the match point and the relationships of each parameter.

4.2 Match at the Stable Design Condition

When the ship is navigation at a stable condition, if the design operation is at MCR (100% rated speed, 100% power output of main engine), when the resistance of the ship (or external load) is changed, see Fig. 4.9, for example the load increased from line 1 to line 2, to keep the same speed of engine, the engine will be overload, so as the limitation of the mechanical load, the operation point will be changed from A to B, the injection racker is still at the rated position, the power supplied by the engine is less than the power needed by the propeller, then the speed of engine drops, and the power output is less than the rated power. When the load decreased from line 1 to line 3, if the injection racker is still at the rated position, the power supplied by the engine will be bigger than the power needed by the propeller, the engine will be over speed. So the injection racker should be adjusted to the direction of less fuel injection

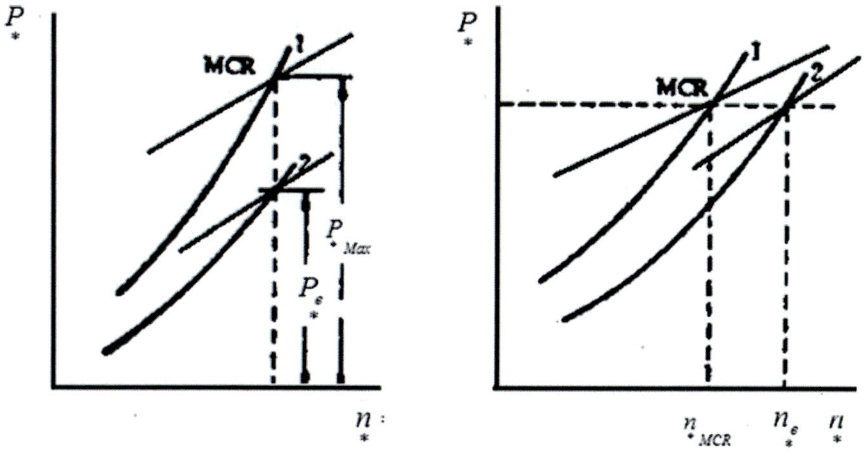

Fig. 4.9 Sea margin

to keep the engine speed at the rated speed, that is the limitation of the engine speed, the operation point changed from A to C, the effective pressure of each cycle is less than the rated one, so the power output drops. In this case, whatever the external load increases or decreases, the operation point always leaves the rated point, the power of the engine is less that the rated power.

When the ship has sailed for some time, the hull and propeller become fouled and the hull's resistance will increase. Consequently, the ship speed will be reduced unless the engine delivers more power to the propeller, i.e. the propeller will be further loaded and will be heavy running.

As modern vessels with a relatively high service speed are prepared with very smooth propeller and hull surfaces, the fouling after sea trial, therefore, will involve a relatively higher resistance and thereby a heavier running propeller. If, at the same time the weather is bad, with head winds, the ship's resistance may increase compared to operating at calm weather conditions.

When determining the necessary engine power, it is therefore normal practice to add an extra power margin, the so-called sea margin, which is traditionally about 15% of the propeller design power. It may considered power reservation: 10—15% or speed reservation: 3.5—5%, see Fig. 4.9.

When determining the necessary engine speed considering the influence of a heavy running propeller for operating at large extra ship resistance, it is recommended - compared to the clean hull and calm weather propeller curve 2 - to choose a heavier propeller curve 1 for engine layout, and the propeller curve for clean hull and calm weather in curve 2 will be said to represent a 'light running' propeller. Compared to the heavy engine layout curve 1 we recommend to use a light running of 3.0–7.0% for design.

Engine margin.

Besides the sea margin, a so-called 'engine margin' of some 10% is frequently added. After a period of time operation, as the change of clearance or worn, the performance of the main engine may decline, to keep the ship at the design velocity, 'engine margin' is necessary. In this case, if there are axial loads which is not over 3% of output of main engine, there is no need to change the main engine, unless the shaft load is big like shaft generator, the extra power demand of the shaft generator must also be considered.

For marine diesel engine, the match point should chosen in the area of $A_1A_2A_3A_4$, see Fig. 4.10, A1 is 100% of rated power and 100% rated speed (MCR). It is known that when the engine is running below MCR, the specific fuel consumption g_e is decreased.

Continuous service rating (SCR).

The Continuous service rating is the power at which the engine is normally assumed to operate, SCR is less than MCR, it could be 90% MCR. When SCR is chosen as the match point, the specific fuel consumption decreased. This makes the engine efficiency higher.

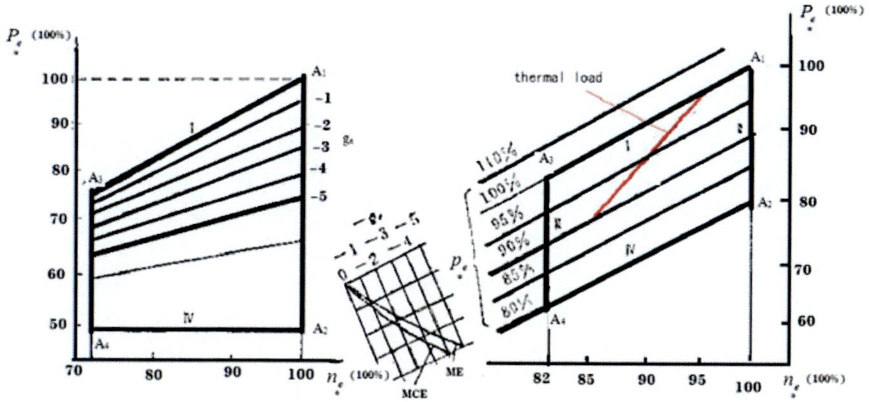

(a) Wasilia RTA engine (b) MAN-B & W L-MC/MCE engine

Fig. 4.10 Engine operation range and performance

4.3 Matching and Operation Characteristic at Stable State for Topical Propulsion Plants

4.3.1 Single Engine and Single Propeller

For a power system with single engine and single propeller, if MCR is the design match point, as it discussed above, when the load changed, the match point changed too (Fig. 4.11):

$$heavy\ load\quad J_1 < J_H$$

$$\underset{*}{p_e} = \underset{*}{p_{eH}} = 1 \quad \underset{*}{M_p} = 1 \quad \underset{*}{n_p} < 1 \quad \underset{*}{P_p} < 1$$

$$light\ load\quad J_2 > J_H$$

$$\underset{*}{n_p} = 1 \quad \underset{*}{p_e} < \underset{*}{p_{eH}} \quad \underset{*}{P_p} < 1$$

$$mooring\quad J = 0 \quad a_R = \infty \quad v_s = 0$$

$$\underset{*}{p_e} = 0.80 - -0.85 < 1 \quad \underset{*}{n_p} = 0.70 - -0.75 < 1$$

$$\underset{*}{P_p} < 1$$

Part load operation condition: when the resistance (or external load) keeps stable, to decrease the speed of the ship manually, the main engine only operated at part load. In this case, the main engine has ability to drive other loads besides the propeller see Fig. 4.12. That is the main engine has excess power: at the part load state, the difference between the max output of main engine and the power needed by the propeller. So if the injection racker is set at the rated position, the effective power of

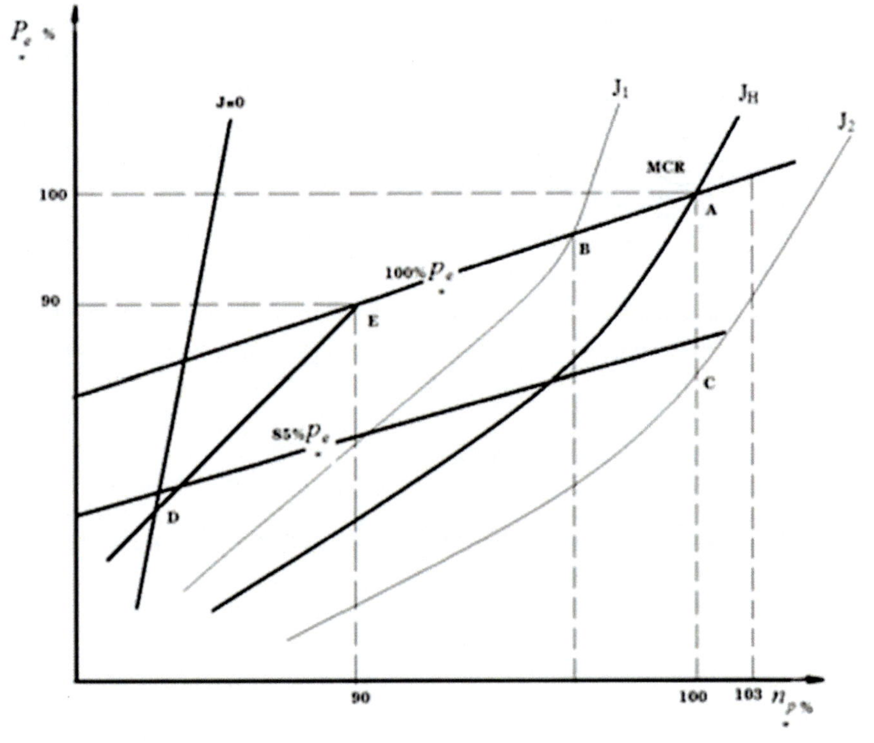

Fig. 4.11 Single propeller with single engine

the main engine is bigger the power needed by the propeller. The excess power can be calculated by following formula.

See Fig. 4.13.

$$\underset{*}{\Delta P} = \underset{*}{P_e^{EA}} - \underset{*}{P_e^{FA}} = \underset{*}{n_p} - \underset{*}{n_p^3}$$

$$\underset{*}{P_e^{EA}} = \underset{*}{n_p} \times \underset{*}{p_e}(= 1) = \underset{*}{n_p} \quad \underset{*}{P_e^{FA}} = \underset{*}{n_p^3}$$

$$\frac{d(\underset{*}{\Delta P})}{d\underset{*}{n_p}} = 0 \quad \underset{*}{n_p} = 0.577$$

$$\underset{*}{\Delta P}_{max} = 0.577 - 0.577^3 = 0.3849$$

For example, if the engine is running at C (Fig. 4.12)

$$\underset{*}{M_p} = \underset{*}{p_e} = 0.8 = \underset{*}{n}^2$$

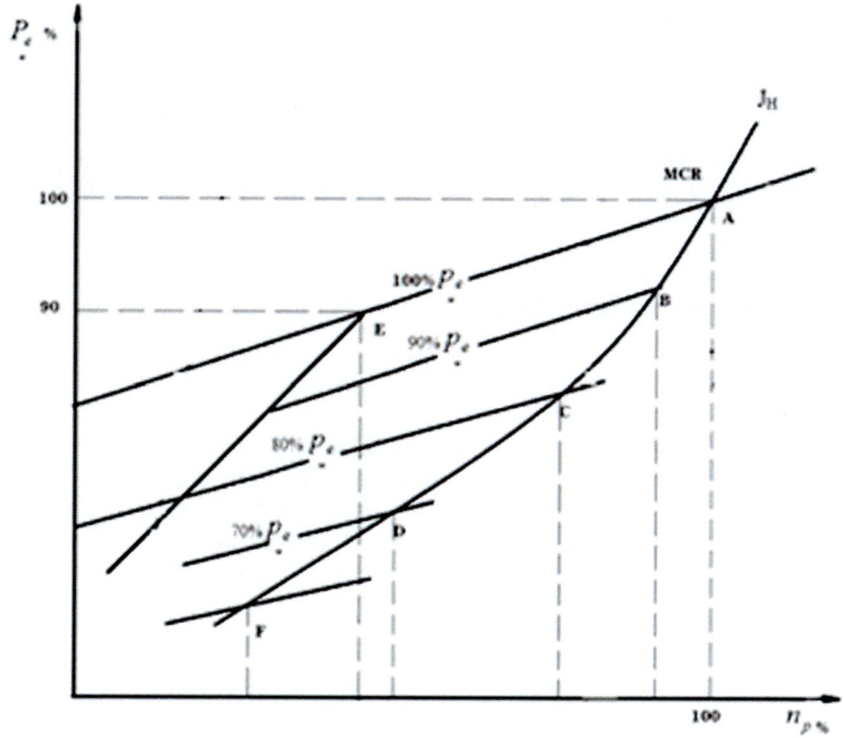

Fig. 4.12 Excess power at port load

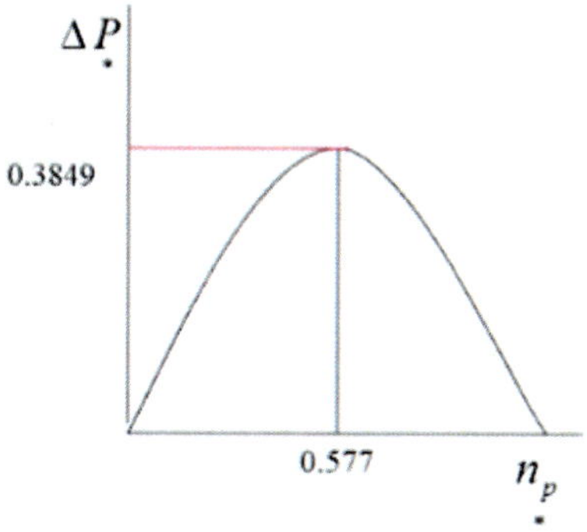

Fig. 4.13 Excess power curve

$$\underset{*}{n} = \sqrt{0.8} = 0.8944$$

$$\Delta \underset{*}{P} = \underset{*}{n_p} - \underset{*}{n_p^3} = 0.8944 - (0.8944)^3 = 0.179$$

This means when the engine is running at 80% of specific effective pressure, speed is at 89.4% of rated speed, and it has 17.9% of excess power to drive other loads.

4.3.2 Multi-Paralleled Propulsion

Double engines drive a propeller system is a typical multi-paralled propulsion, see Fig. 4.14.

Instead of a lower speed engine, two medium speed engines are used to drive a propeller. With the development of engine technology, the heat efficiency of the medium speed engine is nearly the low speed engine, and it has small size and light weight advantage, so it getting more propular in commercial ship. For this system, when the load is less than that of B, there are option to use two engines or only one engine. To get a lower specific fuel consumption, stopping one engine is a good choice. At point B:

$$\underset{*}{M_B} = \underset{*}{p_{eB}} = 0.5 = \underset{*}{n_B^2}$$

$$\underset{*}{n} = \sqrt{0.5} = 0.707$$

$$\underset{*}{p_B} = \underset{*}{n_B^3} = (0.707)^3 \approx 0.35$$

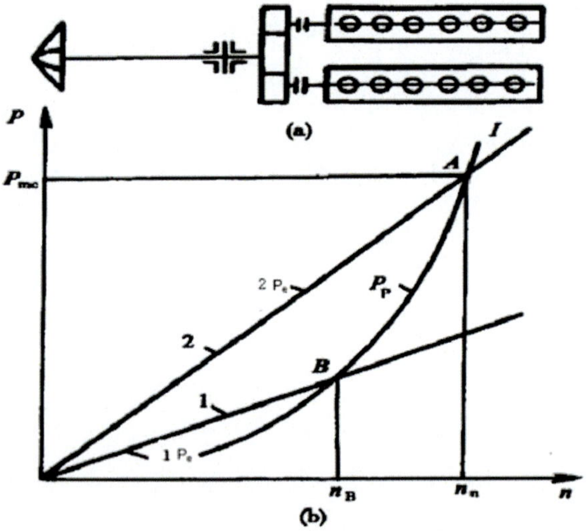

Fig. 4.14 Two engine drive one propeller

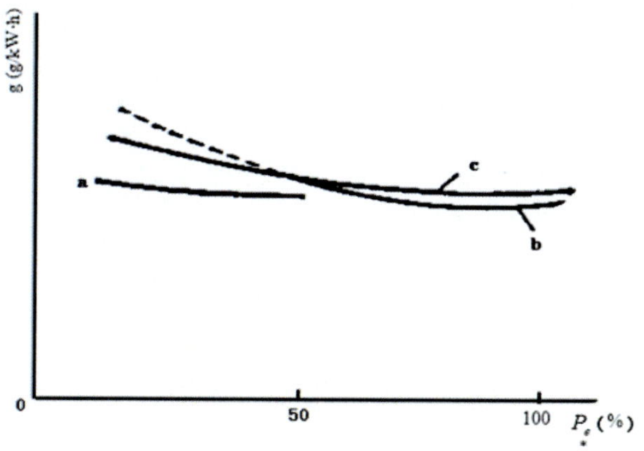

a-one engine running *b*-two engines running *c*-two propellers with two engines

Fig. 4.15 Comparison of specific fuel consumption

It shows that for double engines driving a propeller system, if one engine stopped, the one left could give the power of 35% of total power (compared itself, it is 70% of MCR), and the engine speed is 70% of rated speed.

Compared with one engine one propeller system, there are following advantages, Fig. 4.15 shows comparison of specific fuel consumption.

High efficiency: lower excess power and specific fuel consumption
Convenient to maintain
High flexibility
Good manoeuverability.

4.3.3 Multi-Engines with Multi-Propellers

The power system with three engines and three propellers is a typical multi-engines with multi-propellers, when the power needed is lower, one or two engines could be stopped. The match characteristic shows in Fig. 4.16.

When three engines driving three propeller simultaneously, each engine bears one third of total load, line OP_1 and line $OEFP_1$ are propulsion curve and engine power curve respectively, when one engine is stopped, the propulsion curve is OFP_2', II is additional power needed as one propeller became a resistance, when two engines are stopped, OEP_3' is the propulsion curve, I is additional power needed as two propellers became a resistance. At lower load, stop one or two engines makes the specific fuel consumption lower. Compared with the system of one engine one propeller, there are following advantage.

High efficiency: lower excess power and specific fuel consumption
High flexibility.

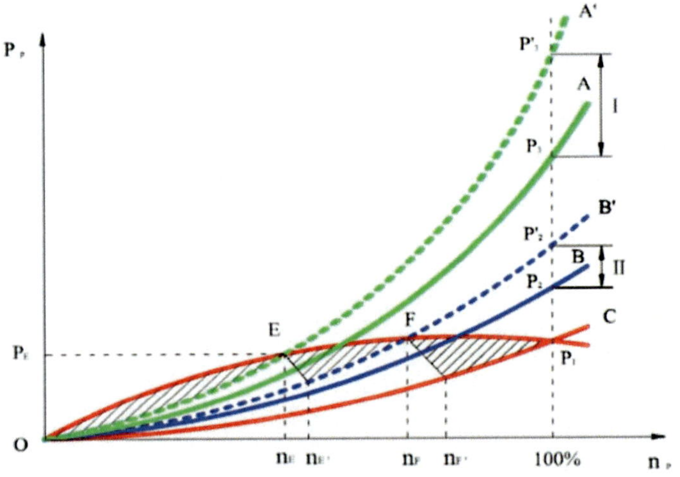

Fig. 4.16 Three propeller with three engines

The system multi-engines with multi-propellers is usually used for warships.

4.4 Controllable Pitch Propeller (CPP)

For a fixed pitch propeller, when external load changes, the power output of main engine is always less than the rated power, because of lower speed or lower mean effective pressure. For the ships which work with variable load, controllable pitch propeller is a good choice. The pitch could be adjusted according to the load or requirement of speed. "Controllable pitch" means a propeller whose pitch varies along the length of the span of the blade, and which can vary the collective pitch of all sections through the blade assembly on an axis. See Fig. 4.17.

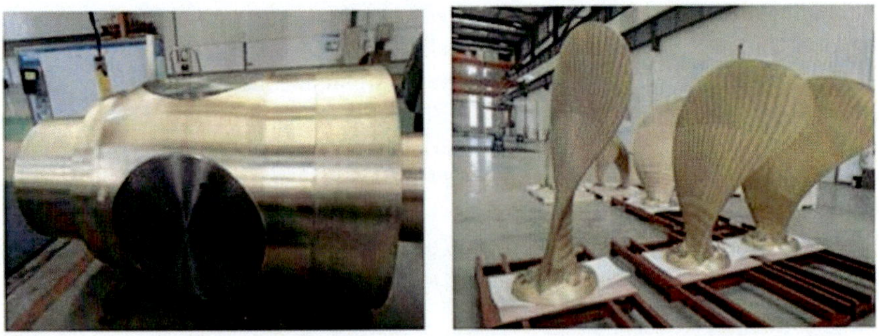

Fig. 4.17 Controllable pitch propeller

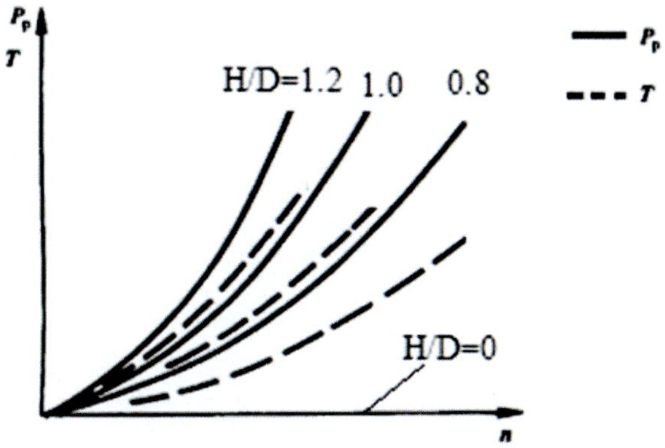

Fig. 4.18 *Propulsion characteristic of* controllable pitch propeller

Controllable Pitch propeller(CCP) systems provide excellent performance and manoeuvrability, and are recommended for vessels with frequent sailing routes that involve multiple operating conditions. These can be, for example, vessels requiring full power in both bollard pull and freesailing conditions, or that make frequent port calls. CCP systems can also be applicable for vessels that encounter varying weather conditions or demanding operational requirements such as dynamic positioning. A controllable pitch propeller is often the optimal choice for installations with a shaft generator operating at constant rotational speed. Full propulsion power is available in both heavy and light conditions through an automatic pitch adjustment. Engine overload is avoided regardless of the conditions.

When H/D increases, the power needed by the propeller increases, the thrust increases too, if the speed of propeller keeps same, the velocity of the ship increases. This means the sailing speed can be changed with change engine speed. When H/D = 0, no thrust, see Fig. 4.18. So if the external load changes, through adjustment of the pitch, the engine can keep running at the design point.

If the load increases, enlarge the pitch, makes K_Q constant, the speed of the propeller will be not changed, the engine speed keeps the same, power of the engine keeps the same, see Fig. 4.19.

If the load decreases, reduce the pitch, makes K_Q constant, the speed of the propeller will be not changed, the engine speed keeps the same, power of the engine keeps the same see Fig. 4.20.

Compared with fixed pitch propeller (FPP), the CPP has following advantages:

The main engine could supply rated power at any operation condition. It just like when the load changes, to adjust the pitch to make the propulsion come back to the original one, see Fig. 4.21.

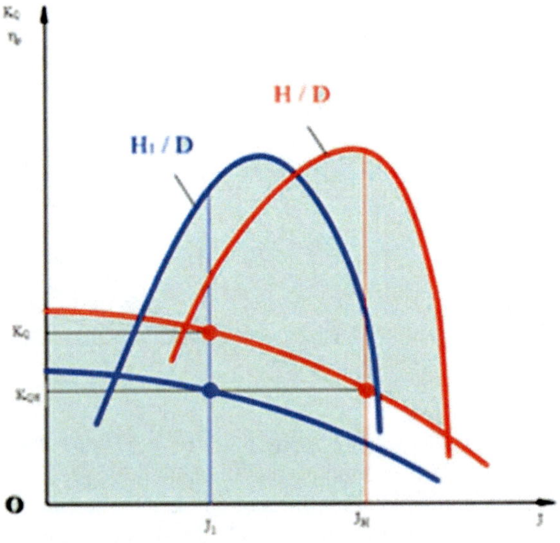

Fig. 4.19 Heavy load

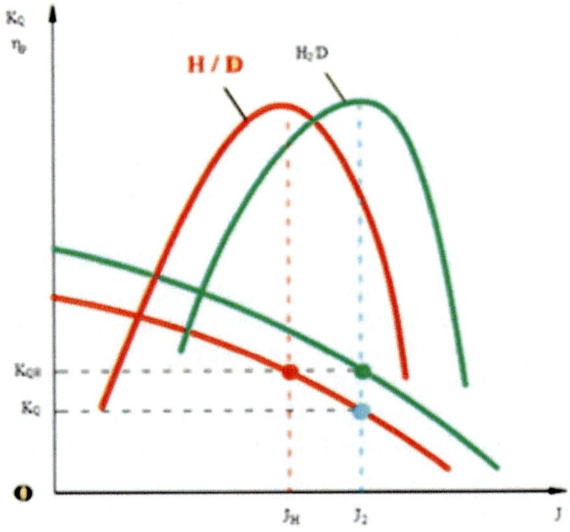

Fig. 4.20 Light load

At any load, once to match speed n_e and H/D, highest efficiency of the unit could be got. That is lower g_{emin} and higher η_{max}, see Fig. 4.22.

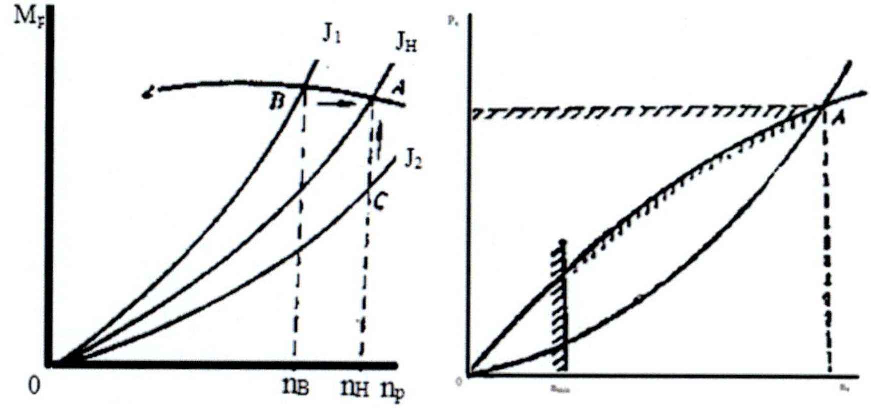

Fig. 4.21 CPP Power Comparison with FPP

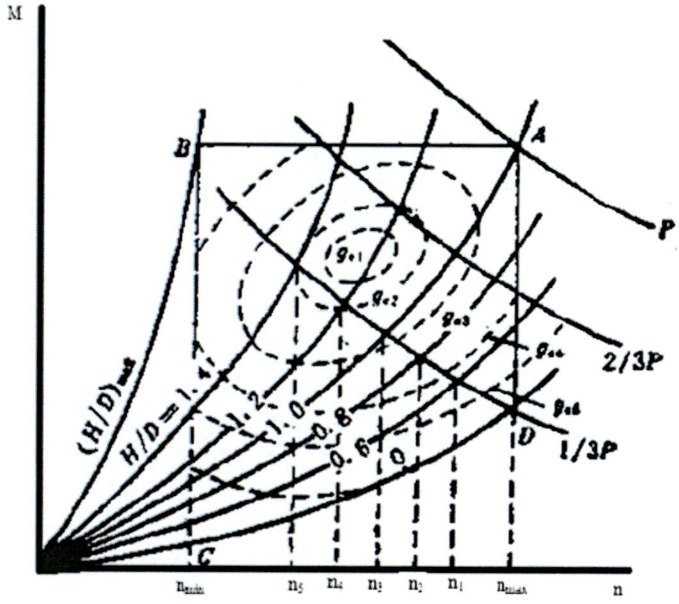

Fig. 4.22 Universal characteristics of diesel engine

For example, if the power needed is 1/3P, the engine speed could be n_1, n_2, n_3, n_4, n_5, with the propeller H/D 0.6, 0.8, 1.0, 1.2, 1.4, engine speed at n_4 and n_5, the specific fuel consumption g_e at the minimum, so it considered to choose engine at speed of n_4 or n_5, with the propeller H/D = 1.2 or 1.4, than to check which one have higher efficiency η_{p}.

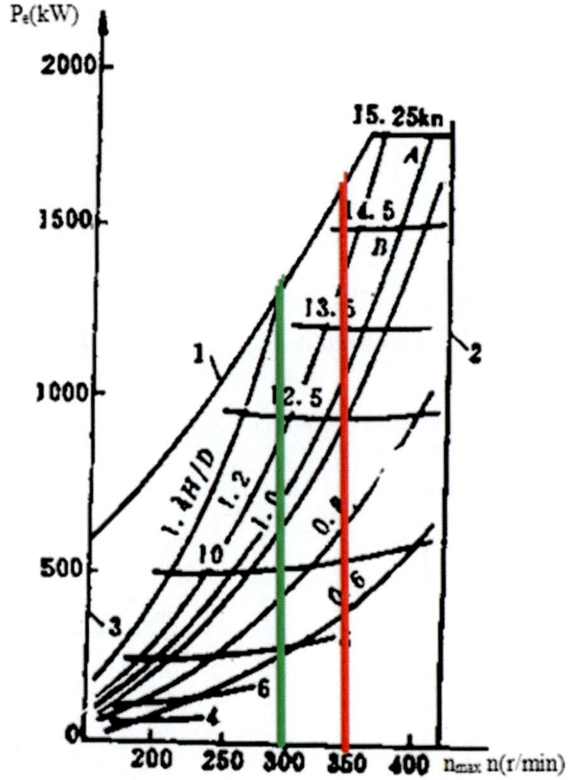

Fig. 4.23 Example of CPP application

Keep the speed of main engine constant, change H/D to get different navigation speed. It is possible to navigate at very low speed, even astern navigation. For example, if the engine speed keeps at 300r/min, when adjust H/D = 0.6, 0.8, 1.0, 1.2, 1.4 the ship velocity is about 8kn, 9.5kn, 11.0kn, 12.5kn, 13.8kn respectively, see Fig. 4.23.

At port load, the controllable pitch propeller has its own advantages of high propeller efficiency and makes main engine running at lower specific fuel consumption. Even it is complex and cost more for manufacture, but it is still good choice for the ships often working at variable external load.

Chapter 5
Marine Steam Turbine Power Plants

5.1 Principal and Thermal Schemes of a Marine Steam Turbine Plant

In a steam turbine plant, the working body continuously circulates along the closed circuit, where it cyclically changes its state (water-steam-water).

The main elements of the simplest steam turbine plant (STP) (Fig. 5.1) are a steam boiler (SB), a steam turbine (ST), a condenser (C) and a feed pump (FP). In the steam boiler, water (referred to as feed water) is heated to the boiling point and boils at a constant temperature. After that, the steam is superheated to a specified temperature $T_1$with the heat of fuel combustion. These processes take place at a constant pressure p_1. The heat consumed by the formation of 1 kg of steam and its superheating is marked as q_1, measured in kJ/kg and is equal to the difference between the enthalpies of the superheated steam and water that enter the boiler.

Further, the superheated steam with the pressure p_1 and temperature T_1 enters the turbine and expands in it. The received mechanical energy is transmitted from the shaft of the turbine T through the reducer R to the propeller, the thrust of which is perceived by the main thrust bearing (MTB) for the movement of the ship.

The exhaust steam is fed to the condenser C. There, at constant pressure and temperature of condensation, it releases heat to seawater and again turns into feed water, which is returned to the steam boiler. Heat q_2, kJ/kg, is transferred to the seawater.

The cycle followed by such an STP is called the Rankine cycle in honor of the famous Scottish scientist. The thermal efficiency of this cycle is:

$$\eta_t = \frac{q_1 - q_2}{q_1}.$$

This efficiency takes into account only the losses related to the heat removal to the cold source (losses in the condenser). At the same time, the real processes of energy conversion in the STP are accompanied by energy losses in its various elements.

Z. Yang et al., *Marine Power Plant*,
https://doi.org/10.1007/978-981-33-4935-3_5

Fig. 5.1 Principal scheme of a simple steam turbine plant

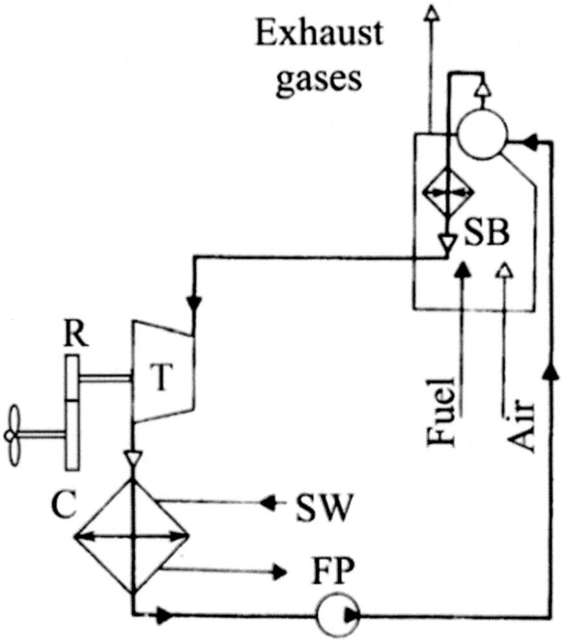

These losses are determined by the corresponding values of the efficiency of the boiler η_b, the main turbo gear unit (MTGU) η_{tg}, the shaft line η_{sl}, the propulsorη_p and the coefficient η_{am}, which takes into account the energy consumption of the drive of auxiliary turbomechanisms (turbopumps and turbogenerators which use the steam from the main SB) and other needs of the STP, as well as leakage, pressure and heat loss in the superheater, steam lines and condensate-feed pipeline.

Considering the above stated, the expression for the efficiency of a propulsive STP will have the following form:

$$\eta = \eta_t \cdot \eta_b \cdot \eta_{tg} \cdot \eta_{sl} \cdot \eta_p \cdot \eta_{am}.$$

In order to increase the efficiency of the energy conversion processes (thermal efficiency), the following measures are taken in modern STPs: increasing the initial steam parameters, regenerative feed water heating, and intermediate steam superheating (ISS).

Concerning the increase of the initial parameters of steam, there are some restrictions that prevent their significant growth. The modern STP is characterized by the following values of initial steam parameters: pressure $p_1 = 10$–15 MPa, temperature $t_1 = 510$–560 °C.

The principle of regenerative heating is that the feed water before being supplied to the boiler is heated in intermediate water heaters (IWH). Thus, the average temperature of heat rises and the thermal efficiency increases in the regenerative cycle when compared to the conventional non-regenerative cycle. Theoretically, feed water can be heated from the condensation temperature to the boiling point with an infinite number of IWHs. In practice, when the number of IWHs is more than three, the growth η_t slows down. In real STPs, no more than five intermediate heaters are usually used.

Feed water is heated by the steam taken from intermediate stages of the main turbines (it is used most often), or the steam used in auxiliary turbine mechanisms and heat exchangers (used in a limited number of STPs).

During intermediate superheating, the steam expanded in several stages of the turbine (usually in the air-gas channel of the high-pressure turbine (HPT)) is fed to the boiler again. There, it is heated at a constant pressure (lower than p_1) to the initial temperature and enters subsequent stages of the turbines for further expansion. The increase in thermal efficiency of the cycle with ISS is ensured by an increase in the average temperature during the supply of heat in the boiler. The increase in the efficiency of the plant is 4–5%. Intermediate steam superheating is used in combination with regenerative heating of feed water in modern STPs.

Figure 5.2 shows the thermal scheme of the STP installed on the tanker "Krym". The steam turbine plant has the following initial steam parameters: pressure – 7.65 MPa, temperature—510 °C, ISS—up to 510 °C. At the pressure of 1.46 MPa, the STP has a developed five-stage system of regenerative feed water heating.

The main equipment of the plant includes the following elements: the main steam boiler with the main and intermediate steam superheaters; MTGU consisting of HPT, MPT, and LPT; reducer (not shown in the scheme); main condenser.

The generator (PG) of the marine power station and the main feed pump are driven by the MPT auxiliary reducer (AR). In the maneuvering modes, the auxiliary steam turbine serves as the drive for the PG and the feed pump; its exhaust steam is diverted to the MC. The RFP has its own turbo drive with the discharge of exhaust steam into the deaerator, and the remainder of steam—into the MC. The PG and RFP turbines operate on cooled steam which comes from the MB steam cooler.

The thermal scheme of the SRP provides for a five-stage regenerative heating of feed water and five steam selections from the turbine casings. Steam selection for the LPT is used for heating water in the feed water heater of the first stage (FWH1) and ensuring the WDP operation. The deaerator is used as the second stage of feed water heating via the steam from the MPT; it also receives condensate from the FWH1 and FWH3. The third (steam from the MPT), the fourth (steam from the HPT), and the fifth (steam from the HPT) stages of heating are made up by the FWH3, FWH4, FWH5 feed water heaters of the surface type. The condensate of the heating steam of the third, fourth and fifth stages is discharged consistently into the lower-lying FWHs, the end point for which is the deaerator. Along with this, the consumer of the fourth selection is the ABP. Under normal conditions (during movement and parking), it operates in the mode of foul condensate evaporator. This mode is provided by feeding the selected steam to the coils located in the steam-water collector of the auxiliary boiler. The selected steam condensate is discharged into the deaerator, and the obtained saturated steam is used for the operation of the ship

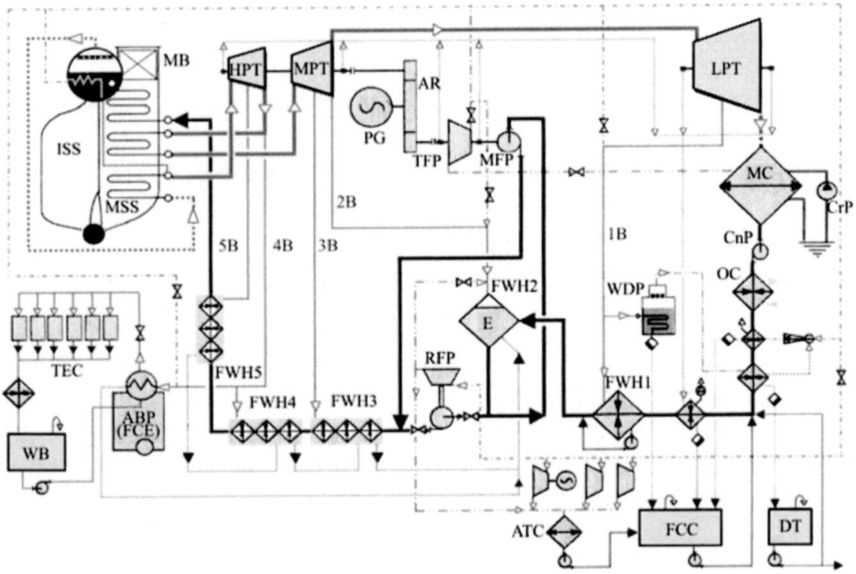

Fig. 5.2 Thermal scheme of the steam turbine plant TS-3 of the tanker "Krym": I, II, III, IV, V—five selections of steam; FCE—foul condensate evaporator; WDP—water desalination plant; MFP—main feed pump; MB—main boiler; MC—main condenser; D—deaerator; ATC—auxiliary turbomechanism condenser; ABP—auxiliary boiler plant; AR—auxiliary reducer; PG—power generator; FCC—foul condensate collector; CnP—condensate pump; OC—oil cooler; MSS—main steam superheater; ISS—intermediate steam superheater; FWH1, FWH2, FWH3, FWH4, FWH5—feed water heaters of the first, second, third, fourth and fifth selections, respectively; RFP—reserve feed pump; TEC—thermal energyconsumers; TFP—turbo feed pump; WB—warm box; DT—distillate tank; CrP—circulation pump

TEC. Condensate from these consumers enters the warm box, from which it is fed into the steam-water collector via the feed pump.

In case of the failure of the MB, the saturated steam of the auxiliary boiler is fed into the HPT. Such a scheme ensures the emergency operation of the vessel.

5.2 General Arrangement of the Main Elements of the Marine Steam Turbine Plant

5.2.1 The Main Turbo Gear Unit

The power complex which consists of turbines, a condenser and a gear is called *the main turbo unit*. The main turbo unit which includes a reduction gear is called *the main turbo gear unit*(MTGU), and the turbo unit with electric transmission is *the main turbo-electric unit* (MTEU).

Turbo-electric units are most often made single-hulled, whereas most modern turbo gear units have two or three hulls. In this case, the first turbine that the steam passes through is the HPT, the second is the MPT, and the third is the LPT.

In a double-hulled unit, the first body in the steam path from the boiler is called a high-pressure turbine, and the second body from which steam enters the condenser is a low-pressure turbine.

A bit of theory on turbines. Marine steam turbines are divided into several types according to the main characteristics.

1. By intention, there are the following turbines:
   - *main* turbines that rotate the propeller shaft (they in turn are divided into the turbines of the forward and reverse stroke);
   - *auxiliary* turbines that activate electric generators and auxiliary mechanisms that serve the main turbines and boilers, as well as mechanisms for general ship use (for example, freight and feed pumps).
2. By the number of hulls:
   - *single-hulled* turbines, which are used as auxiliary or main ones with electric gear (in the form of a turbo-electric unit);
   - *multi-hulled* turbines, the air-gas channel of which is located in several hulls united by a single gear and connected with a continuous flow of steam; they are installed only as the main ones.
3. By the nature of the work process:
   - *active* turbines, in the stages of which full expansion of steam occurs in the nozzles, and the conversion of kinetic energy into mechanical occurs on the working blades of one stage (pressure stage) or several (two or three) rows of working blades (speed stage); such turbines are performed in the multi-staged or single-stage form; the former are turbines of main and auxiliary turbine units where the speed stage is only the first one, and the latter are auxiliary drive turbines where the speed stage is the only one;
   - *reactive* turbines, which consist of a large number of stages where transformation of the potential energy of steam into kinetic energy takes place both in the nozzles and in the channels formed by the working blades.
4. By the steam pressure at the outlet:
   - *condensing* turbines, in which the release of steam occurs in the condenser, where the low residual pressure is maintained (all main turbines are condensing);
   - turbines *with counterpressure*, in which the steam pressure at the outlet is above the atmospheric pressure (these are mainly ship auxiliary turbines).
5. By the method of torque gear:
   - *direct-acting* turbines, which are used only as auxiliary ones;
   - turbines *with gears* (mechanical, electrical, hydraulic).

- By the direction of rotation:
- *reversible*, which have a air-gas channel for the forward and reverse strokes;
- *non-reversible.*

In the structure of a triple-hulled unit, in addition to the HPT and LPT, there is an intermediate turbine called the medium-pressure turbine. Steam is supplied to the MPT from the HPT or the ISS and then flows to the LPT.

The turbine is a non-reversible engine; therefore, a special *reverse turbine* is used to ensure the ship movement in reverse. It consists of several stages and is located in the LPT. Steam flow from one hull to another, as well as from the LPT to the condenser, is provided by the pipelines of a large diameter called *receivers.*

The reduction gear unit (reducer) is designed to reduce the rotation speed when transmitting torque from turbines to the propeller. It is located between the turbines and the shaft line. The *condenser* is usually located under the LPT, it is designed to condense steam and create low residual pressure.

The main thrust bearing, swivel device, oil pumps, oil cooler, air ejector and other units are often assembled together with the MTGU.

Figure 5.3 shows a typical layout of the MTGU of a modern ship on the example of the double-hulled MTGU TS-3 of the tanker "Krym".

The main turbo gear unit consists of the HPT housed in hull *1* together with MPT, LPT *2* connected with a receiver, condenser *3*, reducer *4*, coupler *5* and MTB *6* located in a separate hull and intended to perceive the axial force from the propeller and transmit it through the reducer foundation onto the ship hull. Located across the ship under the LPT, the condenser is designed to condense steam and create vacuum behind the LPT. The coupler transmits the torque from the gearbox to the propeller shaft.

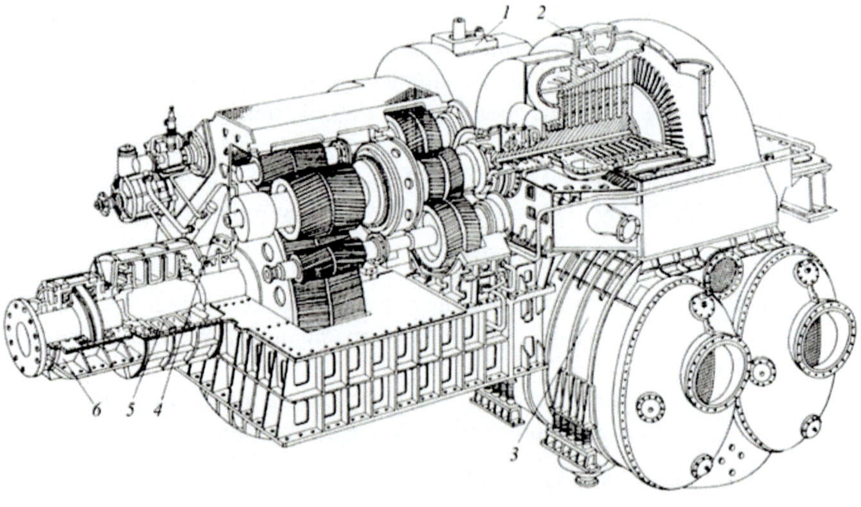

Fig. 5.3 General layout of the main turbo gear unit of the tanker "Krym"

The technical characteristics of the MTGU TS-3 are as follows:

Rated power	22,060 kW

Rotation speed, rpm:

HPT-MPT rotor	5370 rpm
LPT rotor	2850 rpm
Propeller shaft	85 rpm

Steam parameters before the quick-stop valve:

Pressure	7.6 MPa
Temperature	510 °C

Intermediate superheating steam parameters:

Pressure	1.46 MPa
Temperature	510 °C
Condenser pressure	0.0051 MPa
Efficiency coefficient	79%
Weight	300 t

The high- and medium-pressure air-gas channels are located in the HPT housing (Fig. 5.4). Fresh steam with the pressure of 7.6 MPa and the temperature of 510 °C enters the middle part of the HPT housing, passes the air-gas channel of the HPT, in which it expands and, at the pressure of 1.6 MPa, is directed to the ISS of the boiler.

In the ISS, the steam temperature rises to an initial value of 510 °C. After the ISS, the steam with the pressure of 1.46 MPa and the temperature of 510 °C enters the HPT housing again, but only the medium-pressure air-gas channel. After expansion in the medium-pressure stages, the steam is directed to the LPT, and then it enters the condenser. To compensate the axial forces, the steam flows in the high- and medium-pressure stages have opposite directions; there is an intermediate bulkhead with an internal labyrinth seal between the high- and medium-pressure air-gas channels. In case of failure of one of the turbines, the power plant can operate on the other turbine.

The high-pressure turbine is active; the stages are single-crowned. Both high-pressure and medium-pressure air-gas channels have five stages. The turbine housing is cast-welded, made of steel. The rotor is one-piece forged, with a central hole. The disks have a constant thickness; the blades are made with a constant profile along the length, their shanks are T-shaped.

The turbine body rests on a stool with the use of special constructional devices. Aft mount is rigid, and bow mount is flexible. Shaft bearings are self-aligning. The thrust bearing is double-sided and has self-aligning segments.

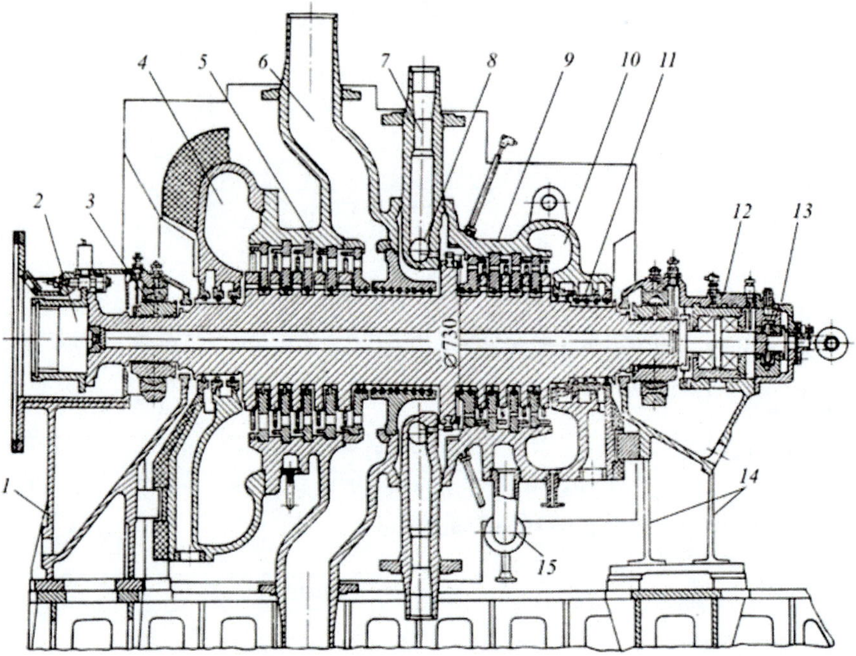

Fig. 5.4 Longitudinal section of HPT-MPT of the main turbo gear unit TS-3: *1*—stern stool; *2*—gear coupling; *3*—shaft bearing; *4*—steam discharge in the LPT; *5*—MPT; *6, 10*—steam supply and discharge from/to the ISS, respectively; *7*—fresh steam supply; *8*—HPT nozzle box; *9*—HPT; *11*—end tightening; *13*—impeller; *14*—flexible support; *15*—steam selection

The low-pressure turbine (Fig. 5.5) is single-flow and has ten active single-crown pressure stages. The steam supply to the turbine is carried out from the aft part of the lower half of the hull. The turbine housing is cast-welded from the thrust bearing side; the rotor is one-piece forged with a central hole. The blades of the first six stages have a constant profile along their entire length; from the seventh to the tenth stage, they have a variable profile along the length to ensure an unstressed steam supply. The hull is mounted to the ship foundation by means of stools. The bow mount is flexible, the stern mount is rigid, and the aft bearing rests on the gear housing.

The following two schemes provide an insight of the layout of the MTGU manufactured by other firms.

Figure 5.6 shows the typical layout of the MTGU of MST type produced by General Electric. The peculiarity of these turbo units is that the LPT is located on the condenser, which is placed across the ship; the HPT and the LPT have active blading. The reduction gear is two-stage with bifurcation of power, and the MTB is placed in a separate housing. The design of the unit allows driving the feed pump and electric generator from the HPT rotor, with their location in the bow from the turbine.

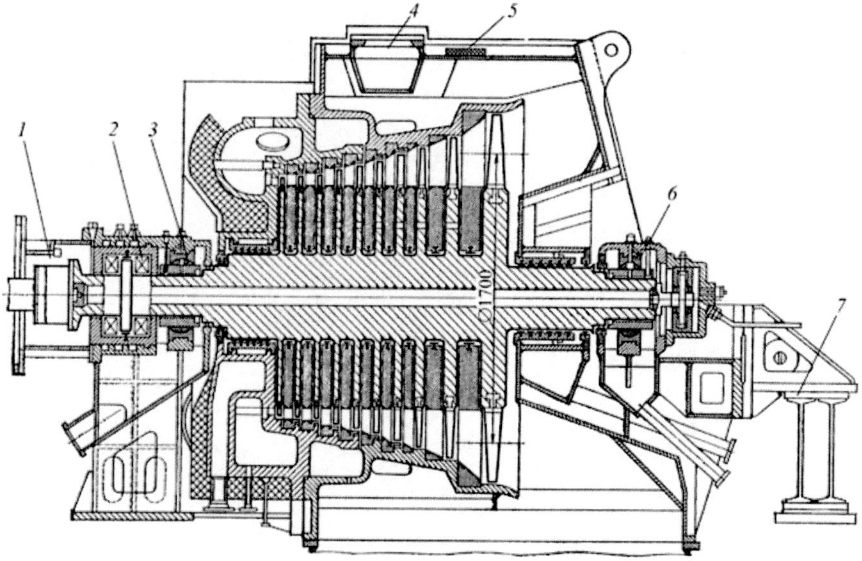

Fig. 5.5 Low-pressure turbine of the turbo unit TS-3: *1*—gear coupling; *2* and *3, 6*—thrust and shaft bearings, respectively; *4*—emergency steam discharge from the HPT; *5*—turbine housing, *7*—flexible mount

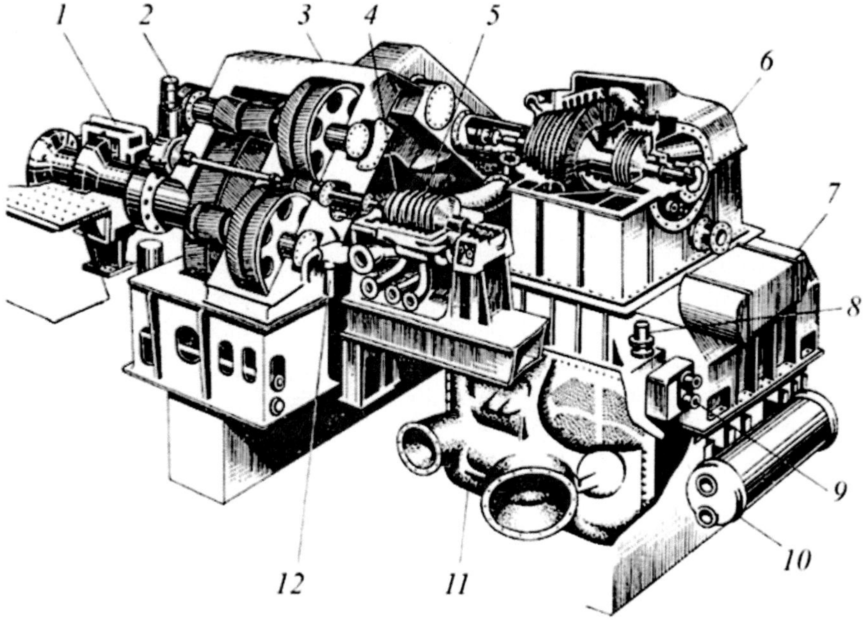

Fig. 5.6 Arrangement of the main turbo gear unit of the MST type with intermediate steam superheating: *1*—MTB; *2*—swivel device; *3*—reducer; *4*—MPT; *5*—HPT; *6*—LPT; *7*—air ejector; *8*—oil pump; *9*—oil filter; *10*—oil cooler; *11*—main condenser; *12*—waste oil tank

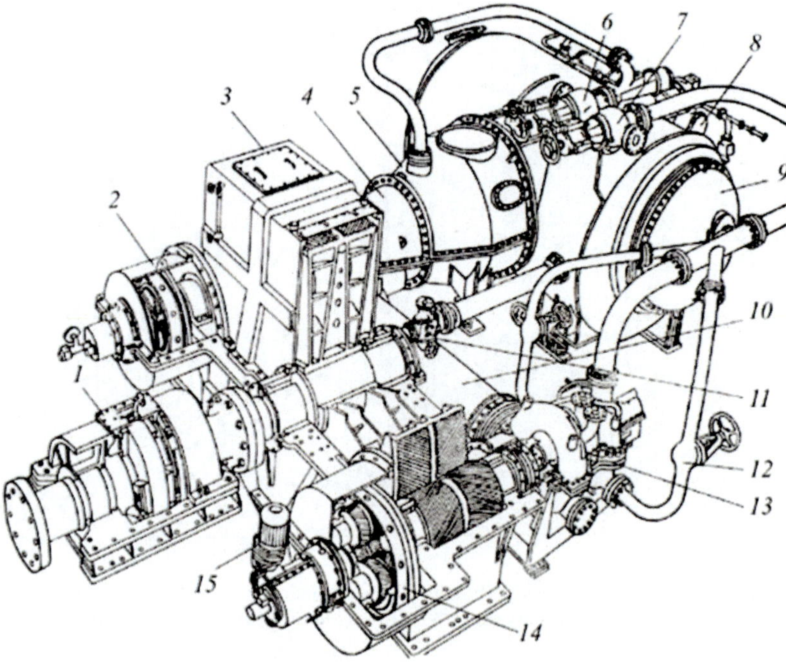

Fig. 5.7 Layout of the Stal Laval main turbo gear unit without intermediate steam superheating: *1*—MTB; *2, 14*—first (planetary) stages of the LPT and HPT reducer, respectively; *3*—second stage of the reducer; *4*—LPT; *5*—reverse turbine; *6, 7*—shunting valves of the reverse turbine and the HPT, respectively; *8*—condenser of the ejector of steam suction from the seals; *9*—main condenser; *10*—receiver of the bypass of the pair on the LPT; *11*—oil pump; *12*—nozzle valve; *13*—nozzle box of the HPT; *15*—swivel device

The MTGU of the AR type made by the Swedish firm Stal Laval (Fig. 5.7) operates without intermediate steam superheating. Its specific features include one-dimensional arrangement of its elements and the presence of a planetary stage in the reducer. The main reducer has three stages on the HPT line (the first two stages are planetary) and two stages on the LPT line (the last stage is cylindrical).

The placement of the condenser with axial steam discharge in the same horizontal plane with the main turbines makes it possible to reduce the height of the MTGU and place the steam generator above the reducer or the condenser. The one-dimensional arrangement of the MTGU makes the foundations lighter and provides easy access to all of its components, which is very important for assembly and repair works.

5.2.2 Main Turbo-Electric Unit

Typical MTEU representatives are the units of power plants installed on nuclear icebreakers "Arktika", "Sibir", "50 Let Pobedy" [1]. Two main turbogenerators with the total power of 55,150 kW are included in each MPP of this type. The turbogenerator consists of a turbine and three alternating current generators connected in series.

The technical characteristics of the main turbogenerators are given below:

Rated power	27,575 kW

Steam parameters before the quick-stop valve:

Pressure	2.94 MPa
Temperature	300 °C
Rotor speed	3500 rpm
Pressure in the condenser	0.0069 MPa
Efficiency coefficient	72%
Weight	2000 t

The turbine of the main turbogenerator is single-hulled (Fig. 5.8). In order to reduce the length of the blades, it is designed to be double-flow, which allows unloading the rotor from axial forces. The first stage is an active radial double-sided stage. Steam supply to the turbine is central. The steam flow that moves to the

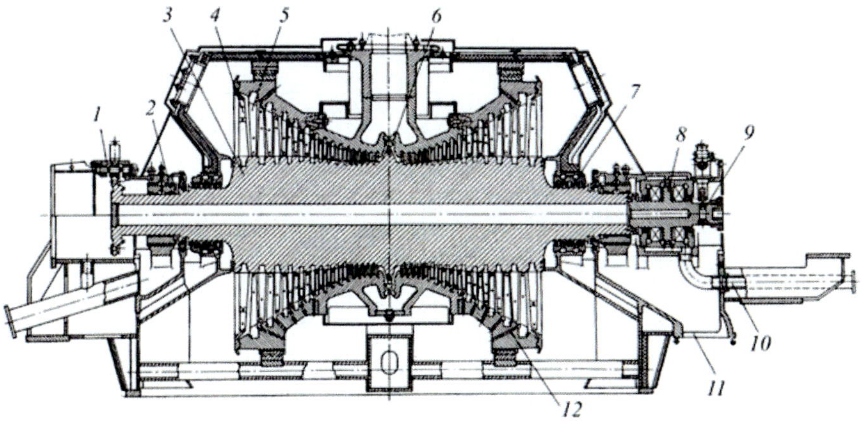

Fig. 5.8 Longitudinal section of a single-hulled turbine of a nuclear icebreaker: *1*—device for axial displacement of the rotor; *2, 8*—shaft and thrust bearings, respectively; *3*—rotor; *4, 5*—working and nozzle blades, respectively; *6*—radial stage; *7*—sealing; *9*—limit regulator; *10, 11*—oil supply and discharge, respectively; *12*—drainage channel

turbine doubles before approaching the nozzles: half of the total amount of steam passes through the bow and the rest goes through the aft groups of the nozzles. After passing through the radial stage, the steam enters the bow and aft air-gas channels, each of which consists of 15 reactive stages.

The turbine hull has both vertical and horizontal connectors. The turbine works on an electric generator, whose load depends on the operation mode of the propulsion electric engines. In the low-load and "propeller stop" modes, the turbine can cool down substantially. To prevent this, an additional housing heated by steam is provided around the main body. Special openings are made to inspect the last stages without opening the hull.

A one-piece forged rotor with a central hole is placed in two rigid mounting bearings. The single-row thrust bearing with an equalizing device is mounted on a separate shaft, which is connected by a flange to the rotor. The bow part of the turbine allocates the swivel device and the regulator of the limiting speed.

5.2.3 *Auxiliary Steam Turbine Generator*

Auxiliary steam turbine generator is most often performed in the form of a turbo block which consists of a multi-stage turbine, a single-stage reducer, a condenser, and other equipment and systems that are compactly located on a common frame.

Let us consider general arrangement of the auxiliary steam TG on the example of the unit TD-600 [1]. The turbogenerator consists of a turbine, a condensing unit and an alternating current generator with the power of 600 kW. The turbine, reducer and generator with exciter, oil cooler and switchboard are mounted on a foundation frame, inside which oil tanks and pumps are located.

Technical characteristics of TD-600:

Power of TG	600 kW
Rotor speed	8500 rpm

Steam parameters before the quick-stop valve:

Pressure	4.05 MPa
Temperature	440 °C
Pressure in the condenser	0.006 MPa
Efficiency coefficient	62%

Figure 5.9 shows a longitudinal section of the turbine. Its air-gas channel consists of a two-crown adjustment stages and six active pressure stages. The first two stages are made with a partial inlet to increase the height of the blades. The single-crown stages are made with a radial seal between shrouds of the working blades and projections of the diaphragms.

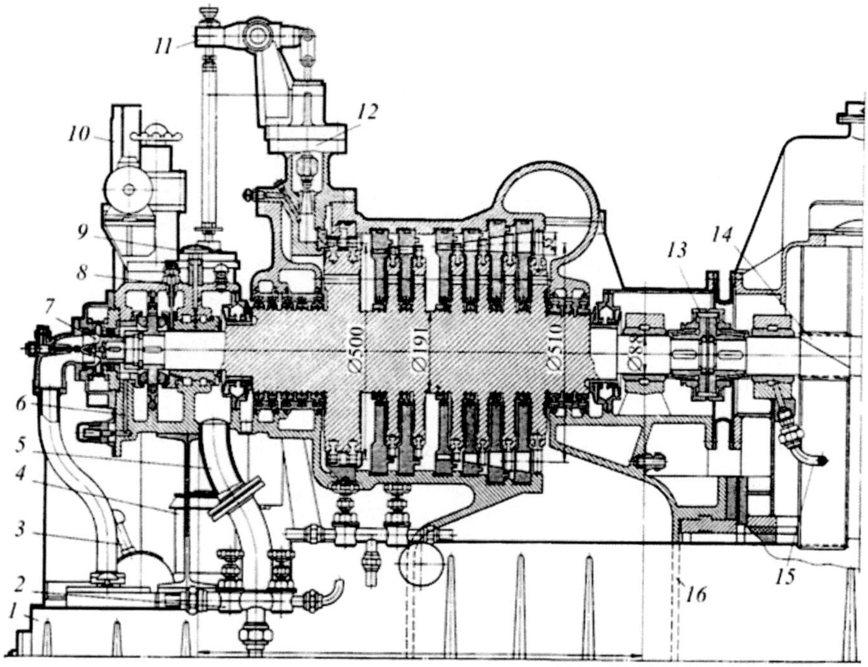

Fig. 5.9 Turbine of the auxiliary turbine unit TD-600: *1*—foundation frame; *2*—valves for regulating pressure in the seals; *3*—manual starting pump; *4*—starting pump; *5*—oil drain pipeline; *6*—oil supply pipeline to the thrust bearing; *7*—impeller; *8*—micrometers; *9*—resistance thermometers; *10*—switchboard; *11*—nozzle valve control mechanism; *12*—nozzle box; *13*—flexible coupling; *14*—reducer; *15*—oil supply to the bearings of the reducer; *16*—outlet fitting

The rotor is one-piece forged and does not have a central hole. The disks of a constant section are made with unloading holes. The rear end of the rotor is connected to the reducer pinion by means of an elastic coupling. The impeller of the oil pump is fixed at the forward end of the rotor; it is also the switch of the limit rotation speed. The turbine housing has a horizontal connector. The turbine stools are cast together with the body, and the front stool is connected to the foundation frame with a flexible mounting. The shaft bearings are equipped with resistance thermometers for monitoring the thermal state and micrometers for determining the radial position of the rotor. The turbogenerator TD-600 has its own condensing unit, which consists of a condenser, two-stage ejector, circulation and condensate electric pumps.

5.2.4 Condensers of the Marine Steam Turbine Plants

The main marine steam turbines and the auxiliary ones for driving electric generators work with condensation of exhaust steam. In the process of condensation, the heat of exhaust steam is absorbed by the cooling seawater supplied to the condenser via the circulation pump. The exhaust steam has the volume that is many times greater than its condensate volume. Thus, at the absolute pressure of 7 kPa, the volume of steam is 20000 times greater than the volume of the condensate. It is obvious that vacuum occurs during steam condensation in a confined space.

Water steam always contains a certain amount of air mixed with it, which does not condense, remains in the condenser and creates some pressure. If one connects pumps to the condenser to evacuate the air released from the steam during condensation, it is possible to maintain a low residual pressure there continuously.

Such maintenance of low residual pressure in the condenser and thus beyond the last stage of steam expansion in the turbine is one of the ways to increase its power and economic efficiency. In the condenser of the main turbo unit, the pressure is usually 5–6 kPa. Its further reduction is not rational, as this would lead to a significant increase in the size of the condenser and the pump power, which is economically impractical.

Ship STPs have made wide use of surface condensers, where condensation of steam occurs on the cold surface of the bundle of tubes with cooling seawater circulating inside. The surface of the tubes (dubbed as the cooling surface) divides the condenser into the steam and water spaces.

Modern condensing units employ centrifugal or axial pumps as circulation and condensate pumps, and jet pumps (ejectors) or vacuum electric pumps as steam pumps.

In a conventional circulation system of the main condenser, the circulation pump receives seawater through the kingston or the onboard ice box and feeds it to the condenser, from which water is removed overboard through the casting pipeline.

On modern ships, scoop circulation systems have been widely used for cooling water of the main condensers. There, dynamic pressure of the oncoming flow arising when the ship moves is used to pump water through the condenser. The circulation pump in this case is used only for pumping water at the reverse, during maneuvering and at a low speed. The supply, charge and power of such a pump are considerably lower than those of a conventional circulation pump. The scoop circulation systems are reliable and easy to operate. Such a system was applied at the large-capacity tanker "Krym" (Fig. 5.10) [2].

Seawater is supplied through the profiled inlet device, then it passes through the inlet pipeline into two parallel bundles of pipes of the main condenser, and then through one common discharge pipeline to the discharge device. The auxiliary circulation electric pump is installed on a separate inlet pipe.

This pump is automatically switched on if the ship speed decreases by the value at which scoop does not occur.

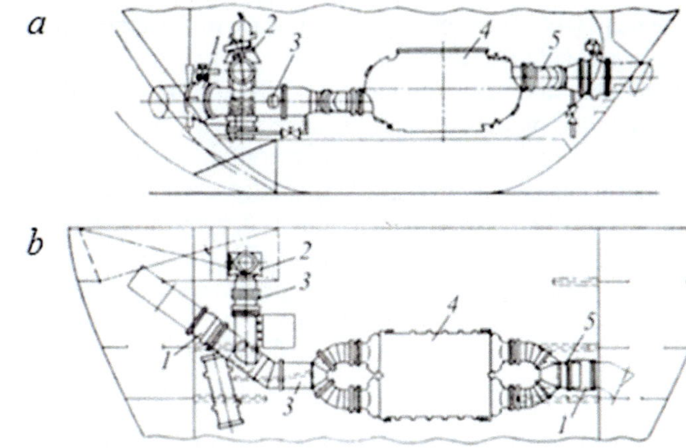

Fig. 5.10 Scheme of the scoop circulation of the tanker "Krym": **a** view of the ship bow; **b** hold layout; *1*—rotary valve; *2*—circulation electric pump; *3, 5*—inlet and outlet pipelines, respectively; *4*—condenser

Design of a typical condenser of the ship STP is represented by the example of the MTGU condenser of the ship "Krym" (Fig. 5.11) [2]. It is a single-pass surface condenser with two parallel flows. The upper part of the steel housing is a neck with rectangular flange *6*, which ensures reception of steam exhausted in the turbines.

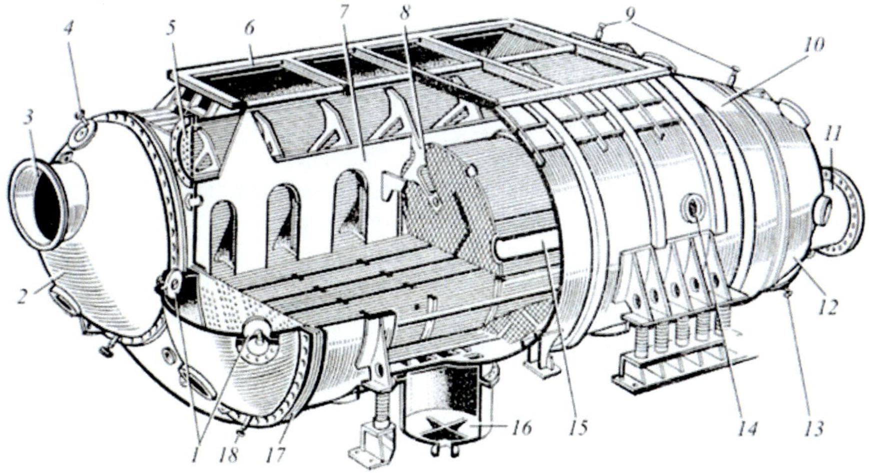

Fig. 5.11 Condenser of the main turbo gear unit of the tanker "Krym": *1*—protectors; *2, 12*—water chamber covers; *3, 11*—outlet and inlet water pipes, respectively; *4, 9* and *13, 18*—air and water valves, respectively; *5*—flange for reception of exhaust steam from turbomechanisms; *6*—steam reception neck; *7*—rigidity sheet; *8*—bulkhead; *10*—compensator of thermal expansions; *14*—opening for air extraction; *15*—condensate tank; *16*—condensate collector; *17*—tube board

The exhaust steam pipeline from the autonomous drive of the turbo block of attached mechanisms is connected to flange *5*.

The melchior tubes are mounted onto brass tube boards *17* by flaring on both sides. Bulkheads *8* prevent sagging of tubes in the casing.

Compensator *10* is provided to compensate the difference in the thermal expansions of the casing and the bundle of tubes. The condensate that drains from the tubes is collected in condensate collector *16* and removed through the pipe by means of the condensate pump. The air is extracted from condensate tank *15* through opening *14*. Seawater enters chamber *12* through inlet pipe *11*, flows along the tube bundle into chamber *2* and is drained through outlet pipe *3*.

5.2.5 Main, Auxiliary and Recovery Boilers of Marine Power Plants

Principle of operation of steam boilers. Steam boilers are regarded as the *main boilers* when they are placed on steam turbine ships and designed to produce steam necessary for operation of the main steam turbines and auxiliary needs.

On ships with ICEs and GTPs, steam boilers are designed to provide steam for some auxiliary mechanisms and meet the needs for thermal energy (premises heating, fuel heating, household needs). These boilers are called *auxiliary boilers* (AB).

Auxiliary boilers which are installed on tankers with the main ICEs often have a high level of steam production due to the high steam consumption for tanks washing, cargo heating and operation of cargo pumps with turbo drive. The ABs installed on steam turbine ships are designed to meet the need for wagering in the parking. The auxiliary boilers that operate on the exhaust gases of an internal combustion engine or gas turbine plant are called *recovery boilers.*

According to their design, ship boilers are divided into two groups: *gas-* and *water-tube* boilers. In a gas-tube boiler, hot gases serve as the main coolant and move inside the tubes, and water surrounds them from the outside. These tubes are called fire tubes. In a water-tube boiler, on the contrary, water and steam-water mixture are inside the tubes, and hot gases flow around them from the outside. These tubes are called water-heating tubes. Currently, only the water-tube boilers are employed as the main marine steam boilers.

Water-tube boilers are divided into boilers with natural and forced circulation. Most ship boilers have *natural circulation*, which takes place due to a difference in density between the water in downcomers (unheated or slightly heated) and the steam-water mixture that forms in risers, which intensively perceive radiation or convective heat. *Forced circulation* is used only in boilers of separate structures. At that, water is pumped through the tubes of evaporation surfaces with the help of circulation pumps. Such boilers are divided into *direct-flowboilers* and *boilers with multiple forced circulation.*

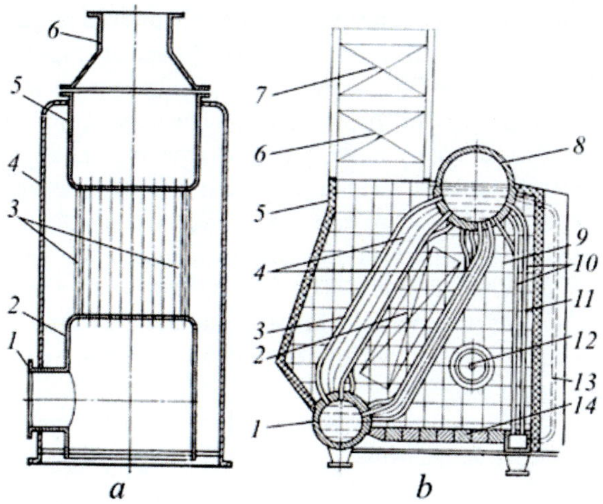

Fig. 5.12 Principal schemes of gas-tube (**a**) and water-tube (**b**) boilers

The circulation ratio is the ratio of the mass of circulating water G_w to the mass of steam G_s produced in the boiler:

$$K = G_w/G_s.$$

The direct-flow boilers have the circulation ratio $K = 1$. The boilers with multiple forced circulation have the ratio of $K = 4$–8 and are widely used as recovery boilers.

A simplified scheme of the gas-tube boiler is shown in Fig. 5.12a. Fuel and air are supplied through inlet pipe *1* from the fuel-injector unit to furnace *2*, where fuel combustion takes place. Combustion products enter boiler intake chamber *5* through fire tubes *3* and then flow into boiler intake *6*. The above structural elements are combined by housing *4*.

The water-tube boiler (see Fig. 5.12b) [2] consists of upper steam-water and lower water drums (*8* and *1*, respectively)connected by water-heating tubes *4*. Bundle of tubes *10* located on the side wall of the furnace is called a *side screen*. Tubes *11* of the second row of the screen, blocked by the first row, and tubes *13* located outside the furnace are called *downcomers*. In individual designs of boilers, tubes *3* also serve as downcomers; they are farthest from the furnace. All other tubes are called risers. Movement of water along the downcomers and steam-water mixture up the risers creates circulation of water in the boiler, which is necessary for its reliable operation.

The furnace of the boiler is formed by screen *10*, prefurnace bundle of tubes *4*, bottom *14* and walls *9*, laid out with refractory bricks. The combustion device consists of fuel nozzle *12* and an air-guiding element, through which the air necessary for combustion is supplied. The gas path separated from the atmosphere by the covering, and outer surfaces of the boiler *5* are thermally insulated. The boiler is equipped with appropriate fittings and instrumentation.

The steam boiler has additional heating surfaces: steam superheater *2*, economizer *6* and air heater *7*, which significantly increase its efficiency. Water economizer and air heater, which make use of the heat of exhaust gases, form the so-called *back-end heating surfaces* of the boiler. Heat transfer from the heating medium to the heated one occurs both in the furnace and in the gas flues.

Depending on their location, the water-heating tubes receive a varying amount of heat. There are two types of heat exchange in the elements of the boiler: radiation and convection. The tubes with gases flowing around them are called convective. At that, the first row of tubes of a convective bundle (counting from the furnace) is the radiation heating surface. The downcomers receive much less heat than other tubes or do not receive it at all.

A steam boiler together with additional heating surfaces is called a *boiler unit*. A boiler or several boilers together with auxiliary mechanisms and devices constitute *a boiler plant*.

The volume of the boiler with water (the lower drum, tubes and part of the upper drum) is called *water space*, and the volume occupied by steam (the other part of the upper drum) is *steam space*. The water-tube boiler is filled with water up to the middle of the steam-water drum. The water surface which forms the interface between the water and steam spaces is called the *evaporation mirror*. The furnace and gas flues belong to the gas space.

The main characteristics of the boiler include:

- steam capacity *D*, t/h, which is the amount of steam produced by the boiler; it is up to 100 t/h for the main boilers on ships with STPs and much lower on ships with other types of main power plants (depends on the purpose of the ship);
- hour fuel consumption *B*, kg/h, which is the amount of fuel burned in the furnace of the boiler; it defines the overall economic efficiency of the power plant and is determined by the fuel quality and operating conditions;
- parameters of steam: working pressure *p*, MPa, and temperature *t*, °C (6–20 MPa and 510–565 °C for the main boilers, 0.50–1.50 MPa and up to 360 °C for the auxiliary boilers, where saturated or slightly superheated steam is generated; the auxiliary boilers on tankers operate on steam of higher parameters).

The efficiency of the STP is also affected by the temperature of the feed water and the air entering the furnace. The former may reach up to 250 °C, and the latter may vary within 200–250 °C, which depends on the type of the air heaters and boiler load.

Main steam boilers. The most advanced, economical and large-capacity steam power plants are performed with intermediate steam superheating. Its implementation leads to specific features of the boiler design. Such boilers are equipped with evaporation elements of radiation type which form a completely screened furnace. The middle screen walls formed of the steam tubes divide the boilers into two parts: a combustion chamber and a chamber with convective heating surfaces of the main and intermediate steam superheaters and the economizer. Arrangement of the convective heating surfaces is called mine arrangement; therefore, such boilers are called mine-type boilers.

Such a boiler unit is used in the power plants of large-capacity tankers similar to "Krym".

Technical characteristics of the main boiler KVG-80/80 are given below:

Steam capacity, t/h	80
Superheated steam pressure, MPa	7.85

Temperature, °C:

Superheated steam	515
Exhaust gases	115
Air fed to furnace	217
Feed water	223
ISS steam	515
Efficiency coefficient, %	96

The principal scheme of the boiler KVG-80/80 is shown in Fig. 5.13, and its cross section is shown in Fig. 5.14 [2] (the numbering of positions is the same). The furnace chamber shaft formed by screens *11, 13, 15* is the right part of the boiler unit. The tubes of the end screens are fixed in drums *9* and *14*. The upper part of furnace chamber *12* locates four steam mechanical nozzles *7*. This arrangement of the nozzles is called *ceiling nozzle arrangement*. Downcomers*10* are the tubes connecting steam-water drum *2* to water drum *16*. The fuel combustion products formed in the combustion chamber pass through sparse tube Sect. 15 and enter the left shaft *18* which accommodates convection coils of the heating surface of the main and intermediate superheaters (*19* and *25*) and economizer *29*, as well as tubular heating surfaces of air heater *1*.

Via steam tubes *8*, steam from the steam-water drum enters the main steam superheater *19*, which consists of two compartments. A portion of the steam after the first compartment can be directed along steam line *17* to main cooler *5* to control the temperature of the superheated steam. From the main steam superheater, the steam goes through steam line *21* to the main steam turbine. There is also a partial discharge of steam through steam line *20* to auxiliary steam cooler *4* and its supply to cooled steam line *3*. The steam partially exhausted in the HPT flows through steam line *27* into ISS *25*, and the newly superheated steam flows from there via steam line *23* to the MPT.

The purpose of bypass steam lines *24* and *22* is as follows: the former is intended for regulating the superheated steam temperature in the ISS, and the latter is for supplying saturated steam to the ISS for burnout protection and discharging through steam line *26* to the steam cooling system.

Feed water is supplied to the economizer via tubes *28*, and to the boiler via tubes *6*.

The air heater is three-pass for gas and one-pass for air. The last bundle of tubes (third pass) on the gas side has a fluoroplastic protective coating against low-temperature chemical corrosion. Air preheating with feed water at low load modes is provided by heater *30*.

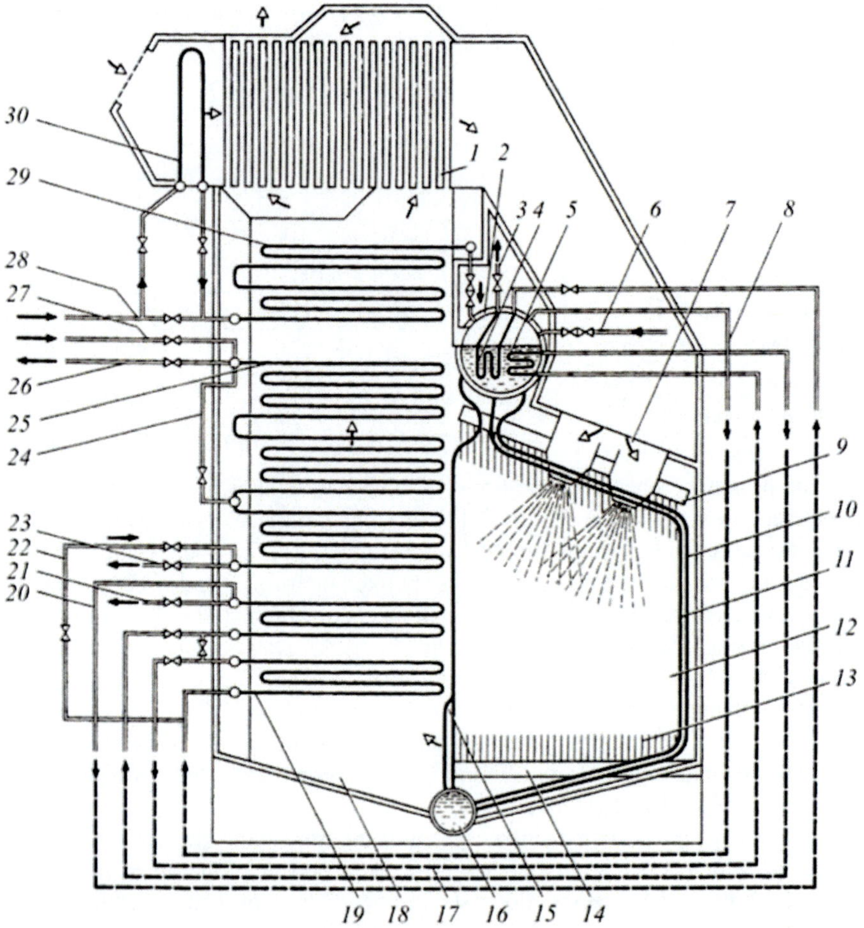

Fig. 5.13 Layout of the boiler KVG-80/80

The system of automatic regulation of fuel combustion allows changing the steam capacity of the unit in the range of 10–90 t/h.

Auxiliary steam boilers of steam turbine plant. Auxiliary boilers are not generally used on steam turbine ships with two main units. At parking, only one boiler operates, and its load depends on the need for steam in the parking modes.

Transport ships (mostly large-capacity tankers) or passenger ships equipped with only one main boiler are provided with an auxiliary boiler plant. Its capacity covers the need for steam at parking, for cargo heating and tanker washing. Along with this, such an auxiliary boiler should provide the work of the MTGU with the power at which the ship would have the speed of 7–8 knots and would be stably controlled.

A typical example of such boilers is the auxiliary boiler KV35/25–1 of the tanker "Kuban" (similar to "Krym"), which is shown in Fig. 5.15 [2].

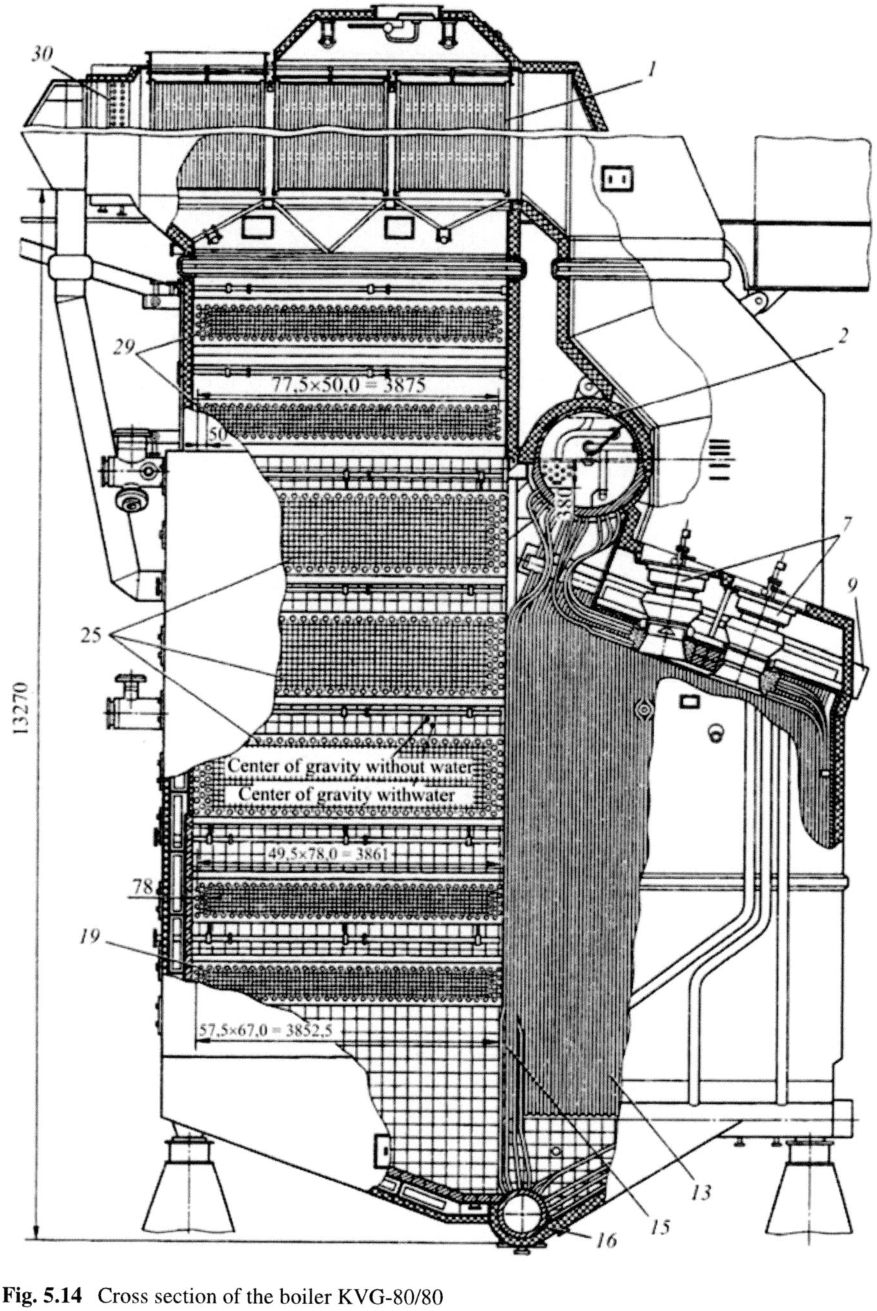

Fig. 5.14 Cross section of the boiler KVG-80/80

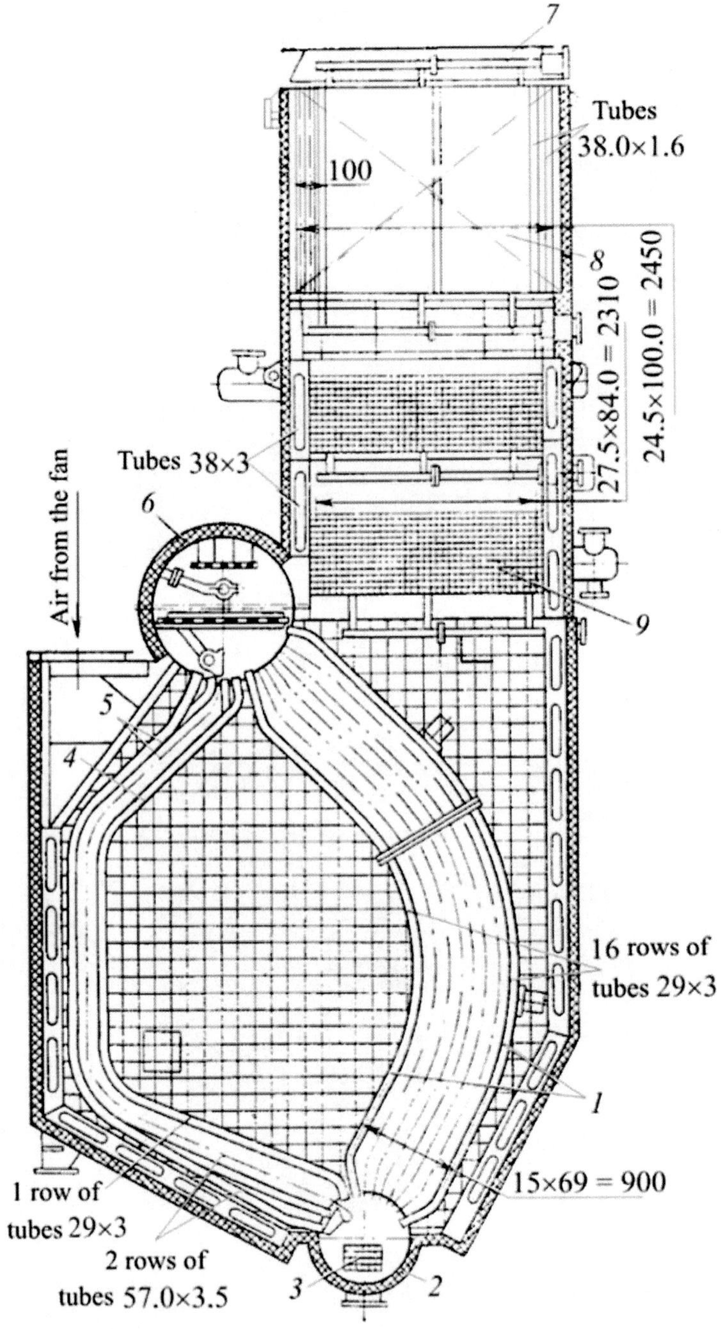

Fig. 5.15 Boiler KV-35/25–1

The water-tube boiler is vertical, single-flow, and two-drum. The steam and air heaters are the back-end heating surfaces of this boiler. The water circulation circuit is formed by the risers of screen *4*, convective bundle of tubes *1* and downcomers*5* located behind the first row of the screen. All the tubes are connected to steam-water and water drums (*6* and *2*, respectively). The boiler is equipped with soot blower *7*, air heater *8* and water heater *3*.

The water heater is installed inside water drum *2*; it provides heating of the boiler water with the heat of the superheated steam. This is necessary to maintain the unit in a hot standby. Superheater *9* consists of two compartments. The tube part consists of 112 twin coils. The air heater is vertical, one-pass for gas and air; its tubes are welded to the grilles. The air from the blower fan is first fed into the interlacing space, and then to the intertube space of the air heater.

Technical characteristics of the auxiliary boiler KV-35/25–1:

Steam capacity, t/h	35
Superheated steam pressure, MPa	2.45

Temperature, °C:

Superheated steam	330
Feed water	60
Efficiency coefficient, %	91

Auxiliary steam boilers for ships with diesel marine power plants. Among all the types of sea ships with the main DPPs, tankers and oil carriers are characterized by the most saturated auxiliary steam power equipment, which is associated with the operational characteristics of these ships. Most marine diesel tankers are multi-tonnage ships with high-capacity power plants. The cargo at the port of destination is pumped out by ship means for 8–10 h, and this is possible only if cargo pumps have the drive of a considerable power. Given that the power of the marine power station is often not enough, auxiliary steam turbines are used as the drive for cargo pumps.

Along with this, tankers are equipped with steam systems for heating cargo, heating water for washing tankers, steaming tankers after pumping out the cargo from them. It is obvious that with such steam power facilities, it is necessary to use high-performance ABs. The products of fuel combustion from the auxiliary boiler plant are used in the system of inert gases, which fill the tanks and tankers in order to reduce the fire risks.

The auxiliary boilers on large-tonnage tankers and oil carriers are similar to the main ones in their size, arrangement of heating surfaces, and parameters. A typical arrangement can be considered on the example of the auxiliary boiler KV1-1 installed on oil carriers of the "Boris Butoma" class (Fig. 5.16) [2].

Technical characteristics of the auxiliary boiler KV1-1:

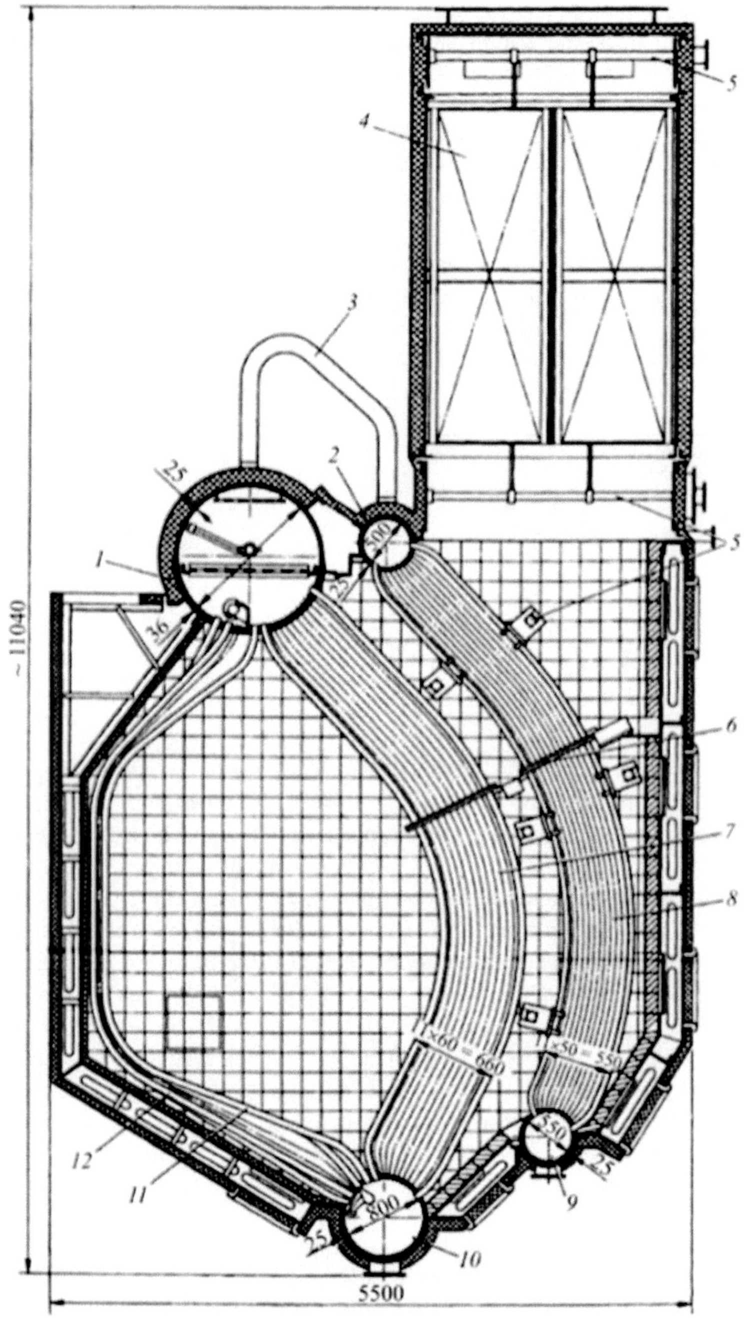

Fig. 5.16 Boiler KV1-1

Steam capacity, t/h	30

Steam parameters:

Pressure, MPa	2.6
Temperature, °C	320
Efficiency coefficient, %	93

A characteristic feature of this boiler is a two-drum steam superheater located not in the gas flue, but directly behind convective beam *7* of steam-generating tubes of the boiler, as well as the absence of the economizer. Four transverse bulkheads are installed in each of steam superheater drums *2* and *9*, so the saturated steam flows from the steam drum of boiler *1* through tube *3* and enters upper drum *2* of the superheater. After eight passes within bundle of tubes *8*, the superheated steam is removed from the upper drum. Reduction in the vibration of bundles of tubes *7* and *8* is facilitated by supporting device *6* made in the form of a comb. The boiler furnace is equipped with two steam-mechanical nozzles. Soot blowers *5* have remote control. Figure 5.16 also has other designations: *4*—air heater; *10*—water drum; *11*—screen tubes; *12*—downcomers.

In 1970–80 s, auxiliary boilers of the KVVA and KAV types were widely spread on dry cargo ships. A standard series of KVVA boilers with the steam capacity of 1–12 t/h and the steam pressure of 0.5–2.8 MPa was put into mass production. These boilers do not have back-end heating surfaces. Their brand is deciphered in the following way. The letter part (in the original Russian brand name) is a shortening, which consists of the initial letters of the words translated as "water-tube auxiliary automated boiler". The number in the numerator determines the steam capacity, t/h, and the number in the denominator stands for the operating pressure, kg/cm^2. The general arrangement of the boilers KVVA-2.5/5 is presented in Fig. 5.17. The steam capacity of the boiler is 2.5 t/h at the operating pressure of 0.5 MPa. The boiler is vertical, two-drum, single-flow; it has natural circulation. Its efficiency coefficient is 80%.

Screen tubes *8* and convective bundle *12* serve as risers, and tubes *7* located behind the screen are downcomers. All the tubes are connected to steam and water drums (*5* and *10*,respectively). Feed water is supplied through inlet pipe *6*, above which there is a soothing hole screen *2*; a similar screen *4* is also provided near steam inlet tube *3*.

The burner is equipped with two nozzles *9* and overview hatch *11* with protective glass. Soot blowers *1* are designed for blasting the surfaces with steam and are also used to wash them with water when cleaning the boiler.

A similar design is also characteristic of boilers of the KAV type. The KAV boilers which found application of ships have the steam capacity of 1.6–6.3 t/h at the pressure of 0.7 MPa and the steam capacity of 10 and 16 t/h at the pressure of 1.8 MPa. These

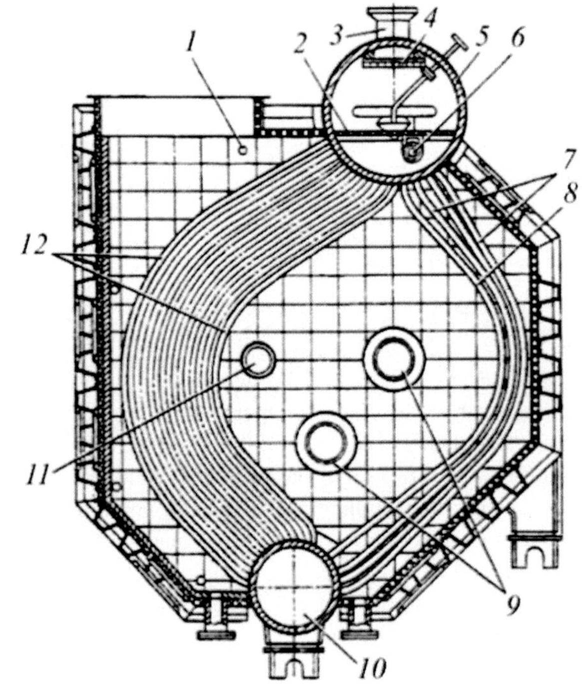

Fig. 5.17 Boiler KVVA-2.5/5

water-tube units are fully automated, two-drum, one-pass for gas; they are designed for operation at unattended maintenance.

Since the beginning of 1990s, major changes have taken place in the power industry of sea ships. The main steam boilers are now used only on gas carriers which employ the main steam turbine units with steam turbines produced by Kawasaki or Mitsubishi. Annually, there are no more than five such ships built annually.

The auxiliary boiler plants used on modern transport ships with diesel power plants are equipped with compact, simple and rational devices with auxiliary steam and hot water boilers. As a rule, these boilers with the steam capacity up to 20 t/h do not have back-end surfaces. They are often made with natural circulation and provide saturated steam to general consumers [3].

More powerful water-tube two-drum steam boilers with the steam capacity exceeding 20 t/h are designed for tankers. They generate both saturated steam for general consumers and superheated steam for driving an electric generator and cargo pumps.

Combined boilers have been widely used in the last decade. They can provide the ship with steam both on the move and at the parking.

Water-tube auxiliary boilers of modern transport ships. Alfa Laval Aalborg Industries (ALAI) occupies the prevailing, almost monopolistic position on the market of modern ship boilers.

The Aalborg D *water-tube auxiliary boiler* (formerly knows as MISSION^{TM}D) is two-drum and has a ceiling arrangement of burners with natural circulation (Fig. 5.18) [3]. The boiler consists of a furnace and convective surfaces. The furnace is formed by gas-tight membrane walls: side wall *10*, middle wall *4* and two front walls *9*. The side and middle walls are connected to drum *15*, from which the steam-water mixture is withdrawn via by-pass tubes *2* to the steam drum. The front walls *9* are made in the form of membrane panels of steam-generating tubes, which are connected to lower

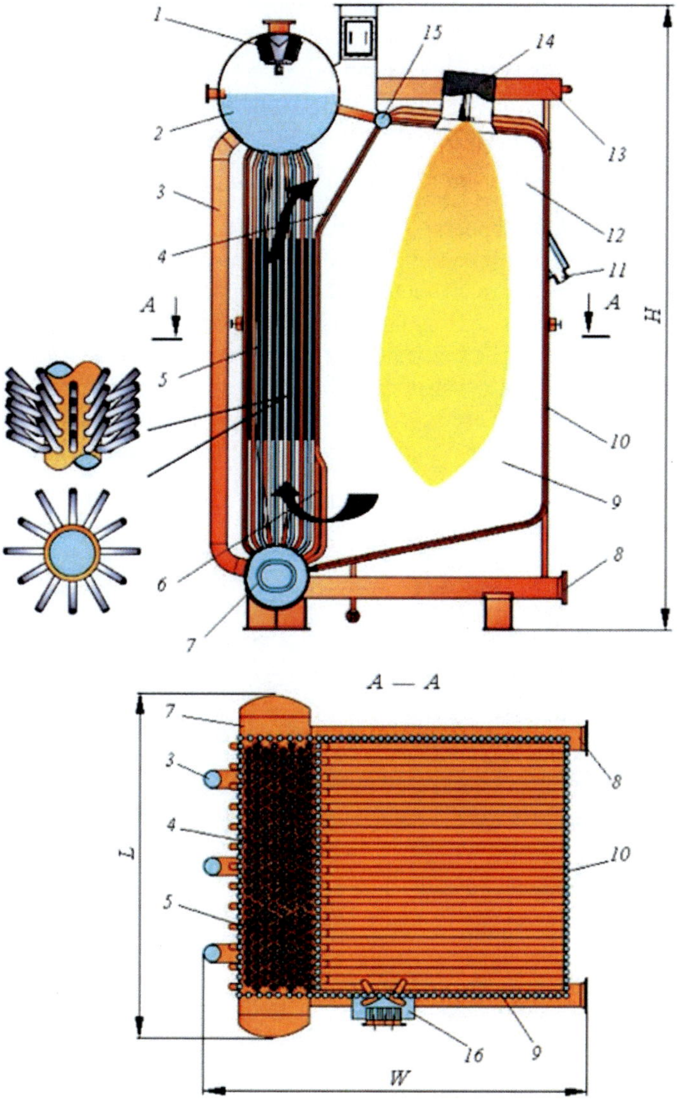

Fig. 5.18 Aalborg D boiler (without fittings and thermal insulation)

dispensing and upper collecting drums (*8* and *13*, respectively). The gases from the furnace are discharged through festoon *6*. The furnace is equipped with viewing device *11* and manhole *16*. Reliable natural circulation is provided by large-diameter downcomers*3* which connect steam and water drums (*2* and *7*) and are located on the outside of the boiler casing. There is separating device *1* in the steam drum.

The whip tube fins of convective steam-generating bundle *5* contribute to intensification of heat transfer in the boiler. Ceiling furnaces *14* with the steam mechanical nozzles provide efficient combustion of both diesel and heavy fuel.

Boilers of the Aalborg D type have the range of steam capacity of 25–120 t/h, saturated steam pressure of 1.8–2.4 MPa, and efficiency coefficient of 84%. They are mostly used on tankers. Boilers can produce slightly superheated steam (at 40 °C above the saturation temperature) for turbo drives of cargo pumps. In this case, they are equipped with a superheater placed in the exhaust gas outlet pipe.

Aalborg AQ-10/12 W water-tube vertical cylindrical steam boiler produced by ALAI.

The company produces boilers of this type with the steam capacity of 0.6 to 6.3 t/h and saturated steam pressure up to 1.0 MPa. In comparison with the previously produced boilers, these are characterized with an increased volume of the furnace for the combustion of waste from the separation of heavy liquid fuel.

Structural simplicity is a key feature of the boilers of this type (Fig. 5.19) [3].

Cylindrical housing *4* consists of a shell, lower and upper flat bottoms. In the middle of the housing, there is cylindrical furnace *3* with coated backstone and convective steam-generating bundle *6* made in the form of inclined tubes arranged

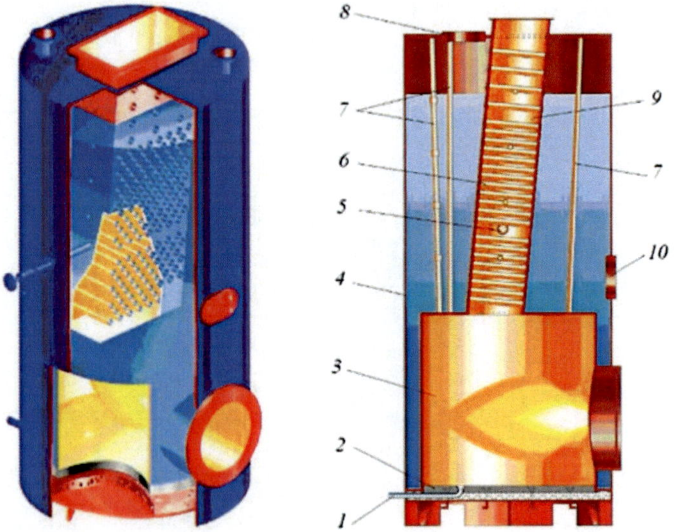

Fig. 5.19 Aalborg AQ-10/12 W boiler

across a rectangular gas flue *9*. To prevent segregation of the steam-water mixture, the angle of inclination of the tubes does not exceed 15°.

Access to the inside of the boiler during the maintenance work is provided by manholes *8* and *10*. The boiler is equipped with stationary soot blower *5*. Drainage tube *1* is intended for discharging water from the furnace while washing the boiler tubes with water.

The burner of the Aalborg KB-W type with a rotary nozzle ensures efficient combustion of both heavy liquid fuel and fuel waste with the maximum water content of 20% and solids content of up to 2 mm (no more than 5%).

Aalborg OL water-tube cylindrical two-drum steam boiler produced by ALAI.

To ensure strength of the boiler construction, power links *7* are mounted between the upper ends of the furnace and the housing.

According to the design type, the Aalborg OL boiler (O stands for oil fired, L stands for large steam capacity) is an auxiliary water-tube two-drum steam boiler with a burner on the ceiling, which has natural circulation and developed convective steam-generating bundle (Fig. 5.20) [3].

The horizontal section of the furnace is an irregular polygon. Enclosing walls of the boiler are gas-tight membrane panels *6* located on the sides of this polygon inscribed in a cylindrical casing. Furnace *5* is separated from the convective steam-generating bundle *12* by gas-tight screen *15*. The floor of furnace *8* is coated with fireproof bricks protecting the water drum from the flame radiation. Tubes of the membrane panels confining the convective steam-generating bundle from all sides and the bundle tubes are connected to water *10* and steam-water *3* drums. Together

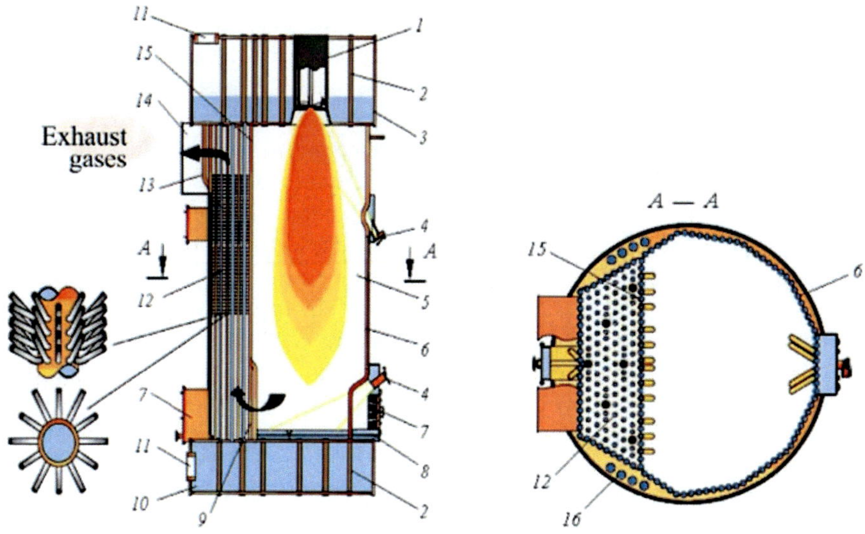

Fig. 5.20 Aalborg OL boiler (without fittings and thermal insulation)

with large-diameter downcomers *16* located outside the walls of the boiler, they form the natural circulation circuit.

Screen tubes *15* in the lower part of the screen are arranged into two-row festoon *9*, and the exhaust gas are discharged to smoke box 14 through two-row festoon *13*. The furnace is serviced by means of viewing devices *4* and windows with covers *7*. The washing of convective bundle tubes is longitudinal, the intensification of heat exchange in the boiler is provided by the whip fins.

Manholes *11* are used for servicing of the steam-water and water drums. The flat bottoms of the drums are supported by links *2*. The boilers are equipped with ceiling-mounted burners, which include Aalborg KBSD steam-mechanical nozzles for burning diesel and heavy fuel.

The steam capacity of the boilers of the considered type is 12.5–55 t/h. The saturated steam pressure is 0.9 MPa with the steam capacity of up to 20 t/h and 1.8 MPa with a higher steam capacity. In addition to being used for steam generation as part of auxiliary plants on tankers, such boilers can function as generators of inert gases.

Water-tube vertical cylindrical steam boilers of the MS series produced by Kangrin Heavy Industries. The steam capacity of the boiler is 0.5 and 1.0 t/h [3], the steam pressure is 0.7 MPa, and the efficiency coefficient is about 85%.

A special feature of the boiler design (Fig. 5.21) [3] is the presence of steam and water drums (*2* and *4*, correspondingly). They have a rectangular cross section and are made in the form of rings connected to each other by vertical tubes *3* arranged in two concentric circles. Tubes *3* form the inner and outer screens. The inner screen forms a cylindrical furnace chamber and includes steam-generating risers. A part of the tubes of the outer sealed screen is steam-generating risers, while the other, slightly warmed, are downcomers.

Burner *1* is mounted at the top of the furnace. The exhaust gases from the furnace chamber are discharged to the boiler intake through the side passage and annular gas flue between the inner and outer screens.

Aalborg OH horizontal cylindrical gas-tube boilers. The OH in the boiler brand name stands for oil-fired, horizontal. These boilers are manufactured by ALAI. The boilers can be made in two versions: a steam version with the steam capacity of 50–1600 kg/h and steam pressure of 1.2 MPa, and a water-heating version with the heat capacity from 500 to 10,700 kW and water pressure of 0.8 MPa [3].

The boiler body (Fig. 5.22) consists of the following elements: shell *13*, front *4* and back *12* bottoms, flame tube *1*, fire chamber *11*, fire tubes *10*, and burner *2*. The fire chamber is completely water-cooled. On the gas side, the boilers are made two-pass: first the gases flow through the flame tube, then make a 180° turn in the fire chamber and enter the fire tubes. The gases are removed from the boiler through smoke outlet pipe *5* via smoke box 3.

In the Aalborg OH boilers, intensification of heat exchange is realized with the help of turbulizers in the form of screw-like ribbon swirlers located inside the fire tubes.

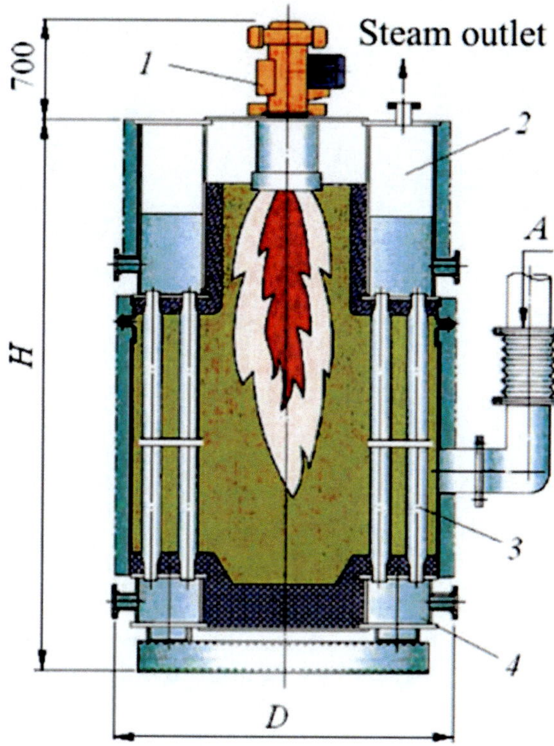

Fig. 5.21 Boiler of the MS series

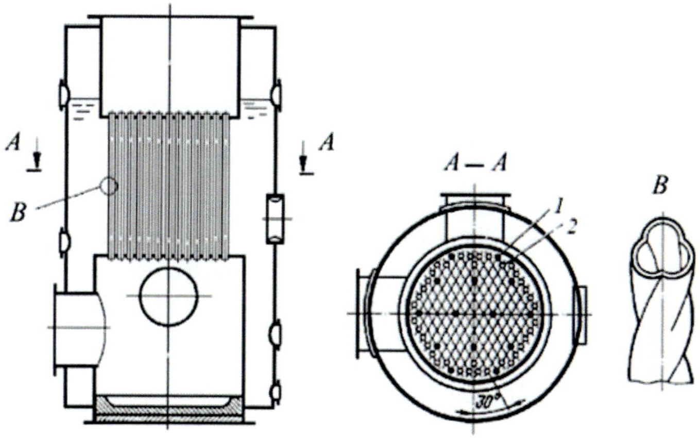

Fig. 5.22 Aalborg OH boiler

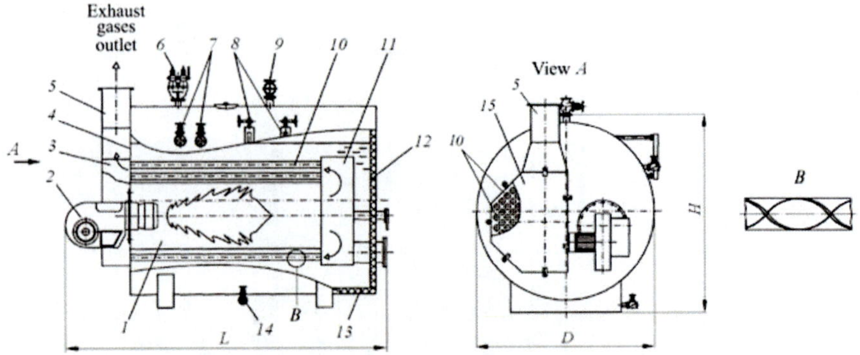

Fig. 5.23 Spanner boiler

Removable cover *15* provides access to the fire tubes for their cleaning. The boiler has the necessary fittings: safety valve *6*, stop valve *9* and feed valve *7*, bottom scavenging valve *14*, and water indicator *8*.

Vertical cylindrical boiler manufactured by Spanner (UK).

The furnace and the bundle of fire tubes are located inside the cylindrical housing of the boiler (Fig. 5.23) [3]. The burner is attached to a flanged pipe welded into the shell.

A special feature of the fire tubes of this boiler is that they are made of screw-like twisted Svirlifor tubes. In Fig. 5.23, 1 stands for the linked fire tubes (50.8 × 9.5 mm), and *2* stands for the Svirlifor fire tubes (50.8 × 3.7 mm). Turbulization of the flow of the gases that pass through these tubes increases the coefficient of heat transfer from the gases to the heating surface.

The steam capacity of the boilers is 1000–6500 kg/h, and the working steam pressure is 0.7–1.2 MPa [3]. Currently, they are produced under the brand Spanner VSF (VSF—vertical side fired).

Vertical cylindrical gas-tube boilers of the Aalborg OM series.

The former name of the boiler is MISSION^{TM}OM. Aalborg OM boilers (O—oil fired, M—medium steam capacity) have the steam capacity of 8–20 t/h at the gas pressure of 1.1 MPa and 14–45 t/h at the steam pressure of 1.8 MPa.

The boilers (Fig. 5.24) [3] consist of two parts: a furnace formed by a gas-tight membrane wall and a housing where the fire tubes are located.

The membrane screen tubes are welded into the screen water drum of the toroidal form with their bottom end. The upper ends of these tubes are welded into the lower bottom of the flat housing.

The peculiarity of this type of boilers is the Sunroad elements placed inside the fire tubes with the diameter of 273 × 12.5 mm. They are fixed in the lower and upper bottoms of the housing. The Sunroad element is a tube, the outer surface of which is covered with welded spikes arranged like sunbeams. These elements are welded

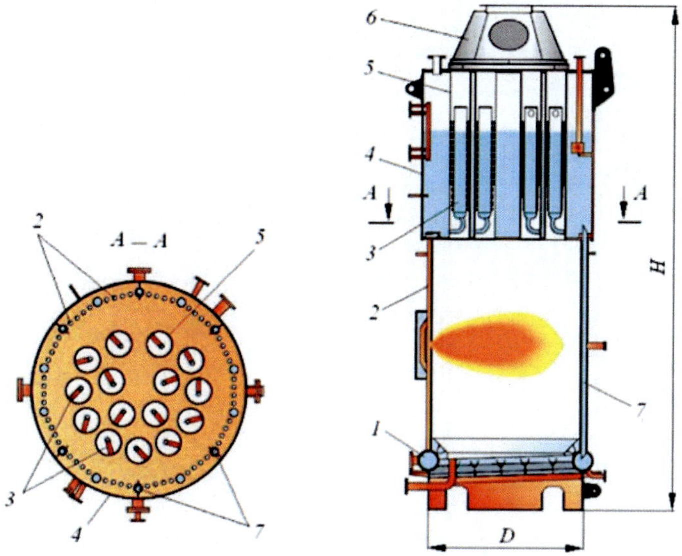

Fig. 5.24 Aalborg OM boiler: *1*—toroidal drum; *2*—steam-generating screen tubes; *3*—Sunroad elements; *4*—housing; *5*—fire tubes; *6*—smoke box; *7*—downcomers

into the part of the fire tube that passes through the steam space of the boiler. As a result, the steam-water mixture formed in the spiked tube enters the steam space.

The number of fire tubes depends on the boiler steam capacity and is between 10 and 30 pcs. The use of Sunroad elements makes it possible to reduce the dimensions of the convective part of the boiler and make it compact.

Recovery boilers of diesel and gas turbine marine power plants can have natural or forced circulation; they are most often made in the water-tube form.

An example of a water-tube recovery boiler with natural water circulation is the KUP-110/5.5 recovery boiler unit, which has only a steam-generating heating surface. The brand name of this boiler in Russian is deciphered as follows: the letter part stands for Steam Recovery Boiler; in the fraction, the numerator is the heating surface area, m^2; the denominator is the steam pressure, kg/cm^2.

The boilers of the KUP type are produced with the steam capacity of 0.175–3.0 t/h at the steam pressure of 0.49 MPa; they have a heating surface area of 15–300 m^2. The KUP 110/5.5 boiler (Fig. 5.25) [2] has the steam capacity of 1.3 t/h, the steam pressure of 0.54 MPa, and the heating surface area of 110 m^2. It has two drums: steam-water drum *9* is connected to water drum *7* by water-heating tubes (risers *8* and downcomers*6*, the latter being heated with a lower intensity). The exhaust gases from the main engine enter inlet pipe *5*, wash the heating surface of the boiler unit and pass through spark chute *10* into boiler intake *1*. Bulkhead *2* installed in bundle of tubes *6* improves washing of the tubes with gases. Shutters *4* and gas inlet guard *3* set the direction of movement of all or part of the engine exhaust gases through the heating surface of the boiler or beyond, depending on the need for steam.

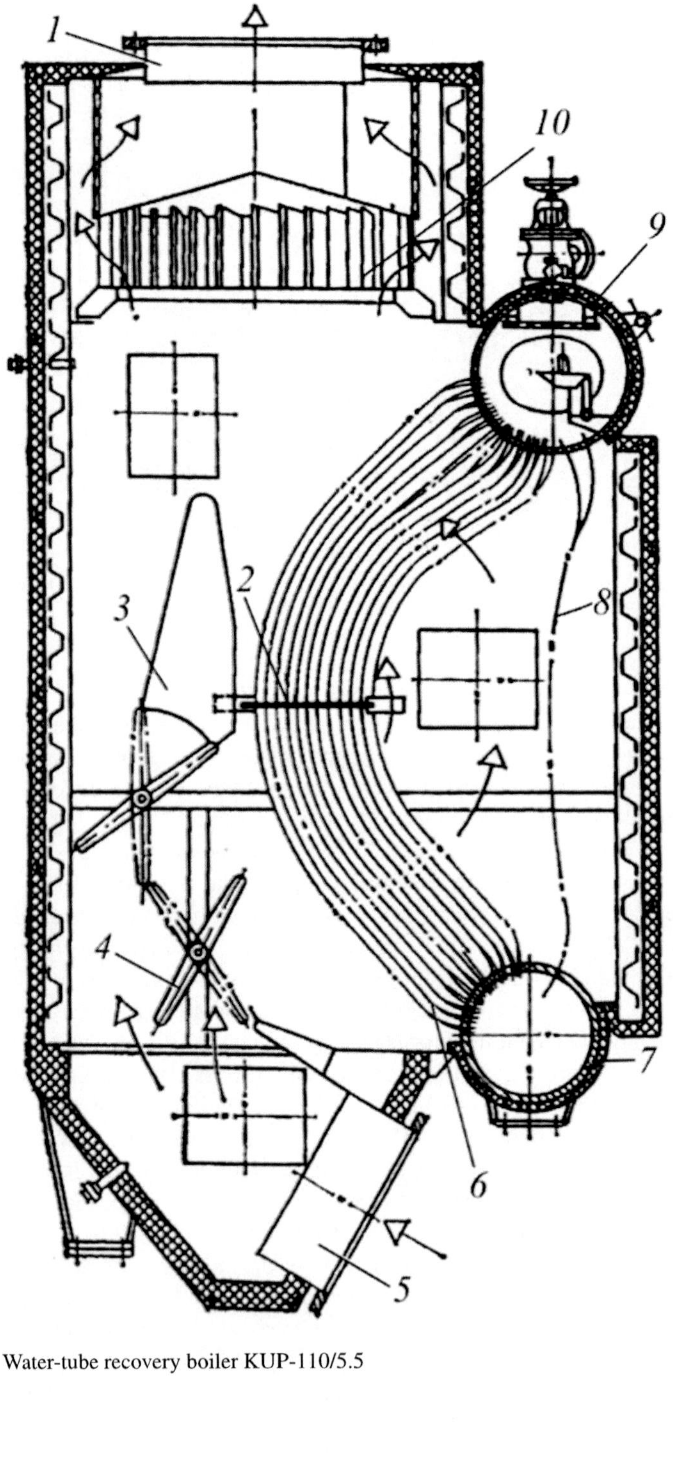

Fig. 5.25 Water-tube recovery boiler KUP-110/5.5

Due to the high humidity of the steam generated by the RB, a *steam separator* is introduced into the boiler plant of medium and large steam capacity. The separator is a steam collector where the steam is separated from the moisture coming in the form of a steam-water mixture from the outlet expansion drum of the recovery boiler. At the same time, it serves as a tank with a water reserve sufficient to compensate the level fluctuations of the boiler pressure and during variable operating modes.

Such boilers are not limited by only one steam-generating surface; along the gas flow, first is a superheater, then there are evaporating surfaces, and at the end there are economizing surfaces. Heating surfaces are a set of spiral horizontal coils with water and steam-water mixture moving inside them. A typical example of such boilers is the boiler KUP-1100 (Fig. 5.26) [2], which was installed in the 1990s on large oil and ore carriers.

Technical characteristics of the recovery boiler KUP-1100 are as follows: steam capacity of up to 9.2 t/h, steam pressure of 0.7 MPa; steam temperature of 270 °C.

Each tube bundle (superheating, evaporating, economizing) consists of two sections that operate in parallel. The feed pump takes water from the warm box and feeds it to the separator, from which it is directed via the circulation pump to coils *2* of the economizer located between upper inlet *1* and the lower outlet *3* drums. Drums *1* are fitted with feed tubes with three rows of openings for even distribution of water over the coils.

From drums *3*, water flows through the bypasses into steam-generating bundles *5* located between lower *6* and upper *4* drums. The generated steam-water mixture is fed to the separator (not shown in the scheme), from which part of the separated saturated steam is distributed to the general consumers, but most of it flows to coils *8* of the superheater located between upper *7* and lower *9* drums.

The resulting superheated steam enters the turbogenerator. The structural elements of the boiler are located in welded gas-tight housing *11*. There are removable screens on its walls that cover the openings made to access the interior of the boiler for inspection, cleaning and repair. The boiler foundation is equipped with four supports *10* for mounting, one of which is stationary. Supports *12* provide vertical detachment of the unit. The spark scrap consists of twisting blades *14*, grid-diaphragm *15* and cone *13*.

The heat recovery boilers of such gas turbine ships as "Kapitan Smirnov" are equipped with the units KUP-3100 (Fig. 5.21) [2].

The KUP-3100 recovery boiler are the following technical characteristics: steam capacity is up to 26.5 t/h; steam pressure is 1.25 MPa for the separator steam and 1.14 MPa for the superheated steam; steam temperature is 310 °C.

A special feature of this unit is the application of fins on all coils; it has the form of a spiral, welded tape. This technical solution ensures a high compactness of the unit with a smaller specific weight of heating surfaces. However, when the boiler is operating, these surfaces are heavily polluted, and their cleaning is rather complicated.

Superheated steam produced by the boiler is used to drive a steam turbine, and an auxiliary turbogenerator with the power of 1000 kW. The former together with

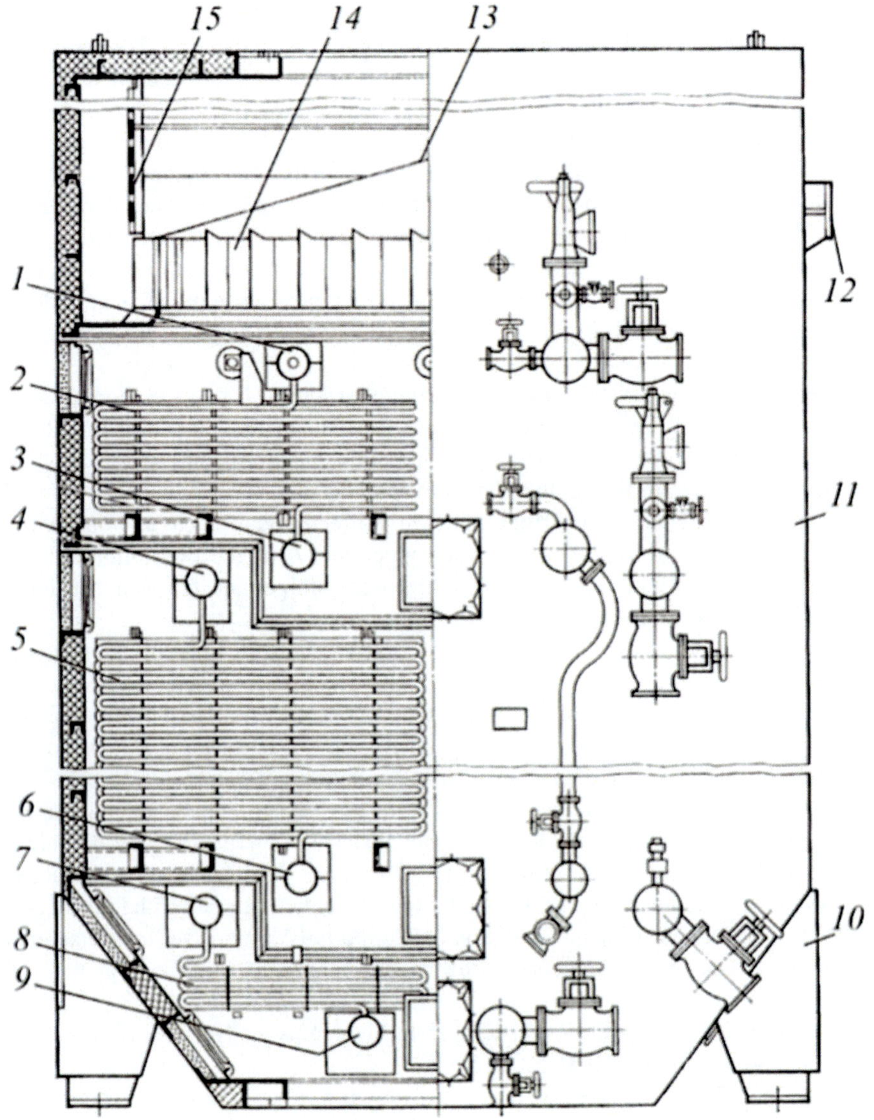

Fig. 5.26 Recovery boiler KUP-1100

the GTE rotates the propeller. In addition, 2.25 t/h of saturated steam is used for the heat supply of the ship (Fig 5.27).

Among the recovery boilers of modern diesel ships, a significant place is occupied by those manufactured by ALAI.

Aalborg P water-tube vertical cylindrical recovery boiler manufactured by ALAI. These boilers (formerly called UNEX^{TM}P) are designed to produce steam or heat

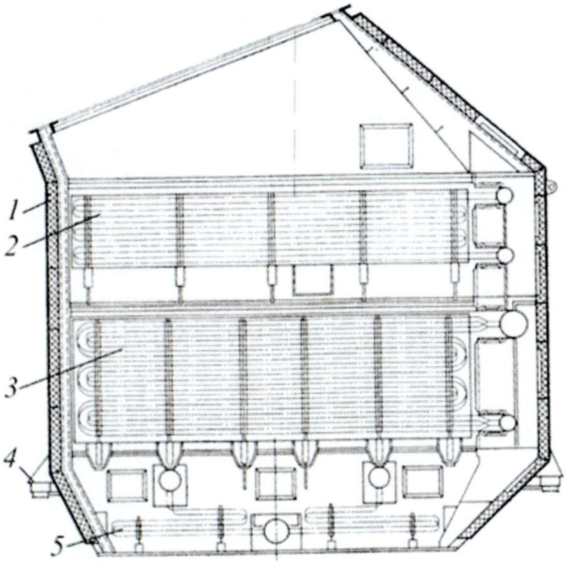

Fig. 5.27 Recovery boiler KUP-3100: *1*—casing; *2, 3, 5*—economizing, steam-generating and steam-superheating bundles of tubes, respectively; *4*—support

water in the ship heat supply systems. The steam capacity is 0.5–5 t/h, and the steam pressure is 1.0–2.4 MPa. By hot water, the heat capacity of the RB is 400–2800 kW. The boilers operate with forced circulation.

The heating surface of the RB (Fig. 5.28) is made in the form of vertical coaxial smooth-tube coils *2* located inside gas-tight casing *1* and connected to horizontal inlet *11* and outlet *8* drums. The coaxial coils are connected in parallel along the gas flow. The bottom drum is equipped with circulating water and drainage valves (*10* and *12*, correspondingly). The upper drum is equipped with the safety valve for steam-water mixture. The boiler housing is equipped with inlet pipe *11* and outlet chamber *5*, a drainage tube with a valve for water draining from the gas flue. Housing supports *4* are used to fasten the RB to ship structures.

Inside the housing, there is a sound absorber and a spark trap. The degree of sound absorption reaches 20 dB. Soot-blowing and water-washing devices are also installed on the boilers.

Similar design is typical for Aalborg XC recovery boilers as well. They generate up to 0.7–7.0 t/h of steam at the pressure of up to 2.4 MPa (Fig. 5.29). The designations in the name of the boiler have the following meaning: X stands for exhaust gas, and C stands for coil tube boiler.

Water-tube steam RB of the SG series with a rectangular shell produced by Osaka Boiler Company.

The RBs of this type are installed after the high-capacity ICE on large, long-distance container ships. They are designed for the production of saturated and superheated steam and provide both the RHG work and heat supply of the ship. The boiler heating surfaces are made of vertical finned coils, which are connected to horizontal drums.

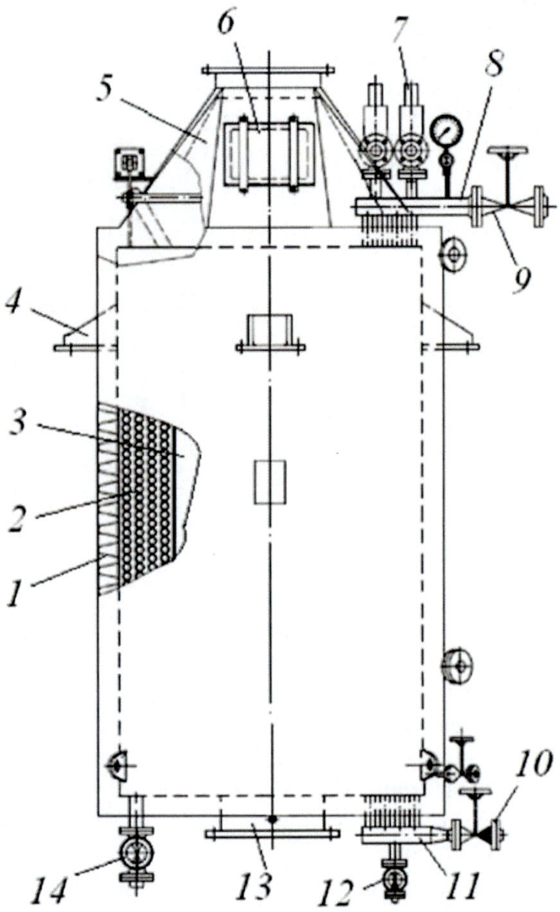

Fig. 5.28 Aalborg P boiler

The finning is transverse, H-shaped, individual for each tube (Fig. 5.30) [3]. The RB includes a steam-generating surface and a superheater, which are connected in a direct and counter flow, respectively. At the generation of steam of two pressures, a low-pressure steam-generating surface is additionally installed and connected to an individual steam separator.

***Boilers with combined heating*.**

Combined boilers or boilers with combined heating (BCH) include the recovery boilers that have a furnace for burning liquid fuel in case of lack of heat from the exhaust gases of the main engine or when it is turned off. As a rule, such boilers have two autonomous parts with heating surfaces: fuel (flame) and recovery ones. Sometimes both these parts have a common heating surface. The BCH can produce steam both during the ship movement and in the parking. In recent years, they are being increasingly used as the only way to produce steam on transport ships.

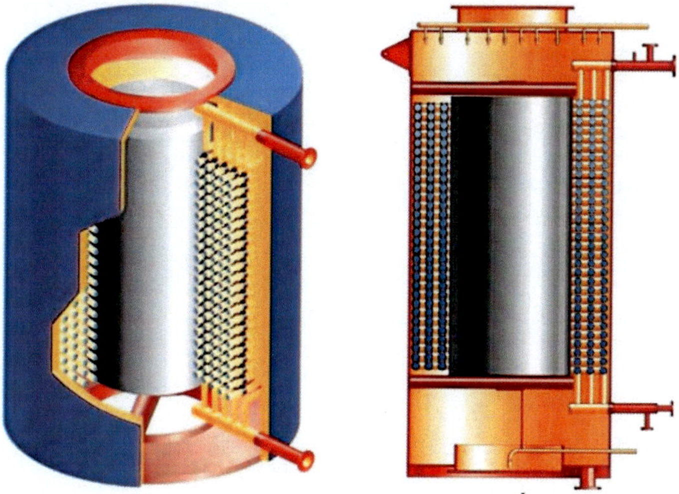

a *b*

Fig. 5.29 Aalborg XC boiler: **a** general view; **b** cross section

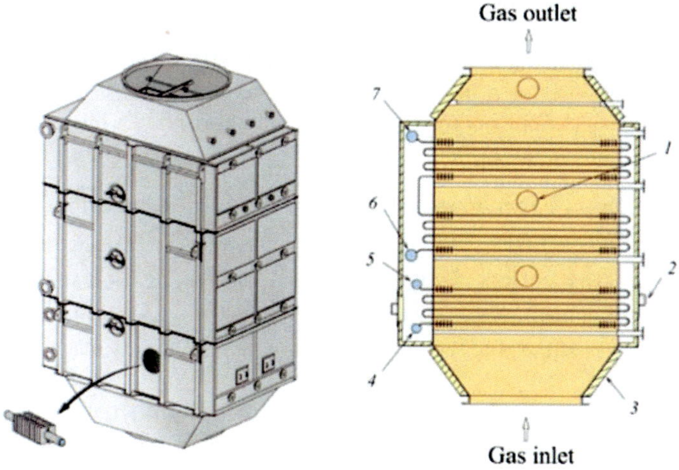

a *b*

Fig. 5.30 Recovery boiler of the SG series: **a** general view; **b** cross section

Aalborg AQ-10/16 vertical gas–water-tube steam boilers with combined heating manufactured by ALAI. Design of the fuel section of the boiler is based on the design of an auxiliary boiler (see Fig. 5.19).

The convective water-tube heating surface is made of inclined steam-generating tubes placed across the rectangular gas flue (Fig. 5.31) [3]. The furnace is completely water-cooled without the use of coating.

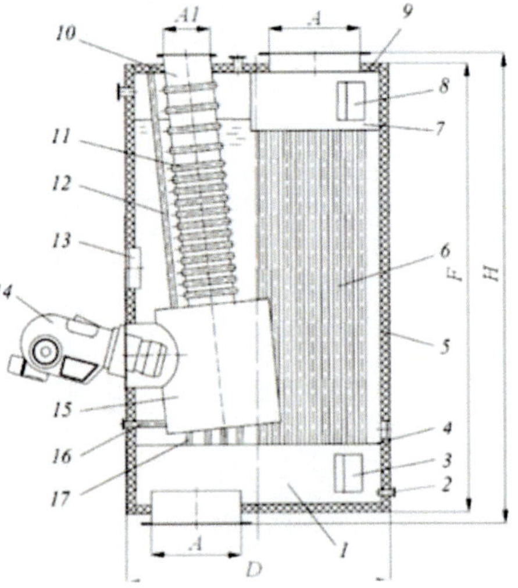

Fig. 5.31 Aalborg AQ-10/16 boiler: *1*—inlet chamber; *2, 16*—drainage tubes; *3,8*—manholes; *4*—lower bottom; *5*—shell; *6*—tubes of the recovery section; *7*—outlet chamber; *9*—upper bottom; *10*—fuel section gas flue; *11*—tubes of the fuel section; *12*—long link; *14*—burner; *15*—furnace; *17*—short links

The vertical fire tubes comprising the gas-tube recovery section of the boiler are fixed from below in the tube board of the boiler housing and from above in the boiler intake of the outlet chamber.

The engine exhaust gases enter the inlet chamber, then they move inside the fire tubes to the outlet chamber, from which they flow into the exhaust gas flue.

The steam capacity of the fuel section of the boiler is 0.7–5.0 t/h, and that of the recovery section is up to 3 t/h.

Vertical cylindrical gas-tube steam boiler with combined heating of the PC series produced by Kangrim Heavy Industries.

Design of the boilers in Fig. 5.32 is similar to the design of the Aalborg OC boilers and consists of a furnace and vertical fire tubes with Sunroad elements inside of them. The burner is fixed at the angle of 15°, which improves the fuel combustion conditions.

The heating surface of the recovery section is made of straight fire tubes. The boiler is equipped with a sound absorber.

The steam capacity of the fuel section is 1–3.5 t/h, and that of the recovery section is up to 3 t/h depending on the parameters of the exhaust gases of the internal combustion engine.

Vertical cylindrical gas boiler with combined heating CMB-VF-LONOX produced by SAACKE.

In the designation of the boiler, CMB stands for Composite Marine Boiler, VF—Vertical with Flame Tube, LONOX—Low NO_x.

The fuel section of the boiler is two-pass: at the beginning, exhaust gases move downward in the flame tube (the first pass), then they make a 180° turn in the fire chamber and move upward in the fire tubes (the second pass) (Fig. 5.33) [3]. After

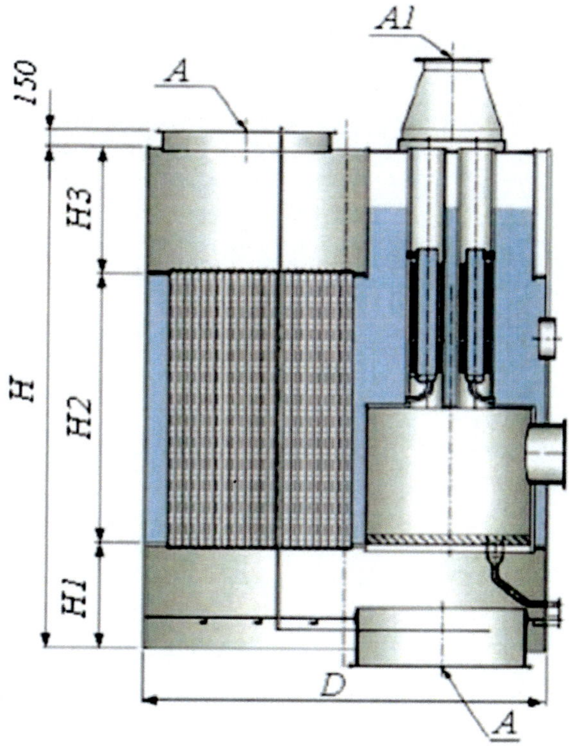

Fig. 5.32 Boiler with combined heating of the PC series

the fire tubes, the gases enter the smoke box, which has windows with covers to clean the tubes from the inside.

Design of the fuel section of the boiler makes use of the combustion scheme with low emissions of nitrogen oxides due to recirculation of exhaust gases in the furnace (Low NO_x Combustion System). The recirculation sets the temperature level in the furnace below 1500 °C, which reduces NO_x emissions. The NO_x emissions do not exceed 0.3 g/Nm^3 when burning light types of liquid fuel, 0.8 g/Nm^3 at heavy fuel and 0.2 g/Nm^3 at gaseous fuel. Another feature of this boiler is that its recovery section made of straight vertical fire tubes can include several independently operating bundles of tubes connected to both the main engine and several (up to three) auxiliary diesel generators.

The steam pressure is 1.0 MPa, the steam capacity of the fuel section is up to 13 t/h with recirculation or up to 18 t/h without recirculation. The boilers are installed on passenger ships, tankers and gas carriers.

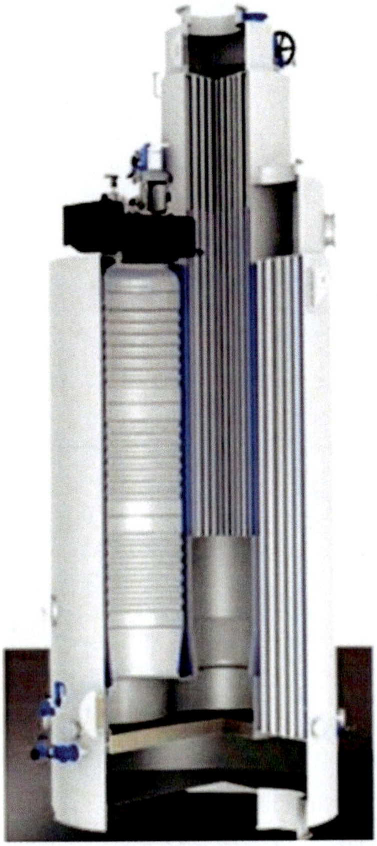

Fig. 5.33 CMB-VF-LONOX boiler with combined heating

5.3 Characteristics of Marine Steam Turbine Units

The number and nomenclature of steam turbine units used on ships have recently been very limited. These are single container carriers and gas carriers equipped with MTGUs of Japanese production, ships with NPPs (mainly icebreakers of Russian production) and recovery steam turbines of GTPs or recovery steam turbine generators of gas turbine or diesel MPPs.

The main reason for such a state of application of steam turbine power technology on ships is its lower fuel efficiency compared to the DPP.

The basic characteristics of the main marine steam turbine units are given in Table 5.1.

Table 5.1 Main characteristics of marine steam turbine units

Model of the main turbine unit, class of ship, year of manufacture	Company, country	Power rating, kW	Parameters		Number of regenerative selections	Structural scheme	Rotation speed, rpm		Pressure in the condenser, kPa
			Fresh steam P, MPa: t, °C	Superheated steam P. MPa; t, °C			Turbine	Propeller	
MTGU TS-1; Dry cargo ships of the "Lenin Komsomol" class (I958-1967)	OJSC Kirovsky Zavod, Russian Federation	9560	4.0 465	–	3	1 HPT 1 LPT	5500 3680	100	4.9
MTGU TS-2: Tankers of (he "Pra;;a" and "Sofia" classs (1959–1970)		13,970	4.0 470	–	3	1 HPT 1 LPT	4780 2860	110	4.9
MTGU TS-3; Tankers of the "Krym" class (1974–1982)		22,060	7.7 510	1.4 510	5	1 HPT + 1 MPT 1 LPT	5370 2850	85	5.1
MTEU; Icebreaker"Lenin™(1959, 1970 — modernization)		8100	2.9 285	–	2	1 HPT	3850	–	10.0
MTEU GTG-642, Icebreakers o f (he "Arkti ka" class (1974–2007)		27,570	2.9 300	–	1	1 HPT	3500	130	6.9
MTEU GTG-M2T: Icebreakers o f ihc "Taimyr" class < 19*9–1990)		18,400	2.9 300	–	1	1 HPT	3000	180	6.9
MTGU 684 OM v Lighter carrier -'ScviiMirput" < 19*S)		29,500	4.0 290	–	3	1 HPT 1 LPT		115	7.2

(continued)

Table 5.1 (continued)

Model of the main turbine unit, class of ship, year of manufacture	Company, country	Power rating, kW	Parameters		Number of regenerative selections	Structural scheme	Rotation speed, rpm		Pressure in the condenser, kPa
			Fresh steam P, MPa: t, °C	Superheated steam P. MPa; t, °C			Turbine	Propeller	
MTOU MST-14; Tanker (1970)	General Electric, USA	23,540	10.1 510	2.0 510	3	1 HPT + 1 MPT 1 LPT	6000 3600	128	5.1
MTGU MST-19; Container ihip 11,971)		44,100	6.1 501	–	1	1 HPT 1 LPT	5040 3370	135	7.0
MTOU MST-21; Tanker (1971)		14,000	5.9 510	–	4	1 HPT 1 LPT	–	–	5.1
MTGU MST-23; Tanker (19*3)		36,800	16.8 566	–	6	1 HPT 1 MPT 1 LPT	–	80/120	5.1
MTGU UR-315. Tanker (1970)	Kawasaki Heavy Industries, Japan	22,000	10.3 520	2.2 520	4	1 HPT + I MPT 1 LPT	7050 3570	90	5.1
MTGU A SPP; Container sli ip (19(0)		58,800	13.8 540	3.1 540	5	1 HPT + 1 MPT 1 LPT	–	–	5.1
MTOU URA-450; LNG earner (2011)		29,890	12.1 565	2.3 565	4	1 HPT + 1 MPT 1 LPT	–	70	5.1
MTGU IJFMCIO: LNG (carrier 2009)		27,600	6.0 520	2.1 520	4	1 HPT + 1MPT 1 LPT	–	82	5.1

(continued)

Table 5.1 (continued)

Model of the main turbine unit, class of ship, year of manufacture	Company, country	Power rating, kW	Parameters		Number of regenerative selections	Structural scheme	Rotation speed, rpm		Pressure in the condenser, kPa
			Fresh steam P, MPa: t, °C	Superheated steam P. MPa; t, °C			Turbine	Propeller	
MR 36–11: LNG carrier(2015)	Mitsubishi Heavy Industries, Japan	3910	10.2 510	2.2 510	4	1 HPT + 1 MPT 1 LPT	–	81	5.1

References

1. Kartsev V. P., Khazanovskiy P. M. *Tysyacheletiyaenergetiki* [Thousands of years of power engineering]. Moscow, Znanie Publ., 1984. 224 p.
2. Horbov V. M. *Entsyklopediiasudnovoienerhetyky: pidruchnyk* [Encyclopedia of marine power engineering: textbook]. Mykolaiv, NUK Publ., 2010. 624 p.
3. Yepifanov O. A. Konstruktsiiasudnovykhkotliv: navch. posibnyk [Construction of ship boilers: textbook]. Mykolaiv, NUK Publ., 2016. 198 p.

Chapter 6
Marine Gas Turbine Power Plants

6.1 The Main Types of Marine Gas Turbine Units and Engines

The simplest scheme of a marine gas turbine unit with fuel combustion at a constant pressure is shown in Fig. 6.1. The compressor sucks air from the atmosphere, compresses it and delivers it to the combustion chamber, which is open both from the air inlet and the combustion products outlet. Fuel, which is sprayed with the help of a nozzle, is supplied to the combustion chamber, mixed with air and then burns at a constant pressure. Combustion products with high temperature enter the gas turbine, where they expand from the initial pressure equal to the pressure behind the compressor to atmospheric pressure. During the expansion of combustion products, their internal energy is converted into mechanical work.

The ratio of pressures in the turbine and compressor is almost the same. Yet, the turbine work exceeds that of the compressor, since it is proportional to the absolute temperature of the working fluid, and the temperature of the gases in the turbine is much higher than the air temperature in the compressor.

Part of the work of the turbine is spent on rotation of the compressor, and the other (redundant) part is transmitted through the gear reducer to the consumer, i.e. to the ship propeller.

The GTE is launched by a starter that rotates the engine shaft until the power that the turbine develops exceeds the power required for the drive of the compressor. The fuel–air mixture in the combustion chamber is ignited by means of starting blocks that combine an electric spark plug and a starting nozzle.

Figure 6.2 shows the scheme of a GTU with an engine in which fuel combustion takes place at a constant volume. Its main difference from the previous scheme is the design of the combustion chamber, which is equipped with valves at the air inlet and at the combustion products outlet. The air that exits compressor *1* passes through open intake valve *2* into combustion chamber *4*; at that, the valve at the outlet of the combustion chamber is closed. After filling the combustion chamber with air, nozzle *3* receives fuel and ignites the combustible mixture with the help of the electric

Z. Yang et al., *Marine Power Plant*,
https://doi.org/10.1007/978-981-33-4935-3_6

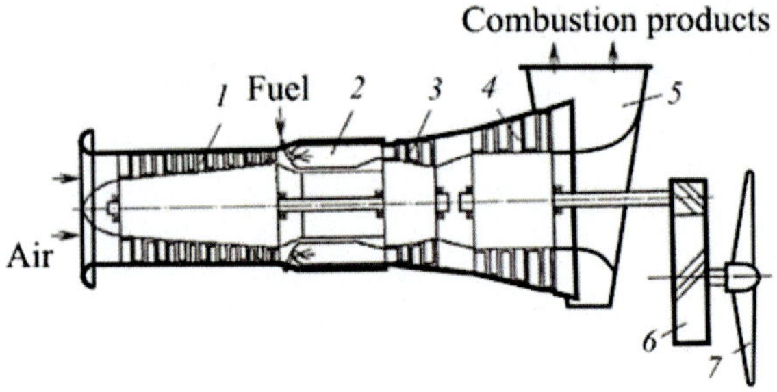

Fig. 6.1 Scheme of a gas turbine unit with combustion at a constant pressure: *1*—compressor; *2*—combustion chamber; *3*—compressor drive turbine; *4*—propeller turbine; *5*—gas outlet device; *6*—reducer; *7*—propeller

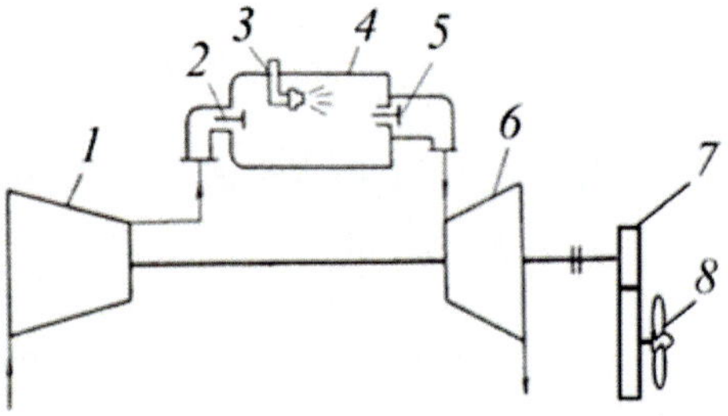

Fig. 6.2 Scheme of a gas turbine unit with combustion at a constant volume

igniter. Thus, fuel combustion occurs in a confined space, which leads to an increase in pressure during combustion. At the end of this process, exhaust valve *5* is slightly opened and the combustion products are fed to turbine *6*. The power is transmitted through gearbox *7* to propeller *8*.

When the gas pressure in the chamber becomes lower than the pressure at the compressor outlet, the air valve opens and air enters the chamber, displacing the combustion products. This air can also provide cooling of the heated elements of the air-gas channel. After closing the intake and exhaust valves, the process is repeated. The use of several combustion chambers for one turbine with particular phases of the processes shifted in time provides a continuous supply of gases to the impeller.

Considering the fact that the pressure rises during fuel combustion in a closed combustion chamber, the GTP could operate even without a compressor or with the use of a low-pressure fan. However, a gas turbine plant with combustion at a constant volume generally includes a compressor as well.

Operation of all modern gas-turbine engines follows a scheme of combustion at a constant pressure. The structural schemes of simple-cycle gas turbine units are shown in Fig. 6.3. This is the case when the GTE includes only one or two compressors, one combustion chamber, and one, two or three turbines.

The block scheme (see Fig. 6.3a) is constructively the simplest. The GTE made according to this scheme includes one turbine, which serves to drive the compressor and the propeller. A characteristic feature of such a gas turbine engine is the dependence of the compressor rotation speed on the propulsion characteristics. When starting and operating in partial modes, the gas turbine engine made according the block scheme requires reduction in the load, which in turn requires the use of electric,

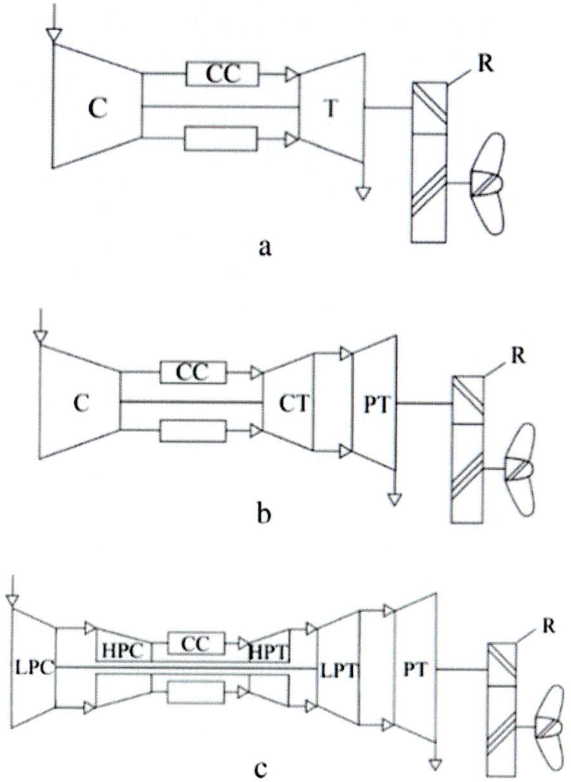

Fig. 6.3 Structural schemes of simple-cycle gas turbine units: C—compressor; CC—combustion chamber; T—turbine; TC—turbine of the compressor; PT—power turbine; R—reducer

hydraulic gears or controllable-pitch propeller. Such engines are used in the units that mainly operate at rated loads, i.e. they are unsuitable for ship conditions. At the same time, the gas turbine engines made in accordance with this scheme found application at the early stages of the development of marine gas turbine engines.

Figure 6.3b shows the scheme of the GTE with a free power turbine, which has only a gas dynamic connection with the rest of the engine path and performs only one function, which is propulsion drive. This scheme eliminates the disadvantages typical for the GTEs performed according to the previous scheme. The GTEs LM1600, LM2500, and MS series manufactured by General Electric were made according to the one-cascade scheme (a *cascade* is a combination of a compressor and a turbine that drives it) with a block power turbine.

The two-cascade GTE scheme with a free power turbine (see Fig. 6.3c) allows obtaining a sufficiently high pressure increase rate corresponding to a high initial temperature of the gases, as well as ensuring the operation of each compressor under optimal conditions. Such a gas turbine engine has two compressors (low- and high-pressure ones) connected in series. LPC and HPC are driven into rotation each by its own turbine (low- and high-pressure turbines, respectively), and the free power turbine produces useful power. The combination of the compressors, turbines that drive them and combustion chamber forms a *gas generator*. The arrangement of both compressors in the same housing, as well as the placement of all turbines in the common housing (direct-flow scheme), allows for the compact design of the gas turbine engine and reduces the pressure losses for air and combustion products. This scheme is typical for all ship engines of Ukrainian production, as well as those manufactured by Rolls-Royce and Pratt & Whitney.

At the existing level of *initial temperatures* in the cycle (initial temperature being the temperature of gases at the outlet from the CC), gas turbine units of the simple scheme have a relatively low thermal efficiency.

The maximum efficiency of such units at initial gas temperatures up to 1250 °C does not exceed 38% [1–3]. In order to increase this value, the circuits and cycles of GTU are made more complex with the use of the following methods [1]:

- intermediate cooling of air during compression;
- regeneration of heat of the GTE exhaust gases;
- intermediate heating of gas during expansion; the use of heat of the GTE exhaust gases in the recovery steam turbine circuit (RTC);
- recovery of heat of the GTE exhaust gases with the organization of its operation according to the contact gas turbine cycle.

Intermediate air cooling in the process of increasing the pressure is provided by using an additional surface heat exchanger pumped by seawater, which is called an intermediate air cooler (IAC) (Fig. 6.4).

After some stages of compressor K1, the air with the temperature increased due to compression enters the IAC, where it is cooled by seawater. Then it enters the second part of compressor K2. In this and the following figures depicting complex GTU schemes, the following designations are adopted: SC—steam condenser; SGC—steam gas condenser; RST—recovery steam turbine; SW—seawater; RG—regenerator.

In the range of initial gas temperatures T_3 = 1100–1400 K, the efficiency coefficient of the unit with intermediate air cooling is 33–38%.

The economic efficiency of the GTU can be improved by introducing an additional heat exchanger into the scheme. It would make use of the heat of the GTE exhaust gases to raise the temperature of the cyclic air after the compressor before it is fed to the combustion chamber (Fig. 6.5). Such a heat exchanger is called a regenerator

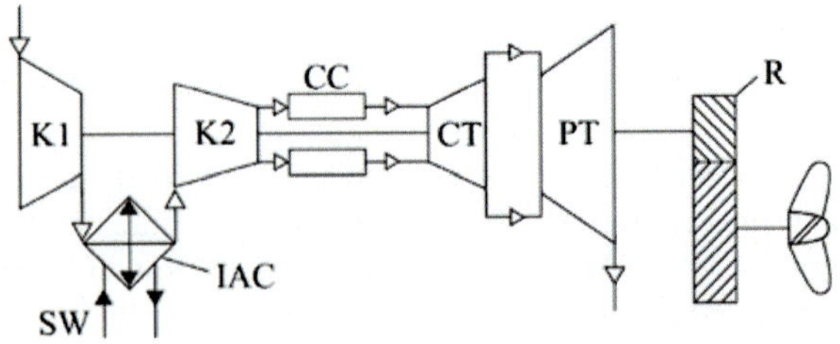

Fig. 6.4 Scheme of a marine gas turbine unit with intermediate air cooling

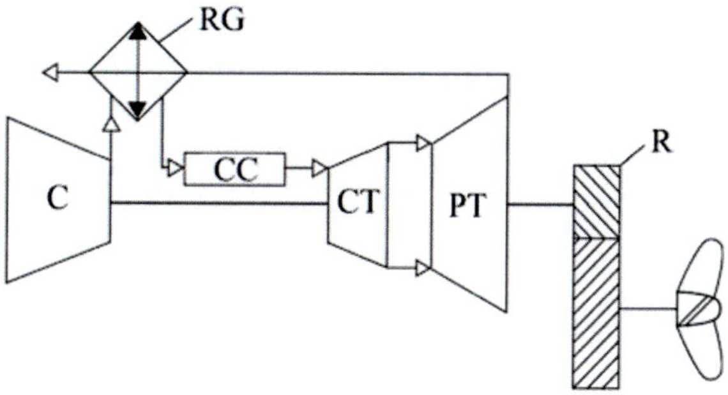

Fig. 6.5 Scheme of a marine gas turbine unit with regeneration of exhaust gas heat

(sometimes referred to as "recuperator", derived from a recuperative heat exchanger), and the GTU is called a *gas turbine unit with regeneration of the exhaust gas heat.*

In the range of initial gas temperatures $T_3 = 1100–1400$ K, the efficiency coefficient of the cycle of GTU with regeneration can reach 34–40%. This increase is explained by saving of a part of the fuel used for heating the cyclic air due to the use of the exhaust gas heat.

Ship practice demonstrates examples of use of the GTU with regeneration of the exhaust gas heat (the GTU of the tankers "Auris", "Shevron Oregon", dry cargo ship "John Sergeant", ferry "Seaway Prince"), but placing both the intermediate air cooling and regeneration in one unit is more common. First, it was implemented in the experimental GTUs in 1950s, then in the GTU installed on the dry cargo ship of Paris Commune Shipyard and the GTU WR-21 jointly produced by Westinhouse and Rolls-Royce. The expediency of such a solution is explained by the fact that the advantages of both ways of increasing the efficiency of ship GTU are most fully manifested in this option.

By including an additional combustion chamber in the scheme of the unit, it is possible to achieve a substantial increase in power and a moderate increase in the efficiency of the unit. Such a chamber is referred to as *combustion chamber for intermediate heating* (CCIH) of gases). In this scheme (Fig. 6.6), the first combustion chamber along the air flow is called the main combustion chamber, and the second combustion chamber, CCIH, is located between the turbines.

The increase in the efficiency of such a scheme makes up 1.0–1.5% in comparison with the GTU of a simple scheme, and the specific power is 15–20% higher [3]. The

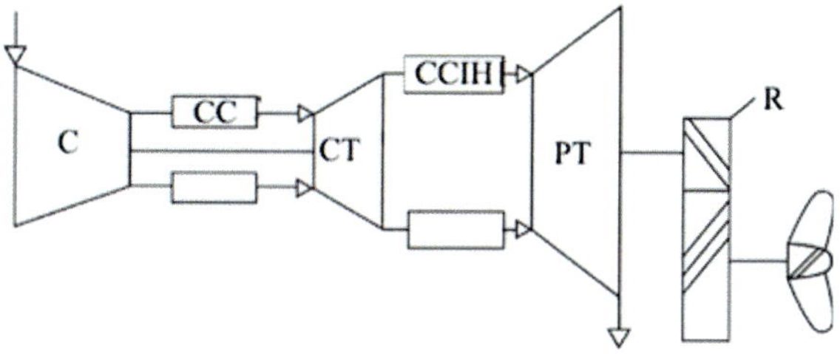

Fig. 6.6 Scheme of a marine gas turbine unit with intermediate heating of gases

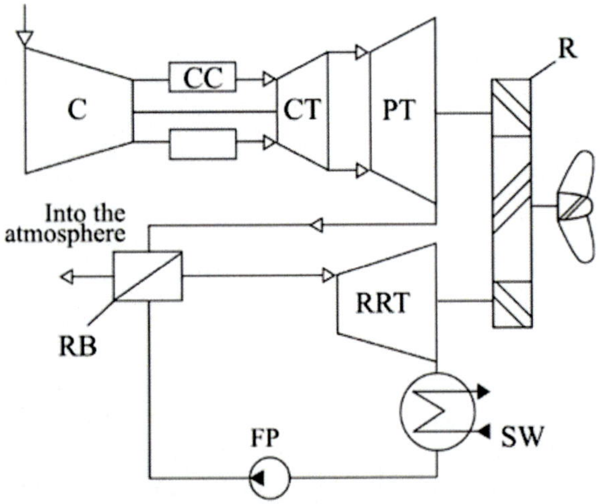

Fig. 6.7 Scheme of a ship unit with a steam turbine heat recovery circuit

GTUs made according to the scheme with intermediate heating of gases have not yet been implemented in ship practice.

The most effective solution for increasing the efficiency of the ship GTU is combination of the Brighton gas turbine cycle and the Rankine steam turbine cycle in a single-unit scheme. This solution is implemented in the GTU using the exhaust gas heat in the steam turbine heat recovery circuit, which includes the recovery boiler and the recovery steam turbine and condenser (Fig. 6.7). The name of the scheme is shortened to *GTU with a steam turbine HRC*.

The structure of the considered unit includes the GTE of a simple scheme; its exhaust gases flow to the recovery boiler (RB), which generates superheated water steam of relatively low parameters. The resulting steam is fed to a recovery steam turbine (RST), which, together with the gas turbine engine, operates on a totalizing reducer. The steam from the RST is discharged into a vacuum steam condenser, cooled by seawater, and consequently condensed. The resulting condensate is fed to the RB by the feed pump.

In the range of initial gas temperatures $T_3 = 1100$–1500 K, the efficiency coefficient for the unit with steam turbine HRC can reach 36–45% [3]. The increase from the simple circuit unit is explained by the fact that up to 25% of the combined GTU power can be obtained without additional fuel consumption, only by using the heat of the GTE exhaust gases in the RST.

This scheme was taken as a basis for GTUs of the power plants of four ships with horizontal cargo handling of the "Atlantika" type, produced in Ukraine [1].

Figure 6.8 demonstrates the scheme of the GTU which provides for combining the Brighton and Rankine cycles without the constructive separation of GTE and HRC. The unit performed in this way is called a *contact GTU with HRC*.

The main element of the unit is a gas turbine engine of a simple scheme; its exhaust gases ensure generation of steam in the RB. The resulting steam is supplied to the combustion chamber. The gas-steam mixture formed there is expanded successively

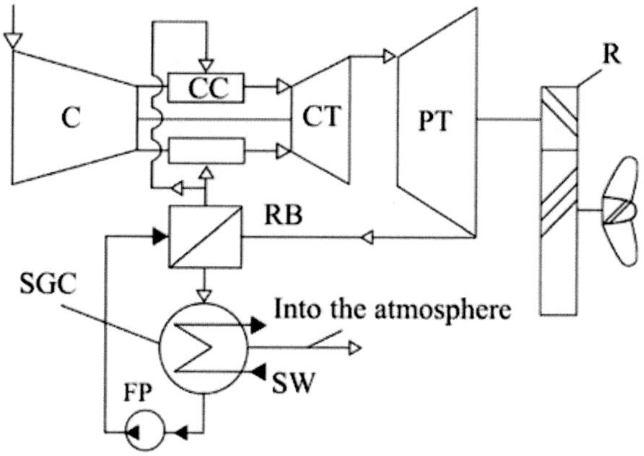

Fig. 6.8 Scheme of a marine gas steam turbine contact unit with a heat recovery circuit

in the turbine of the compressor and in the power turbine. Then it passes through the RB, giving part of the heat to the steam-water working fluid, and enters the steam-gas condenser. This condenser operates at atmospheric pressure and is pumped by seawater. In the condenser, a significant part of the water, both injected into the cycle and formed during fuel combustion, is condensed and fed to the feed pump, and the combustion products are released into the atmosphere almost completely dehydrated.

In the range of initial gas temperatures $T_3 = 1100–1500$ K, the efficiency coefficient of the contact gas turbine unit (CGTU) with the HRC is 35–46%. This high efficiency of the CGTU with HRC is explained by the deep recovery of the heat of the GTE exhaust gases and the high initial steam temperature in the steam-water cycle [1].

The considered scheme of the CGTU with HRC is implemented in the units "Vodoley", which have been developed and put into serial production at the Ukrainian enterprise Zorya-Mashproekt Gas Turbine Research and Development Complex.

6.2 Formation of World and Ukrainian Marine Gas Turbine Construction

Western countries and the USA (1950s). The use of gas turbine engines on ships is associated with specific requirements: the need to regulate its rotation speed for maneuvering the ship and reversing its course. To meet these requirements, the GTP must include astern turbines or stages, reverse gears or controllable-pitch propeller. With the help of the turn of its blades, the latter provides both reverse of the ship and stageless regulation of its speed in the range between full forward and full back.

For certain classes of ships (most often warships), it may be reasonable to use a combined drive comprising a diesel engine and a gas turbine engine. In this case, the GTE is activated only to obtain the maximum speed.

After the World War II, the aircraft gas turbine construction reached a high level and the design of aircraft engines was developing rapidly. As a result, designers of marine power plants got interested in the use of aircraft gas turbine engines on ships. For this to be possible, aircraft engines had to be adapted to operate in ship conditions. The process of adaptation was called conversion. Almost all the first gas turbine engines of foreign production were converted, as was a considerable number of ship engines manufactured both in subsequent years and at present. Most foreign companies with the main focus on creation of aircraft engines have specialized Marine Divisions, which design aircraft engines and develop options for converting them for marine conditions. Taking into account the fact that ship GTEs need a much greater resource than those on aircrafts, the process of conversion includes the following:

- the temperature of the gases in front of the high-pressure turbine is decreased;
- the regulating systems are re-equipped for operation on a more viscous fuel;
- the bearing units are strengthened with account for increased loads during storm and considerable pitching;
- the air purification system is developed (the air under consideration comes from the environment and from water aerosols to the gas turbine engine).

The world's first marine application of a gas turbine engine is associated with a ship that participated in the World War II. In 1947, a GTE was used on a British high-speed ship "MGB 2009" to drive the central line of the shafting system, while the other two shafts were still driven by standard diesel engines.

The use of GTEs as accelerator engines in marine power engineering of the 1950s was justified. It allowed for the rational use of a limited engine life, which at that time did not exceed several hundred hours. Moreover, it was possible for the engine to operate mainly at rated power. In partial modes, the specific fuel consumption of the GTE "Gutrik" with the load of 10% of the rated one was 2.04 kg/(kW h).

At the same time, a number of countries were making attempts to create all-mode gas turbine plants.

The year of 1951 saw the launch of the bench tests of the RM60 gas turbine engine manufactured by the English company Rolls-Royce. This engine was intended for use as the all-mode engine for the gunboat "Gray Goos" with the displacement of 250 tons. In 1955, two RM60 GTPs with the power of 4000 kW were mounted on the ship. These plants operated in a cycle with three compression stages, two intermediate air cooling stages, three expansion stages and recovery of the exhaust gas heat (Fig, 5.9) [1]. All this, along with high cycle parameters, made it possible to ensure a sufficiently high economic efficiency of the plant operating at partial loads. The temperature of the gas before the turbines reached 1100 K, and the pressure increase rate was 18.5. The specific fuel consumption at rated power was 408 g/(kW h), which corresponded to a cycle efficiency coefficient of about 21%. The GTE was launched by an electric starter—a 30 kW engine that spun the HPT shaft via a gear train. The starting time from the cold state to the full stroke made up 30 s. The reverse was provided by a controllable-pitch propeller (CPP). The plant was experimental and had the resource of 1000 h (Fig. 6.9).

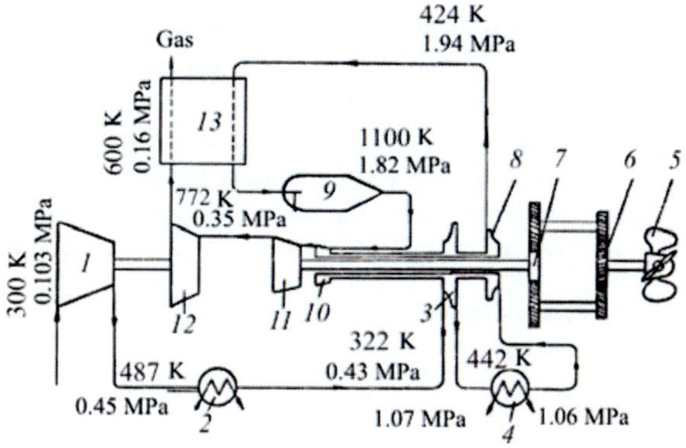

Fig. 6.9 Principal scheme of the RM60 gas turbine plant: *1*—LPC; *2, 4*—low- and high-pressure air coolers, respectively; *3*—medium-pressure compressor; *5*—CPP; *6, 7*—second and first stages of the reducer, respectively; *8*—HPC; *9*—combustion chamber; *10, 11, 12*—HPT, MPT, LPT, respectively; *13*—regenerator

In the 1950s, a widespread application was acquired by the GTE "Protey 1260" created by converting the aircraft turboprop engine "Protey" manufactured by Bristoll Syddly. It was intended for use as the main engine on torpedo and missile boats of the NATO countries. The torpedo boats of the "Breive" class were 30 m long and were equipped with three propellers, each of which was driven by the gas turbine engine "Protey 1260" with the power of 2800 kW. Located in the stern of the ship, three GTEs provided the boat with the speed exceeding 50 knots (Fig. 6.10). The specific weight of "Protey 1260" without the reducer was 0.46 kg/kW, and the operating life exceeded 2000 h. The specific fuel consumption at full load was 370 g/(kW h), which corresponded to the cycle efficiency coefficient of 20.3% [1].

The gas turbine engine "Protey 1260" had a twelve-stage axial compressor, the last stage of which was a centrifugal compressor. Eight tubular combustion chambers were located evenly around the circumference of the engine axis; they wee supplied with air from the last stage of the compressor after the 90° turn.

The compressor turbines and the power turbine were two-stage. The power of the LPT was transmitted to the propeller through the hollow shaft of the high-pressure cascade and the planetary reducer.

The fuel stock placed in fuel tanks with the capacity of 25.4 tons provided the gas turbine torpedo boat with the operating radius of about 750 km at the cruising speed.

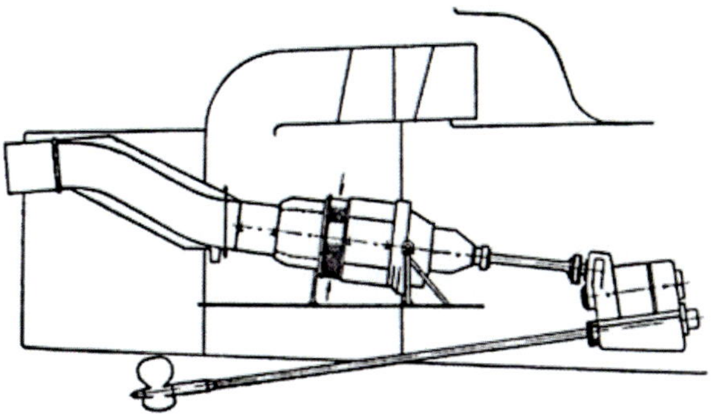

Fig. 6.10 Layout of the gas turbine plant "Protey 1260" on a torpedo boat

The positive results of the tests of the GTEs “Gutrik”, “G2”, and “Protey 1260” resulted in the creation of a more powerful engine that was to become part of the combined marine steam and gas turbine plant. Both the steam turbine and the gas turbine engine transmitted power to the propeller shaft through a common reducer; they were started and stopped independently of each other. This engine was made for use in combined steam and gas turbine plants of destroyers and frigates. The plants of frigates were single-shaft, while the plants on destroyers were two-shaft.

Background information. *Combined marine power plant is a plant that consists of different types of engines or those of the same type, but of different purpose. The appearance and application of combined plants became possible after the creation of efficient GTE designs with high mass-dimensional properties and, unfortunately, insufficient fuel efficiency. For example, if a gas turbine plant and a diesel (of a much lower power than the former) are connected in a power plant of a warship or a high-speed civil ship, the diesel engine will provide the main modes of ship movement at a moderate speed with a moderate specific fuel consumption characteristic of diesel engines. A gas turbine engine is less economical but will ensure the ship movement at high speeds.*

Along with the above, taking into account the insufficient fuel efficiency of both diesel engines and gas turbines at partial loads, it is sometimes advisable to use multi-engine plants. The increase in their power is achieved by connecting individual engines operating at 100% to the propeller shaft.

There is an international classification of combined power plants. The following abbreviations are used in the designations: CO—combine; G—gas turbine engine; S—steam turbine unit; D—diesel.

The letter O (or) is used between the engine designations if only the boost engine is operating in the maximum power mode. The letter A (and) is applied when both sustainer and boost engines operate in the maximum power mode. Thus, COSAG, or COSOG, is a combined steam and gas turbine power plant; CODOG, or CODAG, is a combined diesel and gas turbine plant; COGOG, or COGAG, is a combined power plant with two gas turbines [4].

The engine with the power of 5500 kW was made according to a two-shaft scheme with a free power turbine. The thirteen-stage compressor was driven by a two-stage HPT; the free power turbine also had two stages. The temperature of the gases in front of the turbine was 1066 K. The specific fuel consumption was about 420 g/(kW h), and the efficiency coefficient was about 19%. The gas turbine engine was launched by compressed air at the pressure of about 5 MPa from the ship general system. Expanding to atmospheric pressure, the air drove the starting piston engines, which through the starting planetary reducer rotated the shaft of the turbocharger [1].

At the same time with creation of gas turbine engines for the navy, experimental studies were conducted on the application of such engines on merchant sea ships.

One of the first such ships was the English tanker “Auris” built by Shell in 1949. The tanker with the deadweight of 12,250 tons was equipped with four Sulzer diesel engines, each with the power of 810 kW. They were connected by couplings with four electric generators, and the generators were connected with an electric engine with the power of 2750 kW, which rotated the propeller.

In 1951, one of the diesels of the starboard side of this tanker was replaced with the all-mode GTP manufactured by British TompsonHauston. It had the power of 890 kW and the option of regeneration of the exhaust gas heat.

A structural diagram of such a gas turbine plant is shown in Fig. 6.11. The axial twenty-three-stage compressor compressed atmospheric air and fed it to the upper drum of the regenerator. Then, the air was heated while passing through the heat exchange tubes to two lower drums, where the combustion chambers were located. In the primary combustion zone, the flame tubes had a ceramic lining. In each combustion chamber, there was one main and one starting nozzle with a spark plug. From the combustion chambers, gas flowed first to the compressor HPT, and then to the LPT, which provides the drive of the direct current generator. The exhaust gases from the LPT passed through the regenerator and were discharged from the GTE.

A characteristic feature of the plant is the "heavy" design of multi-stage compressors and turbines with drum rotors, which was borrowed from steam turbine engineering. Along with the presence of a tubular regenerator and the use of power gear, this was the reason for a great weight of the GTP.

The plant was operating on "Auris" for five years over about 20,500 h, including 6650 h on heavy fuel. During this period, the ship covered the distance of about 330,000 km. In 1956, three diesel generators and an experimental GTE were replaced

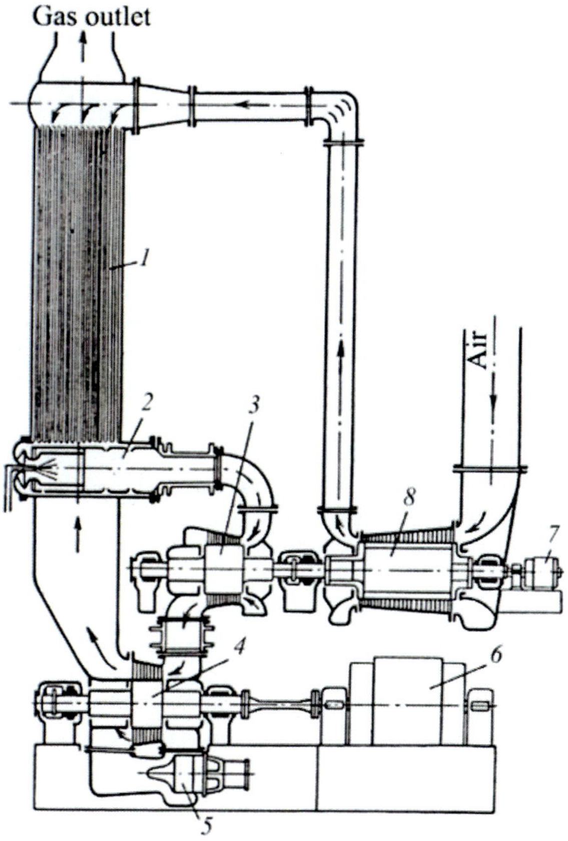

Fig. 6.11 The first gas turbine plant of the tanker "Auris": *1*—regenerator; *2*—combustion chamber; *3*, *4*—high- and low-pressure turbines, respectively; *5*—by-pass valve; *6*—power generator; *7*—starting engine; *8*—compressor

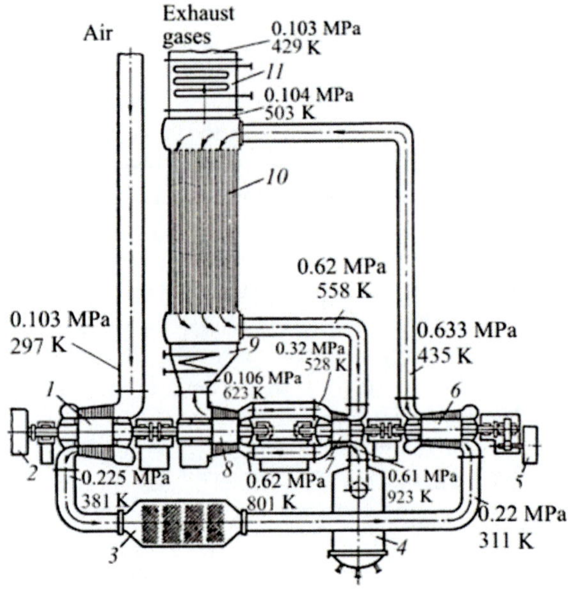

Fig. 6.12 The second gas turbine plant of the tanker "Auris": *1, 6*—LPC and HPC, respectively; *2*—reducer; *3*—intermediate cooler; *4*—combustion chamber; *5*—starting turbine; *7, 8*—HPT and LPT, respectively; *9*—auxiliary evaporator; *10*—regenerator; *11*—RB

with a new, more advanced GTP manufactured by the same company. It had the power of 4050 kW (Fig. 6.12) [1, 4].

The plant operation followed an open cycle with two-stage compression, intermediate air cooling and regeneration. The high-pressure turbine drove the HPC, and the low-pressure turbine drove the LPC and the propeller. In contrast to the previous project, the power was transmitted to the propeller through a two-stage gear reducer with a hydroreversiblePametrade system's gear, which provided the reverse of the ship. At the regenerator outlet, the exhaust gases entered the recovery steam boiler with the capacity of 2.25 t/h and the pressure of 0.35 MPa. There, their temperature was lowered from 503 to 429 K. The steam produced in the boiler was used in a 200 kW turbine generator.

The MPP was provided with a spare engine—a 330 kW steam turbine, which was supplied with steam of the pressure of 1.26 MPa from the auxiliary boiler. The ship was powered by the steam turbine generator, which received steam was supplied from the recovery or auxiliary boiler. Along with this, another steam turbine generator of 125 kW was included to the plant.

The gas turbine engine worked on the tanker for 5240 h; testing and operation demonstrated its compliance with the requirements for SPP. The engine allowed for the use of heavy fuel (according to its characteristics, it corresponded to Soviet fuel—the fleet fuel oil F20) [5, 6].

The tanker had been in operation until July 1960, when it was scrapped due to the economic inexpediency of exploitation.

The middle of 1950s was marked by implementation of a special program by the US Maritime Administration. It was aimed at determining the perfect type of marine power plants for the American merchant fleet. Within this program, in 1956, the General Electric GTP with the power of 4400 kW was installed on the transport ship "John Sergeant" (of the "Liberty" type) with the displacement of 13,825 tons.

The unit was a modification of the plant used for gas pumping stations and power plants. The ship speed was moving at 15.9 knots at full speed.

The gas turbine engine was executed to have a cycle with heat regeneration of exhaust gases in a counterflow tubular regenerator. At the same time, the heat of exhaust gases was also used in the RB. The boiler was located in the gas flue behind the regenerator, and the superheater was in front of it. The steam capacity of the boiler was 2.4 t/h at the steam pressure of 1.6 MPa. Of this amount, 0.9 t/h of steam was supplied for ship service needs, and 1.5 t/h was consumed by a TG with the capacity of 170 HPC. In the marine power station, there was a provision for an auxiliary DG of the same power. The plant was started by a steam turbine with the power of 215 kW; the reverse was performed by means of a controllable-pitch propeller (CPP).

A fourteen-stage axial compressor was driven by a single-stage HPT. A low-pressure power turbine, equipped with rotary nozzles and located in the same housing as the turbine engine, drove the CPP via a two-stage gear reducer. The engine had the gas temperature of 1064 K in front of the HPT, and the air pressure increase in the compressor was $\pi_c = 4.9$. At the air temperature of 32.3 °C, the specific fuel consumption was 322 g/(kW h) (corresponding to the efficiency coefficient of 25%). The gas turbine plant worked for 9700 h, of them 7000 h on heavy fuel.

In 1950, the Soviet Union government adopted a decision to develop a terms of reference for the creation of a ship GTE.

The gas turbine engine was intended for use as the boost one providing the movement of a torpedo boat at maximum speed. The boat was fitted with four diesel engines powering the side propellers and the GTE powering the middle propeller.

The power of the marine gas turbine plant M1 (of the first modification) was 2940 kW (4000 hp); the specific fuel consumption was not higher than 544 g/(kW h), the specific mass was 0.68 kg/kW, and the resource was 100 h.

In 1951, there were factory tests and 100-h commission tests, after which the engine was allowed to operate on an experimental torpedo boat of the 183 TK project. The boat running tests with the GTE began in 1951 and lasted about two years. The planned speed of 50 knots was achieved. On the basis of tests conducted in 1953, the State Commission noted that the engine of S.D. Kolosov's design was the first gas turbine engine in the USSR. It was recommended the M1 gas turbine plant for serial production and installation on torpedo boats.

On May 7, 1954, the Council of Ministers of the USSR adopted the resolution on the establishment of a base for marine gas turbine construction in the city of Mykolaiv. This resolution provided for organization of serial production of marine gas turbines at the Southern Turbine Plant and the creation of a special design bureau for gas turbine plants (SDB GP). The plant was then under construction, and its purpose was the production of marine steam turbine units.

In October 1955, (STP), the first proprietary engine M1 was assembled at the Southern Turbine Plant. In 1956, interdepartmental tests were completed and state tests of the head torpedo boat 183TK with the serial GTE were carried out. Over the period from 1956 to 1958, the plant produced 34 serial gas turbine engines M1. In 1958, their production was discontinued.

The prevalent number of GTEs produced in the former Soviet Union, and then in Ukraine, was intended for use on ships of the naval forces. The main requirements that were put forward to ship GTEs were as follows [2]:

- all-mode, i.e. ability to work on all modes;
- economical efficiency in the entire range of modes;
- substantial resource;
- ability to work on diesel fuel;
- maneuverability;
- adaptability to continuous work in the marine environment;
- provision of the propeller reverse.

These requirements were significantly different from those for aircraft engines. They operated in the narrow range of 0.5–1.0 of full power, had a small resource, and were running on kerosene.

If the gas turbine unit M1 was a kind of transitional model from aircraft to marine gas turbine construction, the next model, M2, was designed as a basic one for different types of ships with the maximum consideration of requirements for marine power engineering.

The unit M2 (Fig. 6.13) was intended for installation on an anti-submarine ship; it worked as a boost engine in a diesel-gas turbine plant. One diesel was placed in the middle of the ship, and two M2 GTUs with right and left rotation of the turbine propellers were placed along the sides, respectively. The power from the free propeller turbine, which was connected to the reducer by means of a spring and couplings, was transmitted to the propeller of the ship.

The first sample of the M2 GTU was manufactured in June 1957, and in 1959 it was installed on a ship with the resource of 750 h. In 1960, the first ship of the 159 project with the M2 GTU was accepted for operation. The interdepartmental tests carried out in 1962 confirmed the M2 resource of 1000 h [2].

Fig. 6.13 General view of the gas turbine unit M2

In 1958, the SDB GP began to develop a new gas turbine unit for anti-submarine ships of the 35 and 204 projects. There is still no analogue of such a GTU in the world practice.

On these ships, the gas turbine unit served as a source of compressed air, which was directed to a hydraulic motor unit whose propeller was driven from the diesel engine. The ship movement was realized by means of traction from air–water mixture ejected from a hydraulic engine. The new unit received the index GTC (gas turbo-compressor) D2. It consisted of a GTE and a separately located compressor as a source of compressed air, which was driven by the engine. The compressor was placed in front of the engine and powered by a low-pressure cascade.

The GTE was made according to the block scheme; the drive power was 11,000 kW. Two D2 GTCs were installed along the sides of the ship with the exhaust gases outlet from the jet nozzle directly through the transom stern. According to some estimates, this added about two knots to the ship speed due to the reactive action of the gases [3]. The year of 1964 marked the beginning of production of the GTC D3 with the power of 13,240 kW [2].

Undoubtedly, the decision of the Mykolaiv gas turbine builders to implement a gas turbine plant on a large anti-submarine ship of the 61 project as the main power plant was crucial for the development of marine gas turbine construction.

In accordance with the terms of reference, the ship had to be equipped with two GTUs, each with the power of 27,500 kW, service life of 3000 h with further improvement to 5000 h and specific fuel consumption of 0.350 kg/(kW h).

In order to increase reliability and reduce fuel consumption, it was decided to install two GTEs on a common frame on each side of the ship; they had the power of 13,750 kW and operated on a total summing reducer. At intermediate and small speeds, only one engine operated on each board. This method of controlling the MPP power was called "quantitative regulation".

The possibility of using the GTE as part of the main power plant of the ship was largely determined by the need to create a propeller reverse system. This problem was solved by the development and manufacture of a 27,500 kW reversible reducer. In the M3 unit (Fig. 6.14), the problem of reverse was solved by including special cam-friction and hydraulic couplings in the kinematic scheme of the reducer. They

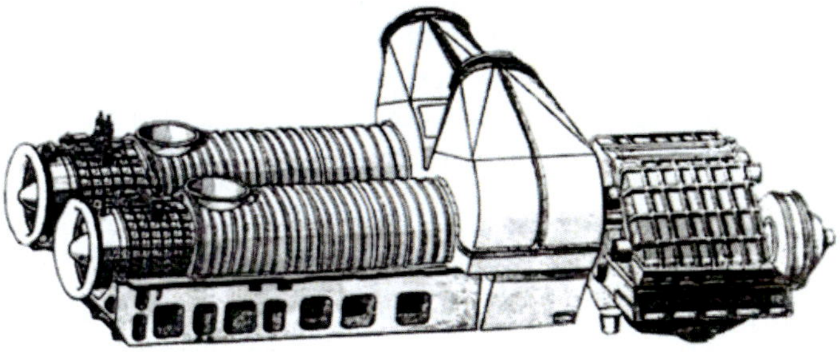

Fig. 6.14 General view of the all-mode main gas turbine unit M3

Fig. 6.15 Gas turbine engines of the M3 unit on the frame

provided the change in the direction of rotation of the propeller and the ability to connect and disconnect one of the engines during ship movement. The forward stroke was provided by the system of friction and cam couplings, and the reverse stroke was provided by the hydraulic coupling.

The first two units manufactured by STP were mounted on a large anti-submarine ship "KomsomoletsUkrainy", which was accepted for operation in 1963 (Fig. 6.15).

Creation of the world's first main all-mode unit M3 of a considerable power and its successful operation on "Komsomolets of Ukraine" finally proved the advantages of the gas turbine power to ship designers. This led to a rapid increase in the number of new ship projects with the GTU. As of 1965, more than 130 GTEs were in operation.

In 1966, the UDB "Mashproekt" started to manufacture the *second generation GTE* and a gas turbine unit on its basis. It was expected to achieve a substantial increase in the efficiency of gas turbine engines (by 15–20%), especially at small speeds and in economic modes, as well as an increase in the service life by 4–5 times and considerable improvement of reliability.

These objectives were realized in the development of the ship M62 GTE with the power of 6.6 MW and the DT59 GTE with the power of 16.5 MW. They were a part of the main M7 and M5 GTUs with their resource reaching 20,000 h. The specific fuel consumption at the nominal mode was 285 g/(KW h). For the first time in the world, gas turbine engines were made reversed. Until now, reverse GTEs have been only produced in the former USSR and Ukraine.

In the middle of 1960s, specialists at the Kirov plant (Leningrad) designed and manufactured the marine gas turbine plant GTU-20 with the power of 8.7 MW.

Structurally, GTU-20 was an alternative to the pattern in marine gas turbine construction typical for the engines manufactured by UDB "Mashproekt". Unfortunately, this path of development of the ship GTE turned out to be a dead-end,

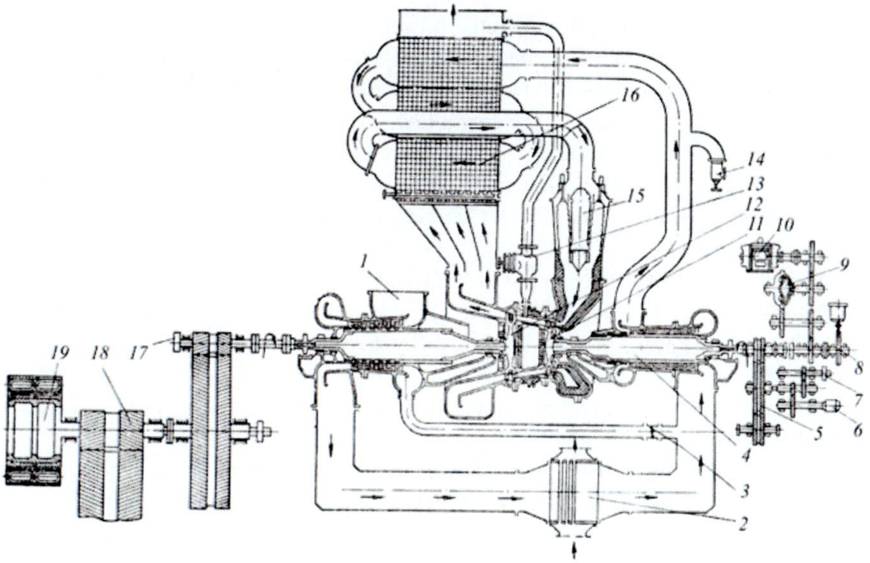

Fig. 6.16 Design scheme of the GTU-10: *1, 4*—LPC and HPC, respectively; *2*—air cooler; *3*—fire-protection device flap; *5*—gear train of the gearbox; *6*—tachometer; *7, 17*—impellers; *8*—shaft-turning gear; *9*—hydrotransformer; *10*—starting electroengine; *11, 12*—HPT and LPT, respectively; *13, 14*—gas and air bypass valves, respectively; *15*—combustion chamber; *16*—regenerator; *18*—reducer; *19*—tire-pneumatic coupling

and GTU-20 became the only one of its kind despite profound support from the government of that time.

GTU-20 consisted of two units GTU-10 with the power of 4350 kW each (Fig. 6.16). These units worked on a total reducer with the gear ratio of 57.5. Structurally, GTU-20 was a typical industrial unit with moderate parameters of the working fluid (initial temperature of gases of 1023 °C, pressure increase rate of 9). The two-cascade engines had a block power turbine and power take-off from the low-pressure cascade. Unfortunately, such a choice of the design scheme for an engine intended for use as the main one on ship was initially erroneous.

The GTU-20 plant was installed on the first Soviet gas turbine ship—dry cargo ship "ParizhskayaKommuna" with the deadweight of 16,000 tons. The reverse of the plant was carried out with the help of the CPP. The ship was in operation as part of the Black Sea Shipping Company until the middle of 1980s.

The period of 1960-70 s is characterized with a growth of interest in the ships with dynamically supported principles: hydrofoil ships (HFS) and air cushion ships (ACS).

From 1969 to 1977, the DT4 and M10 GTUs with the power of 13 and 15 MW, respectively, were developed for dynamically supported ships at the Research and Production Enterprise (RPE) "Mashproekt" (former UDB "Mashproekt"). These GTUs had a complex structure. Thus, the unique DT4 unit of theACS "Jeyran" consisted of two light engines with the power of 13,230 KW each and the mechanical gear, which included 18 planetary and angular reducers of eight types. The latter provided power transmission to four fans and four air CPPs simultaneously. If necessary (in emergency situations), the gear allowed rotating all these units from one GTE to continue the ship movement.

At the beginning of 1970s, an acute need for electricity arose in the remote north-eastern regions of the former Soviet Union due to intensive development of oil and gas fields. To solve this problem, it was decided to build several floating power stations and to deliver them to the destinations by water.

The enterprise "Mashproekt" and the STP have created gas turbine generators for floating power stations of the "SevernoyeSiyaniye" class on the basis of marine gas turbine engines. Two gas turbine generators with the GTE power of 10 MW each were installed at the stations, and more economical engines with the power of 12 MW were also installed in 1974.

In total, six floating power plants were built; they which were installed in the city of Anadyr (in Chukotka near the Schmidt Cape) and Nadym (in the Tyumen region) and in other areas.

The creation of gas turbine engines of the second generation and units on their basis in the Soviet Union was generally completed over the period from 1966 to 1972. These works fully solved the problem of providing power plants for a new type of surface displacement and dynamically supported ships.

Third generationGTEs are characterized with high cycle parameters (the initial gas temperature of 1100–1150 °C with the pressure increase rate of 15–21), high circumferential blading speeds and, as a consequence, reduction in the number of stages of compressors and turbines. Application of a counterflow combustion chamber located around the HPC made it possible to reduce the axial dimensions and mass of the engine [2].

Within the third generation GTE development program, there was created a unified power series range of 3.7, 8.0 and 15 MW engines with the base resource of 20,000 h. They had higher efficiency (of about 35%, which is by 10–15% higher), by 2–4 times smaller mass and 1.5 times smaller dimensions in comparison with the engines of the second generation [2].

A significant achievement of the Soviet and Ukrainian gas turbine construction was the creation of a combined gas-steam turbine plant with a M25 unit with the power of 18 MW for transport ships with a horizontal cargo handling system. The use of the exhaust gas heat in the recovery steam boiler was typical for these units. The boiler provided operation of the recovery steam turbine, which in turn provided additional power for the propeller drive and operation of the auxiliary steam turbine generator.

The structure of each M25 GTU included a reversible unit, a reducer, a recovery boiler, and a steam turbine with a condenser. The main power plant included two M25 GTUs.

The lead ship "Kapitan Smirnov" was built at the Black Sea Shipyard in 1978 in Mykolaiv. In addition, three more ships of this class were built: "KapitanMezentsev"

Fig. 6.17 General view of the UGT15000 gas turbine engine

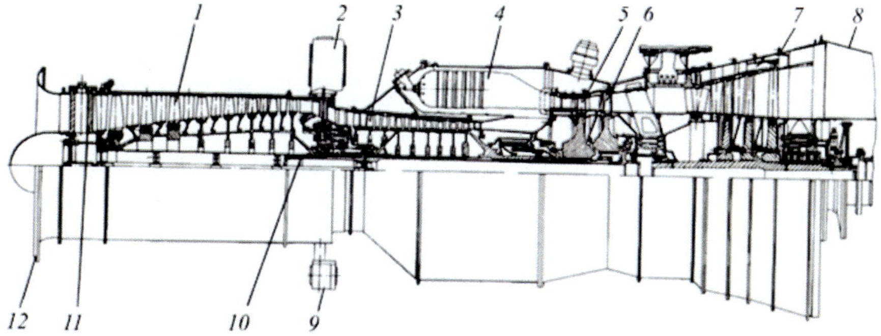

Fig. 6.18 Design scheme of the DO77 (DS77) reverse engine: *1, 3*—LPC and HPC, respectively; *2, 9*—boxes of the main and lower drives, respectively; *4*—combustion chamber; *5, 6*—HPT and LPT, respectively; *7*—power turbine; *8*—gas outlet; *10*—adapter; *11*—forward case; *12*—inlet device

(1980), "InzhenerYermoshkin" (1981), and "Vladimir Vaslyaev" (1987). The experience in operation of these ships demonstrated high reliability, maintainability and economic efficiency of the GTPs with the steam-water HRC.

Serially produced third generation units were taken as a basis for the M37 gas turbine unit with the power of 5200 kW, which was used as part of the MPP on the hydrofoil sea passenger gas turbine ship "Tsyklon". The ship was designed to carry 250 passengers at the speed of 42 knots and with the cruising range of 300 miles. Interdepartmental testing of the unit was completed in 1985, and it was installed on the ship in the same year. The ship mooring tests were completed in December of 1986.

Figures 6.17 and 6.18 shows the general view and the design scheme of the third-generation gas turbine engine of the Ukrainian production.

In 1985, the RPE Mashproekt launched the program of creation of *gas turbine engines of the fourth generation*. It only included a GTE 25,000 engine with the power of 25,000 kW and the efficiency coefficient of 36.8%. The engine has the highest rate of increase in air pressure in the compressor among the existing marine GTEs. It makes up 23.6. The initial gas temperature is 1245 °C.

Four GTE 25,000 engines have been delivered to China for use on warships. Over 90 such engines are being used to drive the superchargers of gas pumping units of compressor stations of main gas pipelines.

To date, the engine UGT 10,000 (DH70) developed by Ukrainian gas turbine designers is at the stage of experimental-industrial operation. It has the power of 10,000 kW, the temperature in front of the HPT of 1270 °C, the pressure increase rate of 22, and the efficiency coefficient of 36% (Fig. 6.19).

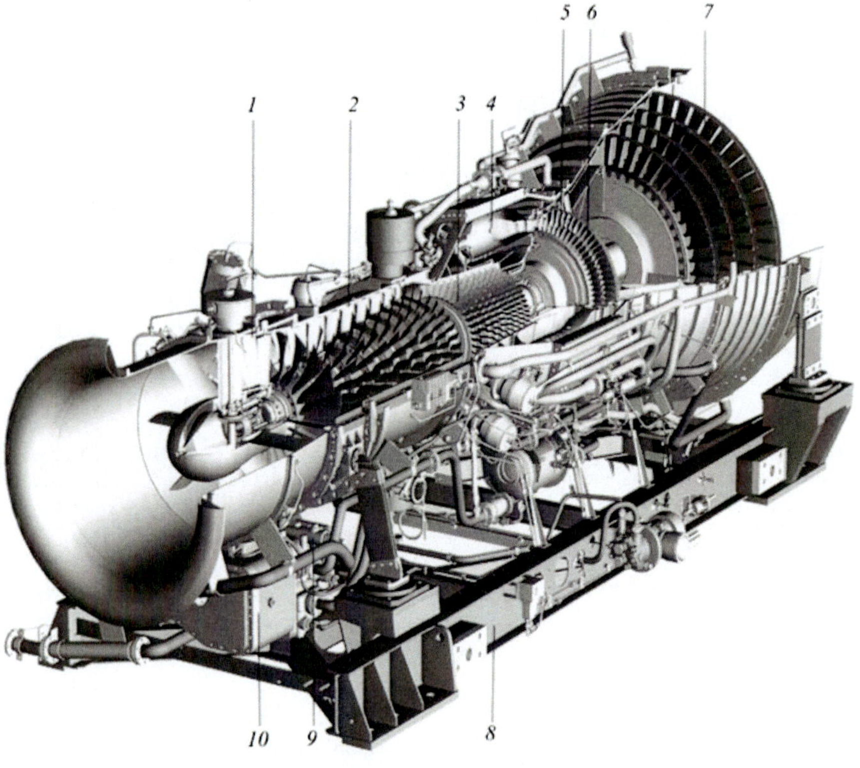

Fig. 6.19 General arrangement of the UGT 10,000 (DH70) engine: *1*—rotary inlet stator; *2, 3*—low- and high-pressure compressors, respectively; *4*—combustion chamber; *5, 6*—HPT and LPT, respectively; *7*—power turbine; *8*—frame; *9*—electric starter; *10*—gearbox

Foreign marine gas turbine construction since 1960s. In 1960s, the most intensive work on creating ship GTEs was carried out by the American companies General Electric and Pratt & Whitney, as well as the English Rolls-Royce.

The first gas turbine for ship application was manufactured by General Electric in 1959. That was the LM100 engine built on the basis of a helicopter gas turbine engine and intended to drive an emergency electric current generator on hydrofoil ships, air cushion ships and vehicles.

The next type of engine was the GTE LM1500, also converted from a serial aircraft engine. The first sample was installed on the air cushion ship "Denison" of the US Navy. In total 182 units had been manufactured until the beginning of 1996. At the beginning of 1990s, production of these engines was discontinued.

In 1961, the American company Pratt & Whitney started works on the development of the ship engine FT4A based on the widely used aircraft turbojet engine J-75. The FT4A GTE has a two-cascade compressor with the pressure increase rate of 12 that consists of an eight-stage LPC and a seven-stage HPC. The combustion chamber is tubular-annular and has eight flame tubes, the HPT is single-stage, the LPT is two-stage, and the power turbine is also two-stage. In the late 1960s, the FT4A GTE became one of the main GTEs that were installed on ships abroad.

In 1967, the transport ship "Admiral V. Callagan" was built in the Federal Republic of Germany. It had the deadweight of 7000 tons and was equipped with a two-shaft power plant with the FT4A engines with the power of 18.7 MW.

A subsequent modification of this engine, FT4A-12, was used in the creation of a class of four "Euroliner" container ships built in the FRG. Those were two-propeller ships with the deadweight of 28,000 tons with the design speed of 26.4 knots, which was provided by two FT4A-12 GTEs with the power of 22.2 MW. The first ship of this class was put into operation in 1971 [7].

A similar power plant was used on a two-shaft Finnish ferry "Finnjet" put into operation in 1977. It was fitted with two FT4S-2 GTEs with the total power of 55.2 MW, which each through its gearbox operated for its own CPP.

The GTE LM2500 has acquired the most widespread use worldwide. It was first released by General Electric in 1969 and is still being manufactured with certain changes at present. The total number of such engines in different versions exceeds 2000 pieces. The gas turbine engine was converted from the TF39 aircraft engine, which had a high efficiency coefficient and a considerable resource. The ship engine was under development simultaneously with its aircraft prototype, and more than 90% of the structural units of the prototype engine were used in the design of the converted engine. The LM2500 is a single-stage, single-cascade engine with a free power turbine.

The GTE gas generator consists of a sixteen-stage compressor with the pressure increase rate of 18.

Air enters the engine through inlet device *1*. The inlet stator and the stators of the first six stages of sixteen-stage compressor *2* are adjustable; the angular position of their blades is determined by the air temperature at the compressor inlet and the

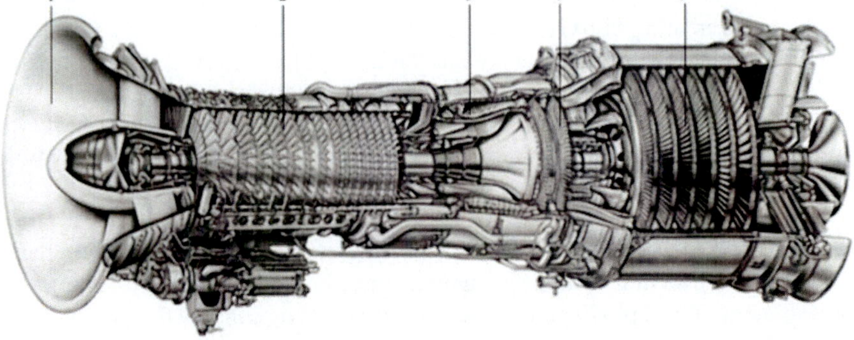

Fig. 6.20 Design scheme of the gas turbine engine LM2500

compressor's rotor speed. Low-emission combustion chamber *3* is annular and direct-flow. Two-stage HPT *4* operates in the range of rotation speed of 4900–9800 rpm. Free power turbine *5* has six stages and the rotation speed of 1000—3000 rpm (Fig. 6.20).

In the first modifications of the LM2500, the initial gas temperature was assumed to be 1100 °C versus 1280 °C in the arcraft prototype.

The LM2500 gas turbine engines were installed for the first time on the "Admiral V. Callagan" ship in 1969, replacing the FT4A GTEs. Then they were used in the US Navy and in other countries [1].

The LM2500 engines for displacement ships are delivered in the form of modules, which greatly simplifies their installation in the engine room (ER), isolates the engine from shock impacts, and helps to reduce the noise and vibration levels (Fig. 6.21). Functionally, the module design provides air supply to the compressor and for the engine cooling, as well as exhaust gas removal, fire protection, ice prevention at the inlet device, and fuel heating. All electronic and control systems of the modules are easily adapted to ship systems.

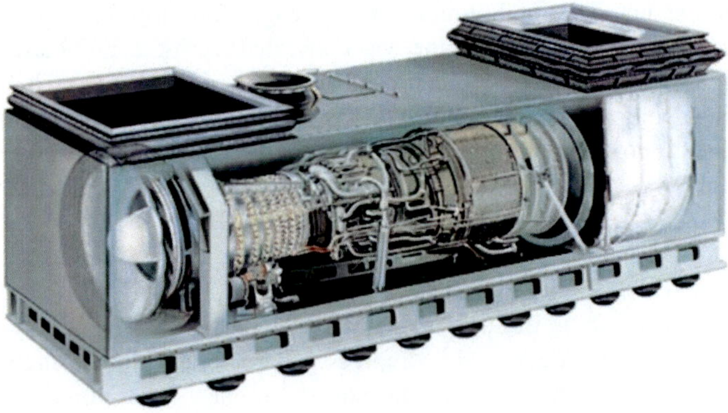

Fig. 6.21 General view of the LM2500 gas turbine engine in a container

Over 1973–1977, there was constructed a series of ships equipped with the GTPs with industrial engines of the MS class manufactured by General Electric. These were the ships with a horizontal cargo handling system: "Iron Monarch", "Iron Duk", "Union Rotoiti", "Seaway Prince", "Seaway Princess", "Shevron Oregon", "Shevron Colorado", "ShevronLuisiana" and others. The power of the gas turbine unit was 9–18 MW. Tankers with the deadweight of 35,000 tons were equipped with gas turbines with electric gear. The main gas turbine generator was located on the main deck, and the propulsive electric engine was directly below it, in the hold. This power plant arrangement significantly increased the ship cargo capacity.

A special feature of the General Electric gas turbine plant MS5212R installed on the gas carrier "Lucaine" is the ability to operate both on oil (distillate or mazut) and on methane, which evaporates from cargo tanks during the voyage.

The gas turbine engine LM500 was developed by General Electric jointly with the Italian company Fiat in 1979. Its design is similar to the design of LM2500. It is a single-cascade gas turbine engine with a fourteen-stage compressor, an annular combustion chamber, a two-stage turbine compressor and a four-stage power turbine. With the power of 3.65 MW, the engine efficiency coefficient is 31%.

The LM500 engine is used as the main engine on dynamically supported ships and as the sustainer engine in combined plants of a considerable power. It is also applied for driving electric generators on large ships.

In 1978, General Electric created the most powerful ship GTE of that time—LM5000 with the power of 37 MW.

The gas turbine engine has two cascades and a free power turbine. The rate of pressure increase is 26.1. The combustion chamber is annular. The high- and low-pressure turbines are made in the single-stage form; the power turbine is three-stage (Fig. 6.22).

The LM5000 GTE has acquired wide application on large high-speed ships, as well as in ground conditions.

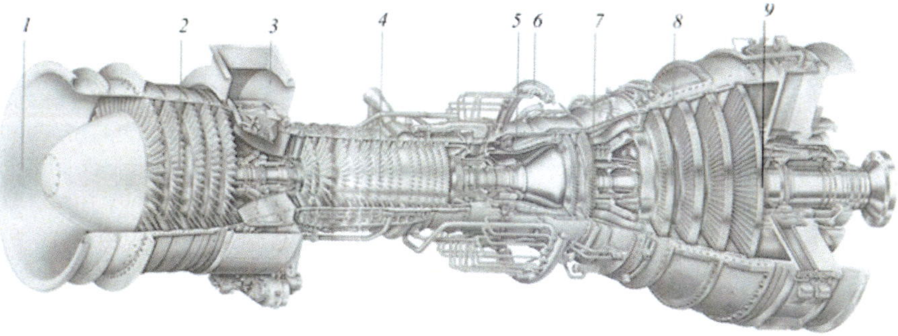

Fig. 6.22 Design scheme of the gas turbine engine GE LM5000: *1*—inlet device; *2*—five-stage LPC; *3*—low-pressure air sampling; *4*—fourteen-stage HPC; *5*—combustion chamber; *6*—fuel tank; *7*—two-stage HPT; *8*—single-stage LPT; *9*—three-stage power turbine

Rolls-Royce has made a substantial contribution to the development of marine gas turbine construction. Thus, more than 1000 engines with the total power of 10 million kW were produced for the naval forces of different countries over the period from 1973 to 2000.

The most popular engine of this company is Olympus TM3B, which was developed on the basis of the turbojet engine Olympus 201.

The Olympus TM3B is made in a direct-flow, two-cascade form with a free power turbine.

In 1983, Rolls-Royce and Allison jointly developed the ship version of the Spey SM1A GTE on the basis of the TF41 aircraft engine. It has a two-cascade gas generator, five-stage LPC, eleven-stage HPC, and two-stage HPT and LPT. The combustion chamber is tubular-annular and has ten flame tubes, and the power turbine has two stages.

Structurally, the gas turbine engine was produced in the form of a gas generator block and a power turbine unit mounted on a common base frame and located in a soundproof container.

The next Rolls-Royce development was a modification of the SMIC GTE of a larger power (19,500 kW) and greater efficiency (37.4%).

The LM1600 engine is of the third generation (the first generation is LM100 and LM1500, while the second one is LM2500 and LM500). For the first time, it was tested in operation on a ship in 1987. Unlike the LM2500, LM500 and LM1500 GTEs, this engine was designed to have two cascades with a free power turbine. The two-stage LPC and seven-stage HPC provide the pressure increase rate of 22, the HPT and LPT are single-stage, and the power turbine is two-stage.

The annular combustion chamber consists of 18 nozzles and two starting blocks. The fuel system can use two types of fuel, and it is possible to switch from one fuel to another at full power. The efficiency coefficient of the LM1600 GTE is 37%.

In 1990, Pratt & Whitney Power Systems created a FT8 GTE. This is a direct-flow, two-cascade gas turbine engine with an eight-stage LPC, seven-stage HPC, tubular-annular combustion chamber with nine flame tubes, and three-stage power turbine. At the power of 25 MW, the engine has the high efficiency coefficient of 38.5%.

The year of 1997 was marked by creation of the LM6000 GTE manufactured by General Electric Marine Engines (Fig. 6.23). This two-cascade engine has a blocked power turbine. Depending on the arrangement of the unit, the power take-off can be carried out both from the air inlet and the gas outlet of the GTE. In marine conditions, the LM6000 GTE with the power of 42.8 MW provides electric power for oil and gas production platforms and can be used as part of the power plants for ships with electric propulsion.

The joint development of Rolls-Royce and Westinhous, the WR-21 GTP, was created in 1997 and became the first ship plant with intermediate cooling and regeneration after 1950–60 s. The GTP design scheme suggests that the air that leaves the LPC is cooled in a heat exchanger, which is an intermediate cooler pumped with fresh water with addition of glycol. The air cooler is inscribed in the engine design, which ensures minimum air pressure losses when air passes through the heat exchange surfaces. Fresh water is in turn cooled by seawater. The air that leaves the HPC is

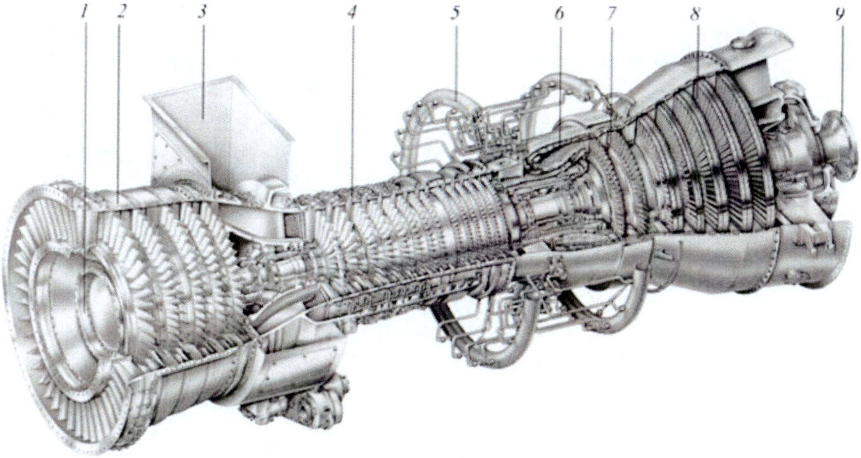

Fig. 6.23 Design scheme of the gas turbine engine LM6000: *1, 9*—power take-off from the cold and hot ends of the GTE, respectively; *2*—five-stage LPC; *3*—low-pressure air sampling; *4*—fourteen-stage HPC; *5*—fuel tank; *6*—combustion chamber; *7*—two-stage HPT; *8*—five-stage LPT

heated in the regenerator, then it enters the combustion chamber, and the combustion products sequentially pass the HPT, LPT and the free power turbine through the first stage with swivel nozzles.

In comparison with that of other GTEs, the combustion chamber design is quite original. It consists of nine radially located flame tubes, each of which is equipped with its fuel nozzle and two igniters.

The radial arrangement is chosen because of aerodynamic considerations, taking into account the fact that the air is fed from the regenerator into the combustion chamber, while it provides combustion with a reduced emission of nitrogen oxides [1].

The regenerator, intermediate cooler and swivel nozzles provide the WR-21 GTP with a low specific fuel consumption (0.200 g/(kW h), which is equal to 42.4% in efficiency).

The intermediate plate-like cooler consists of five heat-exchanging plate-like matrices made of a copper-nickel alloy. The regenerator is also plate-like; it consists of two modules, each including a heat exchanger, inlet and outlet drums, a valve drive, and a housing (Fig. 6.24).

The WR-21 gas turbine plant with an air cooler and a regenerator is delivered to the ship in a container.

The container design is fully adapted for routine repair, maintenance replacement of individual units in the shipboard conditions.

The WR-21 is designed for frigates, amphibious ships, ships of the auxiliary navy, and special high-speed merchant ships.

The LM2500+ engine was created by General Electric in 1998 to become a more advanced GTE in its class. Its main difference from the basic LM2500 model is the

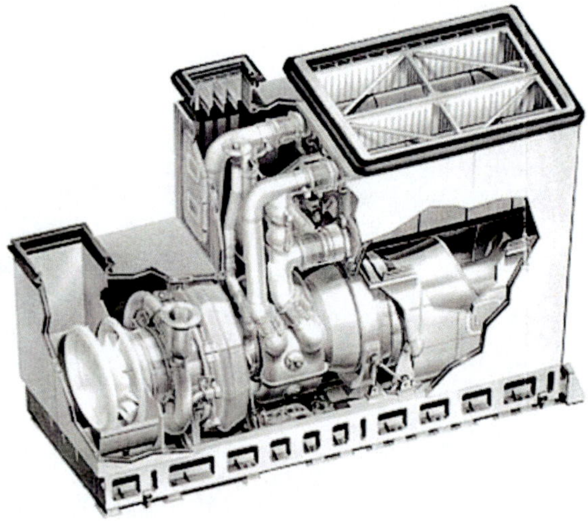

Fig. 6.24 General view of the WR-21 gas turbine unit

additional "zero" stage of the compressor, which allowed increasing air consumption by almost 20%. The rate of pressure increase in the compressor made up 22 (compared to 19.3 for the LM2500), which boosted the GTE efficiency to 37.3%. Over recent years, the LM2500+ GTE has been widely used on large cruise ships.

At the end of twentieth century, Rolls-Royce offered the marine industry a new MT30 GTE designed for use as a propeller or electric generator drive, converted from the aircraft engine Trent 800, which was installed on a Boeing 777. The MT30 is a two-cascade engine with a free power turbine. The combustion chamber is tubular-annular, the LPC is eight-stage, the HPC is six-stage, the HPT and LPT are single-stage, and the power turbine is four-stage.

The rotation speed of the power turbine is 3300 rpm (when used as a mechanical drive) or 3600 rpm (when used as an electric generator drive). The MT30 engine is one of the lightest GTEs, with its dry weight being equal to 6200 kg and its weight with the container making up 22,000 kg (Fig. 6.25).

The power of the MT30 is 36 MW, while the fuel efficiency is 210 g/(kW h), which corresponds to the efficiency coefficient of above 40%.

6.3 General Arrangement of Marine Gas Turbine Engines

The gas turbine engine UGT25000 (DG80) manufactured by Zorya-Mashproekt has a universal purpose. It can be used to drive a propeller with the use of both mechanical and electrical gears, or to drive an electric generator in stationary conditions [8].

Basic technical characteristics. The nominal power at the terminals of the power generator makes up 25,000 kW if ISO conditions are provided. The latter include the atmospheric pressure of 0.9948×10^5 Pa, inlet air temperature of 288 K, relative

Fig. 6.25 General view of the MT30 engine in the container

air humidity of 60%, resistance of the air inlet device to the GTE compressor inlet not exceeding 981 Pa, resistance of the exhaust devices not exceeding 2451 Pa, and the generator efficiency of 98%. The efficiency coefficient of the gas turbine engine is 36.8%. The rate of pressure increase in compressors is 21.2. The initial gas temperature (at the inlet to the HPT) is 1227 °C, the gas consumption is 87.7 kg/s, and the gas temperature at the GTE outlet is 493 °C.

The general view of the UGT25000 GTE placed on the frame is shown in Fig. 6.26.

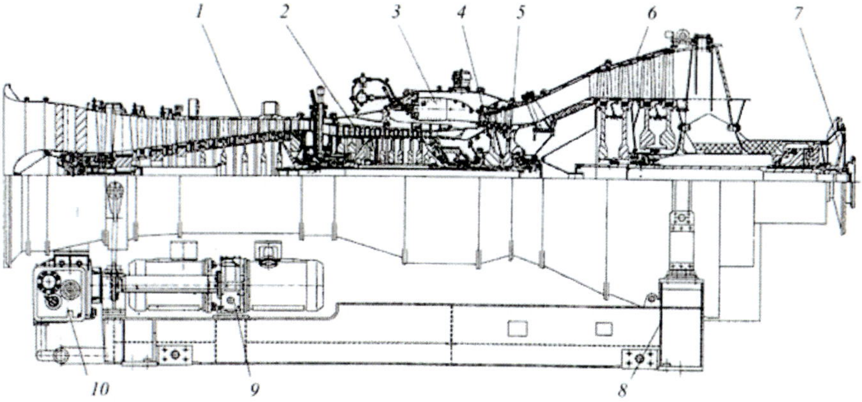

Fig. 6.26 Gas turbine engine on the frame

In this arrangement, the GTE is delivered to the ship and stationary facilities. The engine is two-cascade, with a free power turbine. The low- and high-pressure compressors (*1* and *2*, respectively) are driven by low-pressure turbine (LPT) *5* and high-pressure turbine (HPT) *4*, respectively. The compressors, the turbines that drive them and annular combustion chamber *3* form a gas generator.

The compressors and appropriate turbines form two cascades (a low-pressure cascade and a high-pressure cascade) that have different rotation speeds.

Power turbine *6* transmits a torque through coupling *7* to the consumer—a reducer or an electric current generator. The engine is placed on frame *8*.

The engine is started by spinning the LPC rotor through gear box 10 with the help of AC starters *9* located on the engine frame.

The movement of air and combustion products in the gas turbine engine takes place as follows. Air is fed to the gas turbine engine through LPC inlet device *1*. Before being supplied to the engine, the air must be cleaned from foreign particles and aerosols.

In this engine, a three-stage purification system is applied: moisture removal, pre-filter and fine filter. The latter detects particles of about 1 μm. Next, the clean air is successively compressed in LPC *1* and HPC *2* and enters combustion chamber *3*, where fuel combustion takes place (the above GTE design is meant for operation on natural gas). The primary ignition of the fuel gas takes place with the help of igniters, and its further combustion in the chamber is ensured by continuous fuel and air supply.

The air that is fed into the combustion chamber is distributed as follows. One part directly participates in the fuel combustion, resulting in the formation of combustion products. The second part cools the walls of the flame tubes and, mixing with the combustion products, forms a gaseous working medium, the energy of which is used in turbines. The third part provides cooling of the turbine parts.

The combustion products mixed with the cooling air are supplied to low-pressure turbine *5* and high-pressure turbine *4*, which develop the power required to drive HPC and LPC; in other words, the HPT and LPT power is completely consumed by the respective compressors. The only part of the gas turbine engine that develops the power transmitted to the external consumer is power turbine *6*, which is fed with combustion products that leave LPT *5*. The exhaust gases enter the delivery manifold through the GTE gas outlet.

The low-pressure compressor (Fig. 6.27) [8] is axial and has nine stages. Each stage is formed by one row of working blades placed on the rotor and located next to a number of rectifying blades installed in the housing. The unit is intended to compress air and feed it to the HPC.

The basic elements of the LPC include inlet device *1*, front housing *2*, gear *3*, rotary inlet stator *5*, anti-tamper device *6*, zero-stage rotary stator *7*, rotary stator housing *8*, first-stage rotary device *9*, rotor *10*, LPC housing *12*, front LPC support with ball bearing *15* and thrust sliding bearing *4*, and rear LPC support with roller bearing *14*.

The thrust sliding bearing consists of a housing with eight self-aligning pads *18* being disposed around the circumference; each of them rests on its spring pack *16*

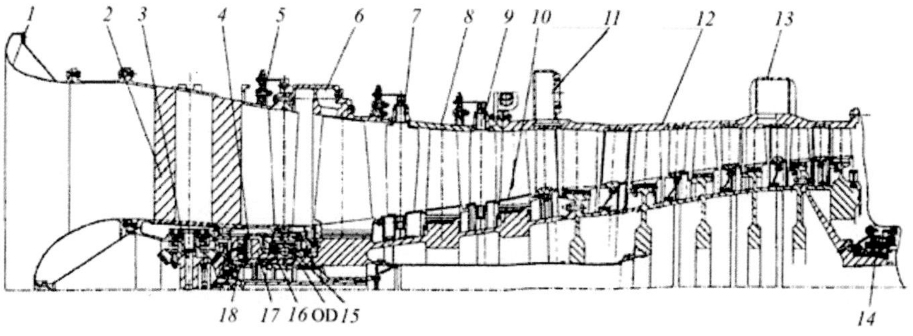

Fig. 6.27 Low-pressure compressor

consisting of two flat springs. Thrust plate 17 is located on the rotor of the LPC. The oil for creating an oil wedge and cooling the bearing is fed through the nozzles located between the shoes.

The peculiarity of this engine's design is that three rotary stators *5, 7, 9* are introduced into its structure in order to ensure constant operation of the compressor in a given range of modes and at the GTE start-up. The blades of these stators are rotated synchronously by a special mechanism depending on the static air pressure in the air-gas channel of the HPC. The purpose of anti-tamper device *6* is to improve gas dynamics of the air flow.

The inlet device is designed to ensure a smooth air supply to the compressor; it consists of internal and external fairings and a spacer. The channel formed by the fairings is the beginning of the air-gas channel of the GTE.

A rotary inlet stator, a front LPC support and a drive are all located in *front housing 6*.

The front support of the LPC rotor is a combined one; it consists of radial ball bearing *15* and thrust sliding bearing *4*.

Drive 3 provides the torque transmission from the electric starter to the LPC rotor at the engine start; it also transmits rotation to all mounted units during the engine operation.

Inlet rotary stator 5 operates synchronously with the zero-stage and first-stage rotary stators (*7* and *9*, respectively).

The LPC housing is a hollow cylinder with flanges of the vertical and horizontal connectors.

Air intakes *11* and *13* are located on its outer surface in the area of the second and sixth stages, symmetrically, on the upper and lower halves. The former is used to supply air for the turbine cooling, and the latter is used to discharge the air through the two valves into the atmosphere in order to increase the LPC stability margin.

The LPC rotor is of a drum-and-disk design. It consists of a drum of the zero to second stage and a drum of the third to seventh stage rigidly connected to each other, a rear trunnion, and a mounted disk of the eighth stage.

The working blades of the zero, first and second stages are secured in the disks with the help of a fir-tree attachment and fixed in the grooves with stoppers. The

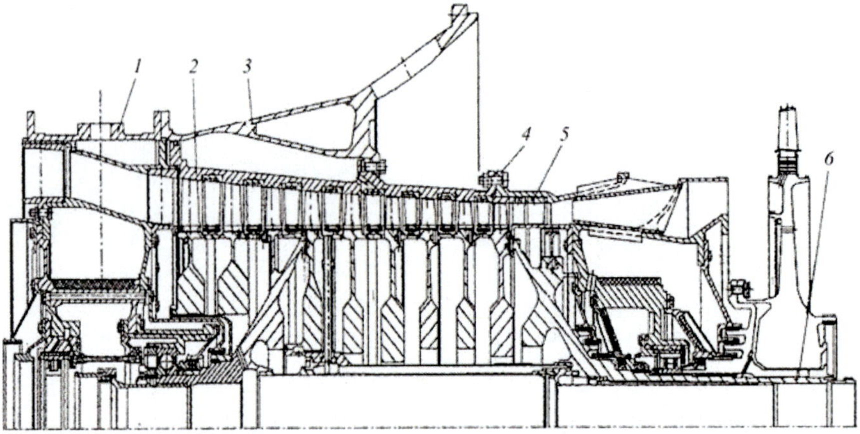

Fig. 6.28 High-pressure compressor

working blades of the third to the eighth stages are fixed with the help of dovetail joints. Axially, the blades sare fixed with plate locks.

A labyrinth belt is placed on the surface of the disk of the eighth stage. It prevents from air access to the discharge cavity of the LPC.

The high-pressure compressor (Fig. 6.28) [8] is designed to compress the air coming from the LPC and supply it to the combustion chamber; it has nine stages. The unit consists of adapter *1*, LPC housing *2*, power casing *3*, rear casing *4*, inlet rectifier *5*, and rotor of the high-pressure turbocharger *6*.

The adapter (Fig. 6.29) [8] provides a smooth air supply from the LPC to the HPC. It accommodates housing of the rear LPC support *2* with roller bearing *26*, housing of the front HPC support *18* with ball bearing *22*, inlet stator of the HPC *14*, two rotation speed sensors *15*, and inlet rectifier of the LPC *7*.

From the side of the HPC, the oil cavity sealing is provided by a contact seal (cover *20* and sealing ring *21*) and a labyrinth seal (labyrinth cover *19* and labyrinth of rotor shield of the HPC).

Lubrication of ball bearing *22* of the front HPC support takes place through pipe *9* and oil supply ring *23*. Direct oil supply to the bearing is performed by nozzle *30*, which feeds the oil through the *D* channels to the bearing.

The HPC housing is a shell with a horizontal connector. The power housing is attached to the middle flange of the housing, and the rear housing is attached to the rear flange.

The purpose of *the HPC power housing* is to strengthen the engine structure in its hottest place and to accommodate the flame tubes, fuel drum and rear housing communications.

The rear housing is made in the form of a diffuser. There, the air velocity decreases and the pressure increases before the combustion chamber inlet. The housing accommodates the rear support of the high-pressure turbocharger and the

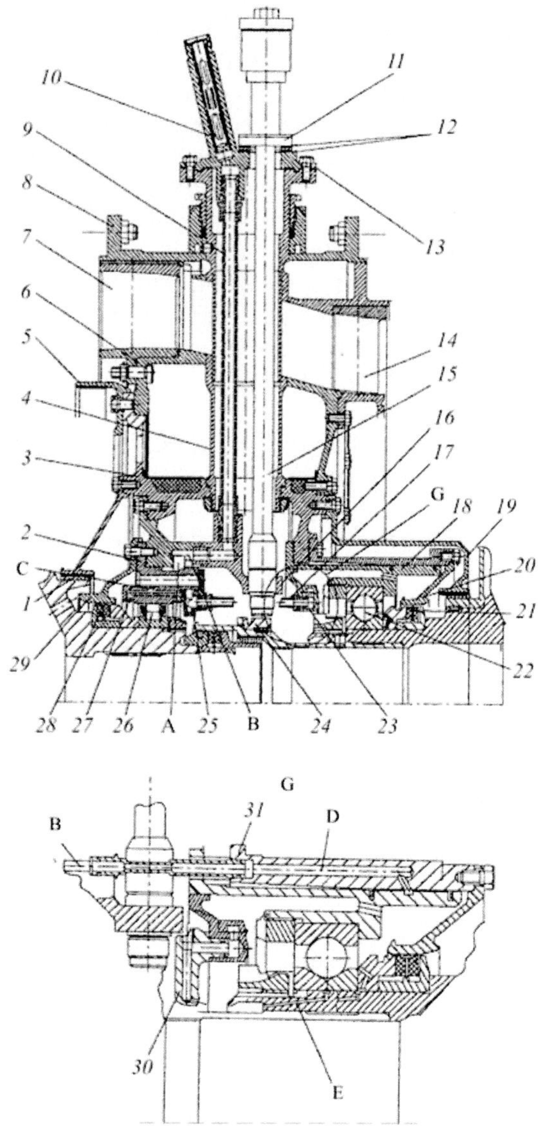

Fig. 6.29 Adapter (cross section through the oil supply to the sensor): *A*—annular oil distribution channel; *B*—oil channel; *C*—annular gap; *D, E*—oil supply channels; *1, 5, 19*—labyrinth covers; *2, 18*—housings of the rear LPC and HPC support, respectively; *3*—power housing; *4*—pipe; *6, 8*—inner and outer walls, respectively; *7*—rectifier; *9*—oil feeding pipe; *10*—filter; *11, 21*—adjusting ring; *28*—sealing rings; *12*—lining; *13*—flange; *14*—stator at the HPC inlet; *15*—rotation speed sensor; *16*—bracket; *17*—oil supply pipe; *20, 29*—covers; *22, 26*—ball and roller bearings, respectively; *23, 25*—oil supply drums; *24*—inductor pinion; *27*—bushing; *30*—nozzle; *31*—limiting cover

air intake for cooling the turbine blades of the HPT through the racks located under each combustion chamber.

The rotor of the high-pressure turbocharger constructively combines the HPC and HPT rotors. It provides conversion of the mechanical energy obtained from the HPT to the kinetic energy of the air flow that passes through the air path of the compressor. The torque transmission from the HPT rotor to the HPC rotor is carried out through a splined connection.

The oil and air cavities are separated by means of contact-labyrinth seals in order to eliminate the possibility of mutual penetration of the working media. The contact seal consists of a bushing, a sealing graphite ring and a restrictive ring.

The purpose of *the combustion chamber* (Fig. 6.30) [8] is to produce the combustion products of the required temperature by burning fuel in the air stream that continuously flows from the HPC and to supply the combustion products to the HPT.

The combustion chamber of the GTE UGT25000 has a tubular-annular form and is made counter-flow, so that the combustion products inside the flame tube and the air outside the flame tube move in the opposite direction. The chamber consists of casings *10* and *19*, sixteen flame tubes *11* located in the annular space between them

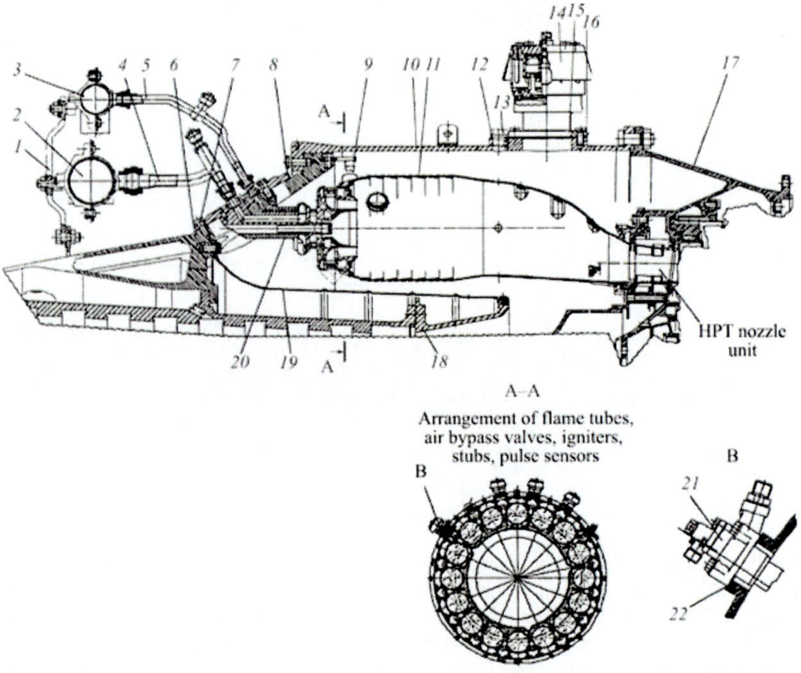

Fig. 6.30 Combustion chamber: *1*—suspension; *2, 3*—drum of the second and the first channels, respectively; *4, 5*—supply tubes of the second and the first channels, respectively; *6*—HPC housing; *7*—propellers; *8, 12, 16, 22*—gaskets; *9*—retainer; *10*—combustion chamber casing; *11*—flame tube; *13*—stubs; *14*—air bypass valve; *15*—glass; *17*—power housing; *18*—diffuser; *19*—inner casing of the chamber; *20*—burner; *21*—igniter

in parallel to the engine's axis, two igniters *21*, sixteen two-channel burners *20*, drums of the first and second channels (*3* and *2*, respectively), sixteen supply pipes of the first and second channels (*5* and *4*, respectively).

Primary ignition of fuel in the combustion chamber during the GTE start-up is performed by two igniters *21*, each of which through the flame-feeding pipe provides ignition of the fuel–air mixture in two flame tubes. The transfer of flame from one flame tube to another occurs through the flame-transferring pipes of the flame tubes.

The flame tube consists of an inlet housing, intermediate cone shells, a mixer, a number of bushings, flame-retarding and flame-feeding pipes.

From inside, the walls of the flame tubes are cooled by means of air film cooling. Outside, the flame tube is cooled by air of the intertube space.

To increase the service life of the flame tube, its inner surface is covered with high-temperature heat-resistant enamel.

Igniter *21* is intended for the formation of the primary flame. Its main elements are a casing, a nozzle, and a plasma-jet plug powered from a pulsed electronic power supply. Periodic electrical discharges lead to the formation of plasma jets that ignite the starting fuel, which is fed through the nozzle. The flame from the igniters (two of them) installed in the flame-feeding pipes of the flame tubes of the combustion chamber is thrown into the flame tube, igniting the fuel that is fed through burner *20*.

The burner is designed to prepare and feed a homogeneous fuel–air mixture into the combustion zone of the flame tube. It is fixed to the flange of the power casing of the HPC.

The engine has four air bypass valves *14*, designed to bleed air from the combustion chamber to the atmosphere during the engine start-up, as well as when the power turbine speed sensor is triggered. The bypass valve opens under the pressure of air from the air cylinder.

The high-pressure turbine (Fig. 5.31) [8] is designed to drive the HPC; it is made single-stage and axial and consists of a nozzle unit and a rotor. Nozzle blades *43* located in power casing *1*, stator *40*, cell sealing inserts *2* located in ring *3*, and distribution shield *4* constitute the nozzle unit of the HPT.

The nozzle blades are coolable, single, tip-shrouded; the upper and lower shelves are sealed with saferite seals *42, 44*. Cooling of the inlet edge of the nozzle blades is film-type, whereas cooling of their middle part and outlet edges is vortex-type. Cooling air is fed to the working blades of the HPT by stator *40*.

The rotor of the HPT is formed by disk *36* which is connected by bolts *38* to trunnion *35*. The assembly is mounted on the trunnion of the HPC and tightened by nut *34*. Working blades *45* are installed in the fir-tree disk grooves and fastened by segments *41*. The scallops made on the protrusions of the disk, trunnions and working blades, together with the seals of the connected parts, minimize the respective leakage of the combustion products.

The axial single-stage *low-pressure turbine* is designed for driving the LPC; it includes a nozzle unit, a LPT rotor and a supporting crown.

The nozzle unit of the LPT is formed by housing *6*, blocks of nozzle blades *5*, diaphragm *39* and ring *8*. The blades are coolable; cooling of the blade profile is vortex-type, and cooling air comes from above.

The *rotor of the LPT* is formed by disk *37* connected to shaft *29* by pins. Working blades *7* installed in the disk grooves are fastened by segments *41*. The scallops made on the protrusions of the disk, the shaft and the working blades ensure minimal flowing of the combustion products.

The *support ring (SR)* of the LPT consists of housing *9*, six power racks *13* connected to the outer housing by means of heat expansion joints *11*, support housing *16* connected to racks *13* by fingers *17* and at the bottom to bearing housing *21* and cone *22* by bolts *20*. Housing *23* is mounted to the bearing housing with propellers *21*. It accommodates the rear support of the LPT.

The air-gas channel of the SR is formed by casings *10, 12*, casing wall *19* and wall *18*. Shield *14* is fixed to support housing 16 with bolts *15*. In order to reduce the temperature expansions of SR housing *9*, the cavity between the casings is filled with the heat-insulating basalt fiber.

Sealing covers *31, 32, 33* are set in the front part of the SR. They form the oil and air cavities of the SR to support the contact seal of the LPT rotor rear support and supply the air to cool the power turbine shaft. Oil supply drum *25* is secured with bolts *24* (Fig. 6.31).

The rear support of the LPT is roller bearing *28* with a damper mounted on its outer casing. The damper is designed to suppress the oscillations of the rotor. The inner casing of the bearing is fastened with nut *27*.

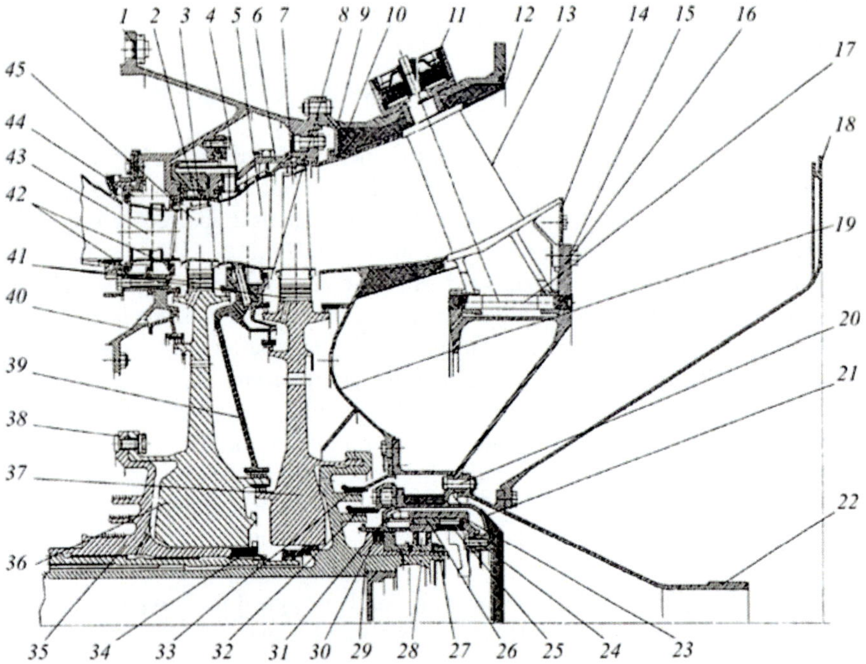

Fig. 6.31 High- pressure and low-pressure turbines

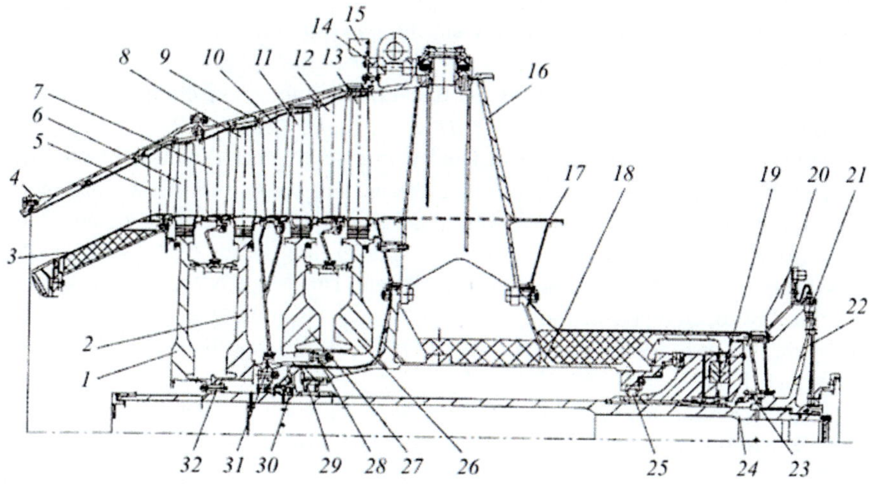

Fig. 6.32 Power turbine

The air and oil cavities of the SR are separated with graphite sealing ring *30*; adjusting ring *26* is located here.

The power turbine is designed to drive the external power consumer (the electric generator) or the propeller through the reducer. It is connected to other GTE elements forming the gas generator only gas-dynamically, along the flow of the combustion products.

The axial power turbine (Fig. 6.32) [8] is has four stages and consists of nozzle units, a rotor and a supporting crown. The air-gas channel from the LPT to the turbine is formed by casings *3* and *4*.

Nozzle unit5 of the first stage is fixed to housing *4*, nozzle units *7, 10, 12* of the other stages are fixed in housing *9*. The lower part of the nozzle unit is equipped with diaphragms with sealing inserts for a radial labyrinth sealing of the cavity.

The rotor of the power turbine includes shaft *24* to which disks *1* and *2, 26* and *27* are fitted in pairs with the help of bolts *28, 31, 32*. The working blades *6, 8, 11, 13* of all four stages are installed in the fir-tree grooves. Roller bearings *25* and *29* serve as the support of the cantilever rotor of the turbine. The axial forces that arise during the turbine operation are perceived by thrust sliding bearing *19*. At the end of the rotor, there is elastic coupling *22*.

The support ring of the power turbine consists of casing *14*, nine racks—fairings *16*, inner casing *17* and housing *18*, which accommodates all the bearings. The oil cavity of the SR is limited on both sides by sealing rings *23* and *30*.Vibration characteristics of this turbine are measured by a vibration sensor mounted on bracket *15*.

Being an element of the diagnostic system, turbine speed-limiting sensors *21* that are attached to bracket *20* prevent the breakdown of the power turbine by signaling to the engine management system when the maximum permissible rotor speed is exceeded.

Reverse marine gas turbine engine. It is important for a ship equipped with a gas turbine plant to have a reliable, fast reverse.

Ship reverse is a maneuver associated with changing the direction of the thrust created by the propeller. The reverse is carried out using one of the elements of the propulsion complex "power turbine—gear—propulsor", which is called *reverse complex* in this case. When using a reverse power turbine, the reverse is called the gas reverse, and the GTE is called the reverse one. The design of the reversible power turbine developed by F.I. Kirzner and V.I. Romanov is shown in Fig. 6.33 [2, 8]. This axial reverse turbine was patented over 1975–1979 in Great Britain, Germany, the USA, and Japan.

The reverse turbine (see Fig. 6.33) has two circuits for gas flow: the internal one and the external one. When the gas passes through the internal circuit, the turbine rotates the propeller shaft forward. If it passes through the external circuit, with the internal circuit closed, reverse is realized.

Gas distribution along the circuits is determined by the position of the rotary blades (forward valve) and the gas bypass belt to the external circuit (reverse valve).

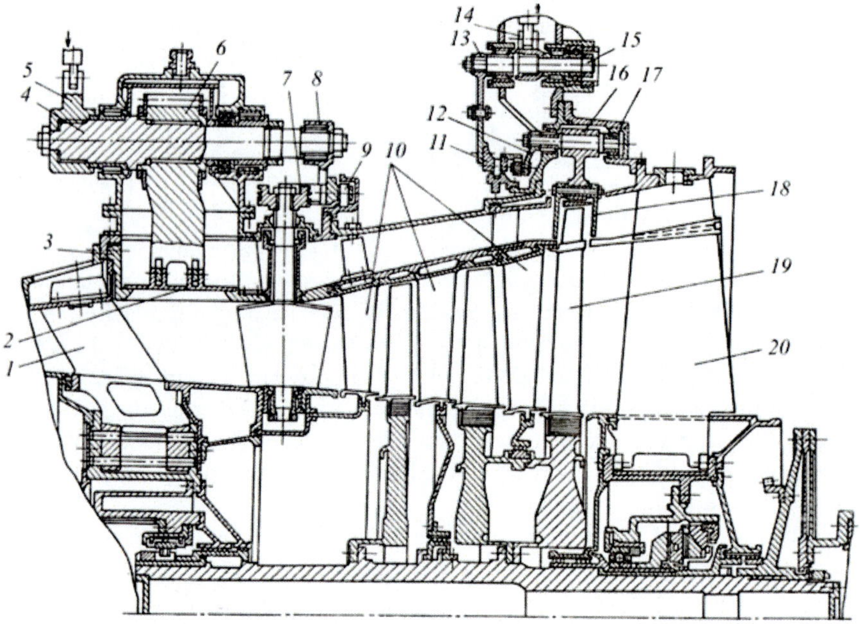

Fig. 6.33 Reverse power turbine: *1*—supporting crown; *2, 3*—gas bypass belt and body, respectively; *4*—belt drive shaft; *5, 12, 13, 14*—levers; *6*—lever of the belt drive; *7*—rotary blade with a pulley; *8*—lever of the rotary ring drive; *9, 11*—rotary rings; *10*—nozzle devices; *15, 16*—shaft and lever of the drive of the U-shaped shields; *17*—housing of the U-shaped shields; *18*—U-shaped shield; *19, 20*—rotor and supporting crown of the power turbine, respectively

The valves are interlocked. Realization of reverse requires simultaneous closing of the forward valve and opening of the reverse valve to direct the gas to the external circuit onto the reverse blades.

It should be noted that the profile of the reverse blades is turned by 180° relative to the profile of the forward blades. Due to this, when the gas hits the reverse blades, the direction of rotation of the power turbine rotor is reversed.

The valves are driven by the pneumatic cylinder through the rotary ring, which is connected to the levers and trunnions of the rotary blades and to the levers of the gas bypass belt. To implement the "propeller stop" mode (when the propeller of the ship does not rotate with the engine running), the bypass belt is slightly opened, the rotor blades are not closed, and the working gas passes both along the forward and reverse circuits. When the rotation moments in the forward and reverse working blades are equal, the rotor speed of the power turbine is equal to zero.

The design of the turbine provides for closing of the reverse working blades of the power turbine rotor with the help of the U-shaped shields when the turbine is operating at the forward stroke. Thus, ventilation losses of the turbine are substantially reduced.

The above mentioned design of the GTE was implemented in the engines produced by Zorya-Mashproekt and has been continuously improved [2]. As for now, there is every reason to assert that Ukraine has an effective reverse GTE of its own production. Other countries that produce ship engines do not have the design of reverse engines.

6.4 Use of Gas Turbine Plants in Marine Transport

6.4.1 Power Plants of Cruise Ships and High-Speed Ferries

Most GTEs used in marine transport are driving gas turbine engines of the main power generators of the propulsive plants on cruise ships. This is explained by the fact that many modern cruise liners are equipped with propeller-steering complexes (such as Azipod), which have the form of an electric engine with a propeller located in a sealed gondola that can make a 360° turn around the vertical axis. This design solution is associated with the need to provide increased maneuverability of cruise ships. For reasons of navigation safety of cruise ships, not less than two main propellers are provided; some ships have three main propulsion-steering complexes, two of which are turning, and the middle one is fixed.

The use of GTE on large cruise liners can significantly reduce the amount of space required for the engine room due to the low mass-dimensional characteristics of the GTE, which allows increasing the number of holds. The high readiness of the GTE for start-up and loading, along with the possibility of the engine's unit replacement, ensures the reliability and survivability of the marine power plant on the cruise liner.

In addition, the use of the GTE in the role of the main engine on such ships allows meeting the requirements of Annex VI of the MARPOL 73/78 Convention, which are particularly stringent for coastal zones, inland seas and other preferential areas for the navigation of cruise ships and high-speed ferry boats. Thus, if comparing a combined gas turbine plant with electric gear installed on a "Millenium" passenger ship with an ICE of the same power, there is a reduction of NO_x emissions from 12 to 4 g/(kW h), SO_x—from 13.6 to 0.9 g/(kW h), other impurities—from 5 to 0.13 g/(kW h). This was the basis for the classification society to assign the Clean Design class to the ship.

Reduction in the fuel costs for propulsive plants with main gas turbine generators is achieved by using heat recovery circuits. Considering that cruise liners have an increased necessity of thermal energy (most of which is consumed by water desalination plants with a daily total capacity of up to 2000 tons), there are two expedient methods of recovering the heat of exhaust gases of gas turbine engines. One of them is production of saturated steam to cover the general demand for heat energy, and the other one is production of superheated steam for use in a steam turbine generator.

The combined gas turbo-electric and diesel-electric plant CODLAG operates according to the first method. It is installed on the cruise liner "Coral Princess" of the "Baby Grand" class with the capacity of 2500 passengers. The abbreviation CODLAG (COmbined Diesel-electric And Gas) denotes the type of plant where electric current generators driven by the ICE and the GTE serve the propeller via the propulsive engine. The unit consists of a 25 MW gas turbine generator located on the 15th deck in a funnel and two main 16.6 MW diesel generators located in the hold in the main engine room [4]. Movement and maneuvers of the ship are provided by two Azipod complexes with the power of 20 MW and six tunnel thrusters (three in the bow and three in the stern) with the power of 1720 kW. The GTE exhaust gas serves as the energy source for the recovery boiler with the steam production of 30 t/h. The main heat consumers are three desalination plants with the capacity of 600 t/h. A similar scheme is typical for CODLAG power plants on cruise ships of the "Vista" class (5 units), larger ships of the "Grand" class (2 units) and "Queen Mary II" (Table 6.1) [4].

On the cruise ships "Millenium" and "Radiance of the Seas", the COGES (COmbined Gas Turbine and Steam turbine integrated Electric drive System) has been used to drive the main power generators. The plant includes two GTEs with heat recovery circuits and a steam turbine generator.

These cruise liners employ gas turbine engines produced by GE Marine. In total, 26 engines of this company are installed on 17 cruise ships: 16 LM2500+ engines are included to the COGES plants, 6 LM2500+ and 4 LM2500 engines are included to the CODLAG plants.

In the world marine transportation market, high-speed cabotage transportation of goods is carried out on the Ro-pax passenger and car ferries. Ro-pax stands for "roll on/roll off passenger" and denotes a ship for transportation of wheeled cargo and passengers. It has the deadweight of 150–1200 tons and the maximum speed of 37–50 knots. High-speed ferries have the form of single-hull displacement ships, as well as multi-hull SWATH ships (Small Waterplane Area Twin Hull—catamarans with a small area of the waterline).

Such ships with the speed above 35 knots usually make use of gas turbine plants or CODAG plants (Table 6.2) [4]. The energy saturation of such ships reaches 68 kW

Table 6.1 The main characteristics of cruise ships with combined power plants

Ship or class name	Vistaclass	BabyGrandclass	Grandclass	QueenMaryII	Millennium	RadianceoftheSeas
Displacement, t	84,000	88,000	113,000	150,000	90,230	90,090
Speed, knots	22	24		28.5	24	
Powerplant type	CODLAG				COGES	
MPParrangement	DG 2 × Sulzer 12ZA40C 3 × Sulzer 16ZA40C GTG 1 LM2500	DG 2 × Wärtsilä 16V46C GTG 1 LM2500+	DG 2 × Wärtsilä 8L46C 2 × Wärtsilä 9L46C GTG 1 LM2500+	DG 4 × Wärtsilä 16V46C GTG 1 LM2500+	GTG 2LM2500 + 1 STG	GTG 2LM2500 + 1 STG
Plantpower, kW	65,340	58,600	60,700	117,200	57,000	59,000

Table 6.2 The main characteristics of high-speed single-deck ships with gas turbine and combined power plants

Ship name	Setfoil	FoilCat	Neptune	FNSLV	NGV Liamone	Corsaire 14,000	MDV 3000	ITLV
Length, m	27.4	35	128	177	134	140	145	121
Width, m	8.5	12.0	19.4	24.0	19.6	21.8	22.0	18.0
Draft, m	5.40	4.70	–	4.15	–	3.62	3.90	3.15
Deadweight, t	100	193	600	4000	900	1300	1200	1100
Powerplanttype	Gasturbine				Combined CODAG			
MPP arrangement	2 × RR Allison 501–KF	2 × GE LM2500	2 × GE LM2500	3 × RR MT30	GTE 2 GE LM2500+ ICE 2 MTU 20V1163	GTE 2 GE LM2500+ ICE 2 Pielstick20PA68	GTE 2 GE LM2500 ICE 4 MTU 20V1163	GTE 1 RR MT30 ICE 2 Pielstick20PA6B
Powerplantpower, kW	6800	8000	42,600	108,000	59,200	66,200	70,000	52,000
Maximumspeed, knots	45	50	42	40	–	–	40	–
Operatingspeed, knots	–	–	–	36	43	42	–	40

per ton of displacement, which corresponds to the energy saturation of dynamically supported ships. A typical example of a single-hull high-speed ship is the "Corsaire 1300" ferry with the length of 134 m and deadweight of 870 tons, which is designed to carry 1116 passengers and 250 cars at the speed of 40 knots. The main power plant is of the CODAG type. It is located in three adjacent compartments: the units that serve water jet propellers are in the stern compartment, while one 25 MW GTE and one 6.5 MW engine are in the two other compartments. On another "Corsaire 14,000" ship with the deadweight of 1300 tons, two GTEs (GE LM2500) and two diesel engines with the total power of 66 MW are installed. They serve four water jet propellers.

Catamarans have been widely used as high-speed ships. Such ships can be also equipped with gas turbine units (Table 6.3) [4].

The power plant with a GTE as the main engine is suitable for assembling in the narrow hulls of Ro-pax catamarans. The concept of a catamaran with recessed hulls of the SWATH type contributes to the reduction of power consumtion when the ship moves at a high speed. Large catamarans with partial application of this concept (semi-SWATH) are produced in Norway, Denmark, Italy, and Australia. The catamarans of the HSS series (Norway) and develop an operational speed up to 40 knots, which is provided by a gas turbine plant of four GTEs with the total power of 68 MW at the deadweight of 1500 tons and two GTEs with the power of 17 MW each at the deadweight of 450 tons.

Recently, large high-speed yachts (more than 35 m long) have become very popular. As of 2015, several dozen ships ranging in length from 80 to 160 m were in operation, and several hundred yachts were under construction, of which 13 megayachts were about 90 m long. Some of these ships have been equipped with gas turbine engines (Table 6.4). One of the fastest is the 42-m megayacht of the "Millennium" project. Its power plant is of the CODAG type: two Paxman 18VP185 engines (4 MW each) each serve its water jet engine, and the TF80 unit (two Lycoming TF40 GTEs with the total power of 6.8 MW) drives the middle water jet propeller. The maximum speed is 68 knots. Most commonly, megayachts are equipped with the TF40 and TF50 engines (already about 20 units).

6.4.2 Power Plants of Dynamically Supported Ships

Power plants of hydrofoil ships. The power plants of non-displacement hydrofoil ships (HFS) and air cushion ships (ACS) should have the minimum weight and dimensions, high acceleration capacity, and develop the necessary power. GTEs of the light aircraft type fully meet these requirements.

Moderate weight of the gas turbine engines allow for the creation of the HFS and ACS power plants with the weight not exceeding 20% of displacement, which can provide a ship with the speed of up to 60 knots. With the same power plant weight/ship displacement ratio, diesel ships can have a speed of no more than 40 knots.

Table 6.3 The main characteristics of high-speed ferries with gas turbine and combined power plants

Ship name	VilljumClausen	HSS 900	AutoExpress 125	Marinteknik 55	HSS 1500	SSCE	HSCE
Typeofhull	Catamaran					Trimaran	
Length, m	86.6	89.8	125.0	55.0	126.6	155.0	214.0
Width, m	24.0	30.5	31.7	15.0	40.0	32.2	48.0
Draft, m	3.2	3.9	4.9	4.5	4.8	4.5	4.9
Deadweight, t	485	450	1200	105	1500	1500	2500
Powerplantpower, kW	36,000	33,500	44,000	9700	68,000	54,000	104,000
Powerplant type	Gasturbine			CODAG	COGAG	CODAG	Gasturbine
MPP arrangement	2 × GE LM2500	2 × ABB STAL GT35	2 × GE LM2500	GTE 2 TF40 ICE 2 MTU 16V396	GTE 2(GE LM2500+ GTE 2 GE LM1600	GTE 2 GE LM2500 ICE 2 MTU 16V1163	GTE 4 GE LM2500+
Maximumspeed, knots	41	38	40.5	38	40	42	45

Table 6.4 The main characteristics of megayachts with gas turbine and combined power plants

Ship name	Fortuna	Moon-raker	Pershing 115	Millen-nium 140	EnigmaEco	EctaSea	Mangusta 108	WalleyPower 118	Diestriero
Length, m	41.5	35.4	36.1	42.0	74.1	86.0	33.5	36.0	67.7
Displacement, t	140	90	114	120	860	1000	94	95	1000
Maximumspeed, knots	68.9	67.7	42	68	38	33	51	60	63
Powerplant type	CODOG	CODAG					Gas turbine		
MPP arrangement	GTE 1 TF40 ICE 2 MTU 16V396	GTE 1 TF40 ICE 2 MTU 16V396	GTE 1 TF40 ICE 2 MTU 16V4000	GTE 2 TF40 ICE 2 Paxman 18VP185	GTE 1 LM1600 ICE 2 Deutz MWM	GTE 1 LM2500 ICE 4 MTU 16V4000	2 × TF50	3 × TF50	3 × LM1600
Plantpower, kW	15,000	11,000	9600	21,600	28,000	31,000	6800	12,400	38,500

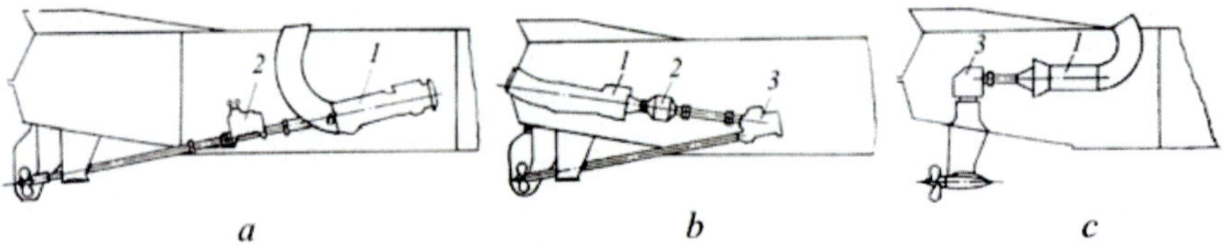

Fig. 6.34 Design schemes of propulsion systems on hydrofoil ships: **a**, **c** with linear and angular transmission, respectively; **b** with a broken shaft line; *1*—TE; *2*—reducer; *3*—angular reducer

The choice of the MPP layout and the technology of power transmission to the propulsive unit are important issues in the design of HFS. This is due to the fact that the ship hull rises above the water surface by 4–5 m when this ship moves with the help of foil (foil mode). Therefore, the propellers must be at a considerable depth in the displacement mode, so that the axis of the propellers would be immersed in the water to a depth of not less than the diameter of the propeller at a full speed on foil.

At linear power transmission to the propeller, the required angle of inclination of the shaft line can be ensured only when the main engine is located in the bow of the ship (Fig. 6.34a). This substantially extends the shaft line, complicating the placement of passenger compartments.

Transmission with a broken (V-shaped) shaft line (see Fig. 6.34b) allows placing the GTE in the stern and using the ship bow for the passenger compartment. This transmission requires the use of an angular reducer that makes the plant more complicated. Considering the fact that the shafting inclination reduces the propeller thrust and propulsion efficiency, the layouts shown in Fig. 6.34a and b are mainly used for HFS with a relatively small height of the hull above the water. On modern HFS and ACS with high seaworthiness, power transmission to the propeller is mainly carried out by means of mechanical angular gears (see Fig. 6.34c).

Let us consider the marine passenger hydrofoil ship "Cyclone-1" as an example of an HFS. It is designed for high-speed passenger transportation on sea lines with the one-way voyage duration of up to 8 h [1, 2]. The length of the ship is 44 m, displacement is 137 tons, the number of passengers on board is 250 people, the speed on foil is 42 knots, and the cruising range is 300 miles.

The main element of the "Cyclone" MPP is the main GTU M37.

The unit (Fig. 6.35) comprises:

- reversible gas turbine engine DO37 with gas outlet and protective casing of the hot parts of the engine and gas outlet;
- angular two-stage reducer RO37 with integrated MTB which perceives axial forces from the propeller;
- automatic control, protection, and emergency signaling stations.

The main characteristics of the M37 unit are as follows. The maximum power at the forward stroke is 5880 kW; the propeller rotation speed is 800 rpm; the power at the reverse stroke is 600 kW; the specific fuel consumption at the maximum mode is 0.295 kg/(kW h). The unit is 6350 mm long, 1900 mm wide, and 2200 mm high;

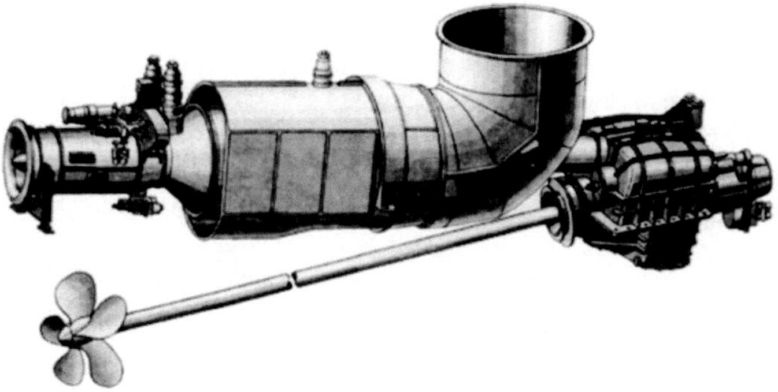

Fig. 6.35 General view of the main gas turbine unit M37

its weight is 7 tons. Its operational resource before factory repair is 8000 h, and total service life makes up 12 years.

The main engine is located at the angle of 4° into the bow relative to the base plane. The angular reducer is placed on a separate foundation. The TG16M gas turbine generator of the marine power station is designed to power the DC starter, which ensures the GTE launch by spinning up the HPC. In the middle part of the ship in a special room there are two auxiliary diesel generators DGA 50M2-9, an electrocompressor and a stack of compressed air cylinders.

The reverse and "propeller stop" modes are provided by the main GTU M37. Relocation of the reversing devices takes place at the mode no higher than 0.5 of the nominal power.

This unit was manufactured by UDBMashproekt (Ukraine) in cooperation with PA Zorya (Ukraine) and delivered in 1985 to the PA More (Feodosia, Crimea). The mooring and sea trials were completed in December 1986. After the experimental operation and commercial voyages in the Black Sea region, the first passenger gas turbine ship "Cyclone-1" sailed around Europe to the port of Tallinn. From there, "Cyclone-1" made daily voyages across the Baltic Sea to Helsinki. Later the ship was sold to a Greek company, which actively used it for regular voyages between the islands of Greece.

According to the results of tests and operation of the ship, it was decided to use not one but two gas turbine units to increase the reliability of the future ships, i.e. make a two-shaft ship.

The UDBMashproekt designed and manufactured an M39 unit with a non-reversing DE76 GTE with the power of 2940 kW to be used for the "Cyclone-2" project. In order to increase the economic efficiency of the unit, reverse was to be performed by the reducer, and the engine was non-reversing.

A special feature of the main GTU M39 is a two-stage two-speed reverse reducer. Connection of the engine to the reducer, operation of the unit at the forward stroke, and maneuvering from the forward to the backward stroke and vice versa are provided by the friction claw coupling of the forward stroke and the friction coupling of the reverse stroke.

Preference is as well given to the propulsion system with the GTE in creating high-speed dynamically supported ferries. The two-hull hydrofoil ships manufactured in South Korea are fitted with the power plants with four gas turbine engines. The ships are intended for transportation of 630 passengers and 160 cars at the speed of 45 knots.

The high speed and maneuverability of the HFS, as well as their high mass-dimensional characteristics resulting from their use of GTE, attracted the attention of specialists in the design of naval ships.

There were several dozens types of HFS created worldwide; this process was especially intensive in the 1970s and 1980s.

On the HFS with a significant propulsion capacity, only corner columns are used. They are able to transmit power up to 15,000 kW and can be made liftable to ensure safety when the ship approaches the shore and convenience of its inspection and repair. Let us consider two most typical power plants with the GTE of Ukrainian production, which are intended for use on the HFS.

The UDB Mashproekt created the M10 main gas turbine unit (Fig. 6.36) for the "Hurricane" hydrofoil ship; two units were installed on the ship.

Each main GTU M10 consists of the following elements:

- all-mode non-reverse DO50 GTE with a free power turbine;
- vertical gear 8 m long, which connects the upper and lower reducers;
- lower reducer located in a detachable streamlined gondola.

The length of the vertical gear is chosen in such a way that when the ship is lifted above the water by 5 m, the propellers of the front reducer of the gondola do not leave the water. In the upper reducer, there is a quick-release coupling, which allows lifting the ship rear stern complex at the angle of 120° [2].

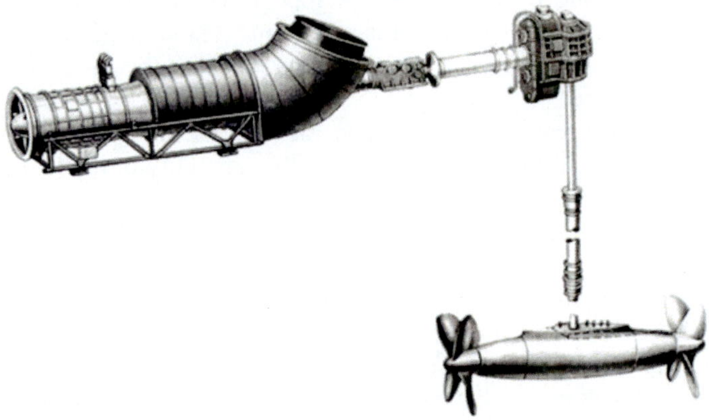

Fig. 6.36 Layout of the M10 unit

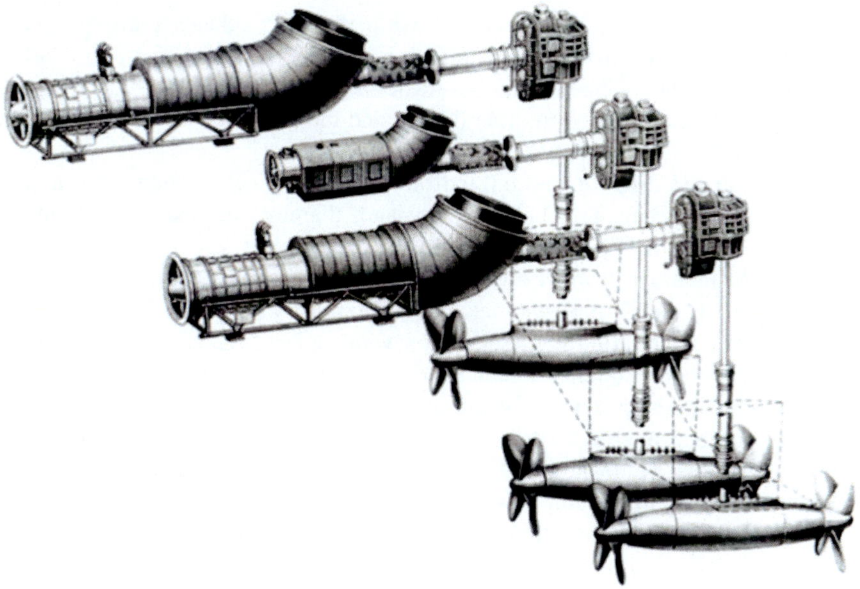

Fig. 6.37 Layout of the M10 and M16 units of the power plant of the ship "Sokol"

The main characteristics of the M10 unit are as follows: total power of 13,200 kW, specific fuel consumption of 285 g/(kW h), resource of 2000 h, specific weight of 0.68 kg/kW.The mooring tests of the ship were completed in 1977.

In 1986, the UDBMashproekt provided the PA More with the main gas turbine unit M16 intended for use as part of the power plant of the HFS "Sokol" [2]. The power plant of the ship includes two main M10 GTUs located along the sides and one M16 unit located in the middle (Fig. 6.37).

The M16 unit consists of the all-mode reverse GTE DS71 and the angular reducing gear, which includes an upper angle reducer, a lower reducer, and vertical and horizontal transmissions.

Each of the three main GTUs serves two coaxial fixed-pitch propellers mounted on the shafts of the lower reducer.

Basic characteristics of the M16 unit are the following: maximum power of 7400 kW, specific fuel consumption of 285 g/(kW h), reverse power of 740 kW, resource of 10,000 h. The total power of the three main GTUs is 33800 kW. The ship movement in the displacement mode is provided by the M16 unit, while the movement in the foil mode is provided by operation of all the main GTUs.

Power plants of air cushion ships consist of two parts: a lifting part which creates the air cushion and a propulsive part that provides the ship movement.

A high speed of the modern ACSs leads to a high specific power of their power plants (50–100 kW/t). Only a light aircraft GTE provides such a high energy saturation.

Depending on the mass and speed of the ship, the MPP includes one to four GTEs, which can be placed both in the ship hull and in the superstructure. Figure 6.38 shows the options for placing the GTE on the ACS.

The engines can be used to drive the fan and the propeller simultaneously both simultaneously and separately.

The combined drive of the fan and the propeller allows reducing the number of engines, increasing their unit capacity and, thereby, increasing the economic efficiency of the unit [1].

A significant progress in creation of the ACS with gas turbine propulsion systems was achieved by the UDB Mashproekt in collaboration with the PA Zorya (Ukraine).

In the 1970s and 1980s, there were several dozens ACSs built for different purposes; they had different size and energy saturation.

The air cushion ship "Jeyran" was developed by the Central Marine Design Bureau "Almaz" and became the first ship of this type in the USSR. Displacement of the ship was 400 tons, while the speed was 50 knots. For this ship, the UDB Mashproekt designed the gas turbine unit DT-4 with the total capacity of 26,500 HPC; the PA Zorya produced it. The unit includes two DO75 (UGT16000) gas turbine engines and 18 reducers of various types (Fig. 6.39).

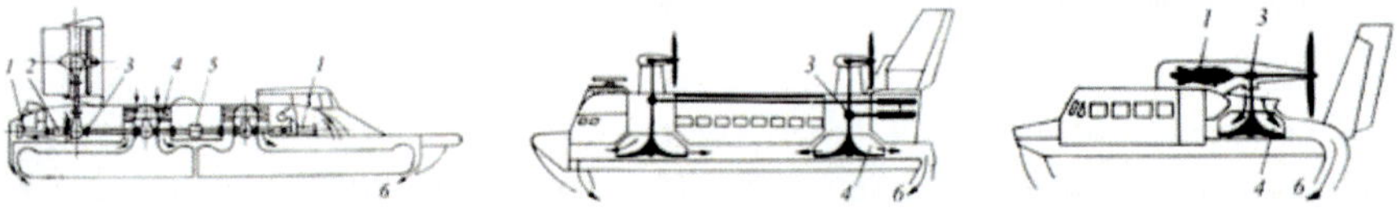

a *b* *c*

Fig. 6.38 Options for arrangement of gas turbine engines in the hull of the air cushion ships: **a**, **b** in the ship hull; **c** in the superstructure; *1*—GTE; *2*—coupling; *3*—reducer; *4*—centrifugal fan; *5*—support bearing; *6*—flexible nozzle

Fig. 6.39 DT-4 Gas turbine unit

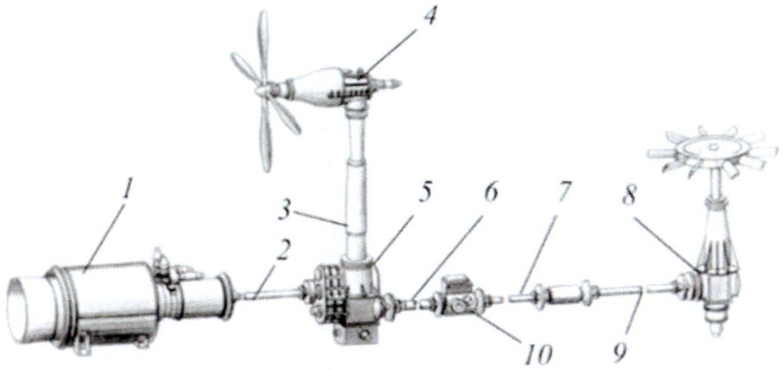

Fig. 6.40 Main gas turbine unit MT70K of the "Kalmar" ship: *1*—M70 GTE; *2, 6, 7, 9*—connecting shafts of horizontal gear; *3*—connecting shaft of vertical gear; *4*—air propeller reducer; *5*—distribution reducer; *8*—fan reducer; *10*—reducer of ship units

Torque transmission to the reducers is performed through flexible disk couplings and torsion shafts. Power is taken off from the front of the engine from the LPC.

The gas turbine engine has a blocked power turbine. The power turbine and the low-pressure turbine are combined into one three-stage turbine. Each GTE drives two air cushion fans and two air traction propellers. The propulsive complex design suggests that in the event of failure of one of the engines, the power of the other one is transferred to all the propellers and fans through two torsion shafts.

Basic parameters of the DO75 engine are as follows: power of 13,250 kW, specific fuel consumption of 285 g/(kW h), gas temperature in front of the HPT of 850 °C, weight of 6150 kg, and resource of 2000 h.

The manufacture of the elements of the main GTUs was completed in 1969, and the ACS was introduced to the navy fleet in 1976.

The main gas turbine unit MT70K (Fig. 6.40) was created by UDB Mashproekt and STP Zorya in 1975 and later built at PA More in Feodosia. It was intended for the air cushion ship "Kalmar" with the displacement of 130 tons and speed of 60 knots.

Two units were installed on each ship. The power on the GTE output flange is 7400 kW, the specific fuel consumption is 258 g/(kW h), engine resource is 500 h, and mechanical gear resource is 1000 h.

The main GTU M34 was developed by the UDB Mashproekt for the air cushion ship "Omar" with the displacement of 50 tons and the speed of 60 knots (Fig. 6.41).

The structure of this unit includes the gas turbine engine DM71 (UGT6000), a gear reducer, and two reducers of the propeller of the right and the left board.

The power from the gas turbine engine is transmitted through elastic couplings and springs to the gear reducer, then to the reducers of the air propellers and to the reducers of the superchargers. The design of the gear reducer, propeller reducers, and supercharger reducer is similar to that of the corresponding M70 elements.

The main characteristics of the main GTU M34 are as follows: power of 4400 HPC, specific fuel consumption of 295 g/(kW h), power turbine rotation speed of 8500 rpm, resource of 1000 h.

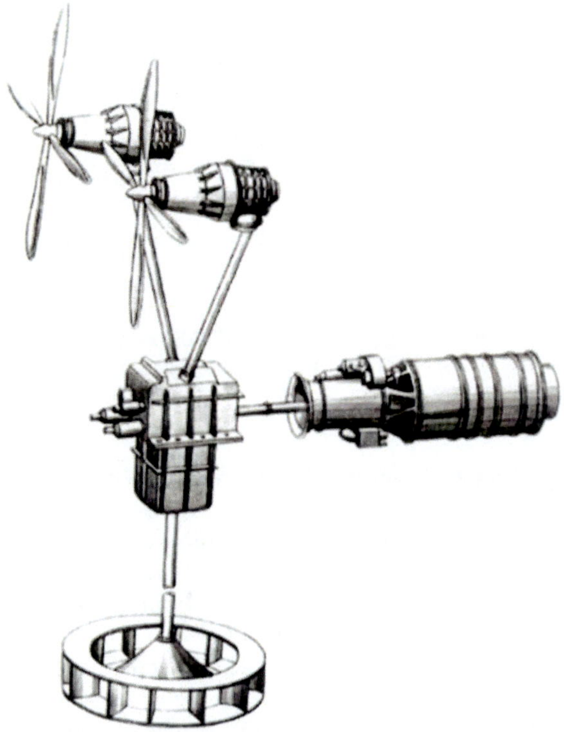

Fig. 6.41 Main gas turbine unit M34

The lead ship was put in operation to the fleet in 1980.

The most advanced product of the Soviet shipbuilding of the Ukrainian production was the ACS "Zubr". Its state tests were completed in August 1988.

The M35 power plant of the "Zubr" ship (Fig. 6.42) is divided into independent gas turbine units: three traction units M35-1 and two injection units M35-2 [2]. The power plant has a separate drive for superchargers and air propellers, which greatly simplifies its design and requires fewer reducers.

The structure of the M35-1 unit includes the engine DP71 (UGT6000) and the air propellers reducer. The M35-2 unit comprises the DP71 engine, bow and stern reducers to drive the superchargers, and transmission shafts.

The basic characteristics of the M35 power plant have the following values: total power (5×DP71) of 36,800 kW (5×7360), specific fuel consumption of 258 g/(kW h), GTE weight of 2500 kg, and specific weight of 0.680 kg/kW.

The air cushion ship "Zubr" can well be used as a basis for the development of a high-speed cargo and passenger ship. There have been developed projects of its multiple-version modernization.

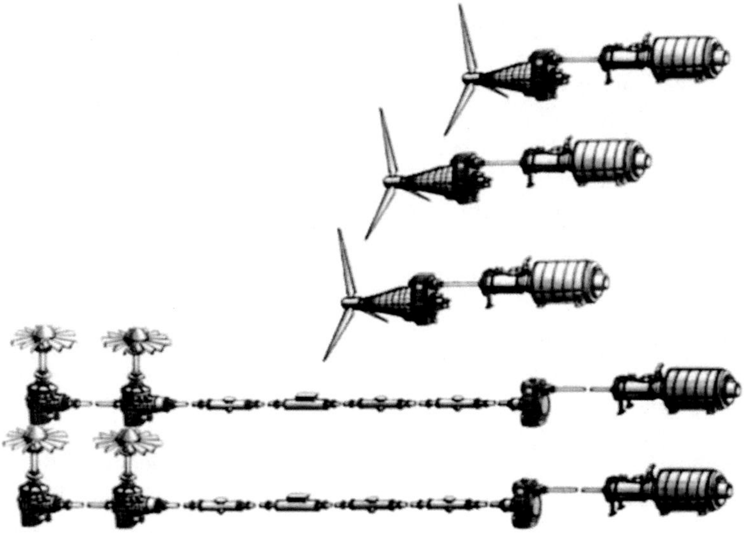

Fig. 6.42 Power plant M35 of the "Zubr" ship

6.4.3 Gas Turbine Plants of Transport Displacement Ships

Application of gas turbine plants on transport ships over the 1960s and 1970s made it possible to elaborate design solutions for the marine power plant elements associated with the GTE use as a heat engine, as well as to accumulate experience in their operation.

After a drastic increase in fuel prices, displacement ships with the main GTEs of the efficiency not exceeding 30% became unprofitable. At the end of the 1970s, a number of decisions were made regarding the replacement of GTPs with more economical plants with ICE on many ships, such as "Euroliner", "Finnjet", and a series of bulk carriers, tankers, ferries, and ro-ro ships with the regenerative GTE MS-3002R produced by General Electric [1].

One of the most effective ways to improve the efficiency of the MPP with GTE is to recover the exhaust gas heat in the steam-water circuit.

The exhaust gases of a simple-cycle GTE have the temperature of 400–520 °C, which is close to the initial temperature of steam in the marine steam turbine plant, which allows obtaining steam of high parameters in the RB.

The use of the exhaust gas heat in the recovery boiler is more effective in a GTP than in a diesel plant. The plant power being equal, the exhaust gas temperature is much higher and its amount is much larger in the gas turbine engine than in the ICE. The excess air factor is 2.0–3.2 for the combustion engine and 3–5 for the GTE.

As a rule, the amount of steam generated in the engine's HRC is sufficient to drive the recovery power steam turbine, which serves the propeller through a reducer common with the GTE, as well as to drive the auxiliary steam turbine generator and meet ship service needs. Marine power plants with a gas turbine engine and a steam turbine engine serving a propulsive unit are called gas-steam turbine plants (GSTP).

The power capacity of the recovery power turbine can reach up to 25% of the GTE capacity without additional fuel consumption [1, 3].

Gas turbine plant of the ro-ro ships "Atlantic". Ships with the displacement of 35,000 tons and the speed of 25 knots (the "Kapitan Smirnov" lead ship) have been equipped with the GSTP where the GTE and the steam turbine transmitted power to the propeller through a combining reducer. The two-shaft plant with two identical GSTPs allows for their simultaneous or autonomous use, making it possible to increase the survivability and economic efficiency of the MPP. For example, there is an operating mode when the GTE and RB work only on one side of ship, and the steam is fed to the RST of the other side, which rotates a separate propeller. At this mode, fuel consumption is reduced by 35–40%, while the ship speed is only decreased by 20–25% [1].

In the power plants installed on the ro-ro ships of the "Atlantic" class, the main gas-steam turbine unit M-25 serves two controllable-pitch propeller shafts (Fig. 6.43).

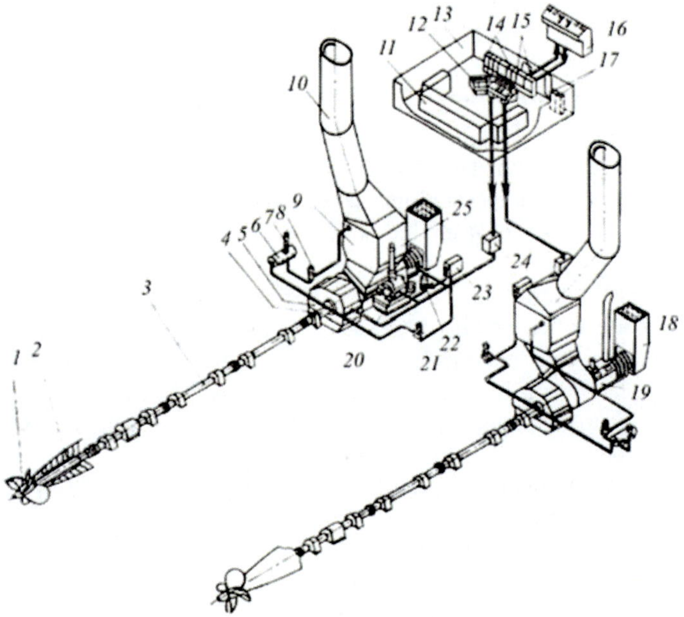

Fig. 6.43 Layout of the propulsive gas turbine plant of gas turbine ships of the "Kapitan Smirnov" class: *1*—propeller; *2*—stern tube; *3*—shafting; *4*—thrust bearing; *5*—reducer; *6*—steam separator; *7, 25*—steam supply lines to the RB and the ST, respectively; *8, 21, 22*—circulating, feed, condensate pumps, respectively; *9*—recovery boiler KUP-3100; *10*—gas outlet pipe; *11*—main shield; *12, 16*—control panels in the central control panel (CCP) and in the wheelhouse, respectively; *13*—CCP; *14, 15*—boards of the automated control and monitoring systems, respectively; *17*—control complex for ship service systems; *18*—air intake shaft; *19*—GTE D59; *20*—recovery steam turbine STP-2; *23*—warm box; *24*—GTE control station

This unit includes a GTE D59, a heat recovery boiler KUP-3100, a steam turbine STP-2 with a condenser and an ejector, and a main reducer.

The steam that is generated in the recovery boiler is used in the propulsive recovery steam turbine and the UTG-1000 recovery steam turbine generator. It amounts to 6 t/h of superheated steam with the pressure of 1.2 MPa and the temperature of 310 °C and 2.25 t/h of saturated steam with the pressure of 1.2 MPa. Then it is directed to cover the needs of general ship services.

The main characteristics of the gas-steam turbine plant of the ro-ro ship "Kapitan Smirnov" are listed further on in Table 6.5 [1].

The DI59 gas turbine engine is direct-flow, two-cascade, with a free power turbine. It consists of two consecutive axial compressors (a seven-stage low-pressure and a nine-stage high-pressure compressor), a tubular-annular combustion chamber with 10 flame tubes, and turbines (two-stage high- and low- pressure turbines, a reverse three-stage power turbine). The high-pressure turbine drives the HPC, the low-pressure turbine drives the LPC, and the power turbine transmits the torque to the reducer. The main three-stage reducer totalizes the power transferred from the GTE and STP-2.

The GTE power is split at the first stage of the reducer and is further divided into four flows at the second stage. The power of the steam turbine is split in the second stage. The engine is connected to the reducer via an elastic disconnection coupling. The steam turbine is connected to the reducer via a friction claw coupling, which is automatically disconnected when the ST is withdrawn from operation. The torque from the GTE and RST is transmitted via two paths of six driving gears to the driven reducer wheel, which is connected to the ship shafting. Above the GTE gas outlet, there is a rectangular water-tube recovery boiler with multiple forced circulation.

The boiler includes an economizer, an evaporative bundle of tubes, and a superheater. All bundles of tubes of the RB are finned. A gas-air ejector located behind the boiler is used to remove the air coming under the heat-insulating casing of the gas turbine engine, cooling it and flowing through the inter-casing space of the boiler. The recovery boiler is provided with a horizontal separator, which at the same time serves as the HRC water tank.

The steam turbine plant STP-2 (Fig. 6.44) is arranged as a single unit. The steam turbine housing is mounted on a two-pass condenser, which serves as a frame and accommodates auxiliary equipment of the STP. The rotor of the steam turbine is made with two speed stages and seven pressure stages.

The auxiliary electric power plant (Fig. 6.45) includes two RTGs with the capacity of 1000 kW each (operating on superheated steam from the recovery boilers of HRC) and three automated generators DGR1000/750 with the power of 1000 kW at 750 rpm (driven by four-stroke diesel engines 6ChN30/38). The system of marine powerstation management provides for the parallel operation of diesel and turbogenerators for any arrangement of the unit.

The emergency power source on the ship is a DGRA-200/1500G emergency diesel generator with the power of 200 kW at 1500 rpm.

During the ship movement, general ship consumers are provided with steam with the help of the recovery boiler; during parking, the auxiliary boiler serves for this

Table 6.5 Characteristics of the M-25 unit

Characteristics	Mode	
	Maximum (without steam extraction)	Nominal (with steam extraction for the turbogenerator and ship service needs)
Forward stroke power, MW, incl.:	18.39	17.28
GTE	14.12	
RTG	4.27	3.16
Specific fuel consumption of GTE, g/(kW h)	238	256
Rotation speed of the output shaft of the reducer, rpm	130	128
Maximumreversepower, MW	–	5.74
Steam capacity of RB, t/h, incl.: superheatedsteam saturatedsteam	26.2 24.0 2.2	
Steam pressure in RB, MPa	1.5	
Temperature, °C:		
steam in RB	310	
exhaust gases before RB	390	
gases behind RB	179	
Gas consumption in RB, kg/s	91.8	
Weight in operating condition, t, incl.: GTE with frame, gas outlet pipe, oil tank Reducer with coupling RB	149 19 55 45	
Overall length (from the GTE inlet device to the output flange of the reducer), m	12,245	
Width, m	5155	
Height, m, incl.: with RB with outlet device	– 10,875 12,525	

purpose. Its steam production capacity makes up 6 t/h; saturated steam is generated at the pressure of 0.5 MPa.

The adiabatic desalination plant M3 has the capacity of 60 t/day and works on the steam from the recovery or auxiliary boiler.

The compressor station of compressed air includes two EKSA-7.5–4 automated high-pressure compressors with the 13 l/min supply of compressed air at the pressure of 20 MPa, two EPA 70/25 medium-pressure electrocompressors with the supply

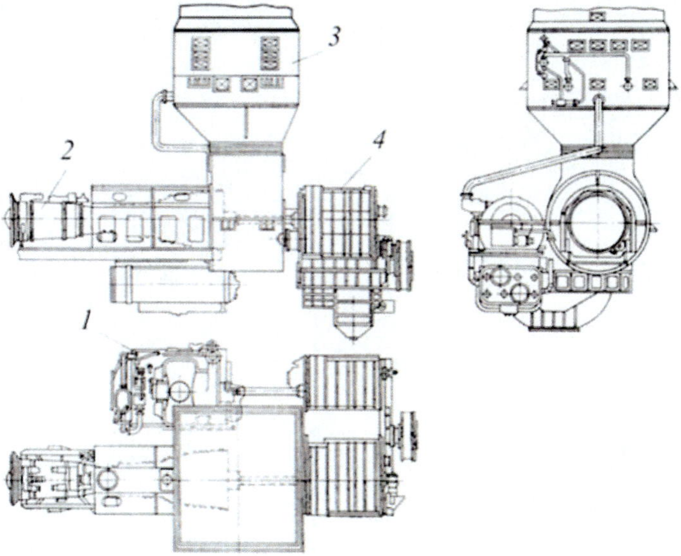

Fig. 6.44 Layout of the gas turbine unit M-25: *1*—STP with a condenser; *2*—GTE DI-59; *3*—recovery boiler KUP-3100; *4*—main reducer

of 70 m^3/h at the air pressure of 2.5 MPa, the required number of compressed air cylinders of high- and medium-pressure, and high-pressure air cleaning and dehumidifying devices.

The power plant is fitted with appropriate mechanisms and equipment that ensures normal operation of the gas turbine unit, diesel and turbine generators, auxiliary boiler, oil and fuel cooling and cleaning systems [1].

Combined marine power plants with an afterburner. Among combined MPPs, a widespread application was acquired by diesel-gas turbine plants. This is explained by the fact that diesel engines are the most economical in terms of fuel consumption in a wide range of loads, and their use as sustainer engines allows covering a considerable range of navigation.

Power plant of icebreakers of the "Polar Star" class. Typical examples of such plants are the power plants of the US Coast Guard icebreakers of the "Polar Star" class with the length of 126.6 m and displacement of 10,800 tons (Fig. 6.46) [1].

The three-propeller MPP of the icebreaker consists of six sustainer diesel generators with the total power of 14,700 kW and three FT4A-2 afterburner GTEs with the total power of 44,000 kW. Each propeller can be powered by two diesel generators of the basic plant producing alternating current, which is supplied to the propulsive engine through a thyristor rectifier or to one afterburner GTE via a gear train.

During electric propulsion (only diesel engines work), the icebreaker passes through an ice field with the thickness of up to 1.2 m. When overcoming ice fields of a greater thickness, GTEs are turned on, and diesel generators are turned off. In this case, at the speed of 3 knots, the icebreaker overcomes ice 1.8 m of ice thickness with

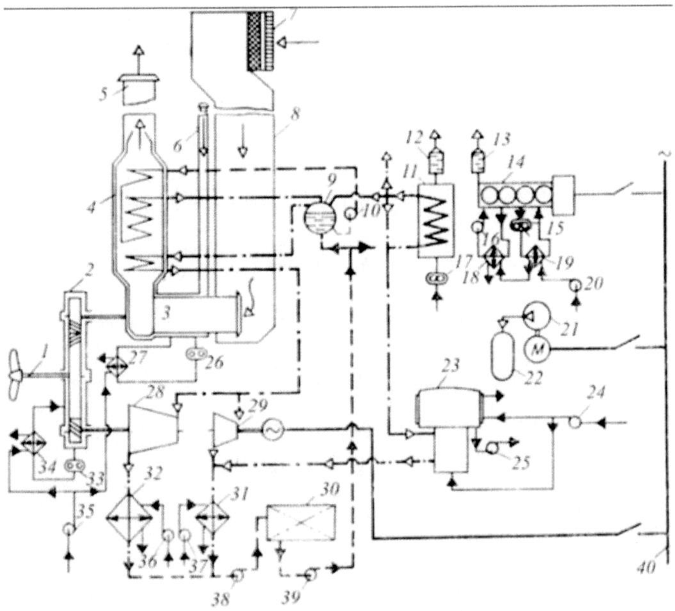

Fig. 6.45 Principal scheme of a marine gas-steam turbine plant: *1*—ship shafting; *2*—main reducer; *3*—GTE; *4*—RB; *5*—gas discharge; *6*—air supply for cooling the GTE and RB casings; *7, 8*—air inlet device and duct; *9*—steam separator; *10*—circulating pump for the boiler water; *11*—AB; *12*—AB spark arrestor;*13*—DE noise suppressor; *14*—DE; *15, 26, 33*—oil pumps of the DE, GTE and reducer; *16*—cooling fresh water pump of DE; *17*—AB fuel pump; *18, 19*—fresh water and oil cooler, respectively; *20, 35*—cooling seawater pumps; *21*—compressor; *22*—compressed air cylinder; *23*—water desalination plant; *24*—seawater pump; *25*—condensate pump of the water desalination plant; *27*—GTE oil cooler; *28*—STP; *29*—turbogenerator; *30*—warm box; *31, 32*—condensers of the TG and STP, respectively; *34*—cooler of the reducer oil; *36, 37*—circulating pumps of the STP and TG condensers; *38*—condensate pump; *39*—feed pump; *40*—power supply

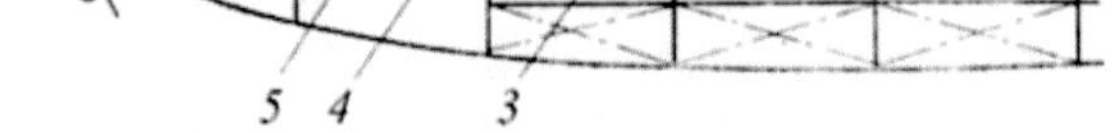

Fig. 6.46 Layout of the power plant of the "Polar Star" icebreaker: *1*—main diesel generators of the sustainer plant; *2*—auxiliary DGs; *3*—afterburner GTE; *4*—reducers of the afterburner GTEs; *5*—propulsive electric engines of the sustainer plant; *6*—fuel tanks

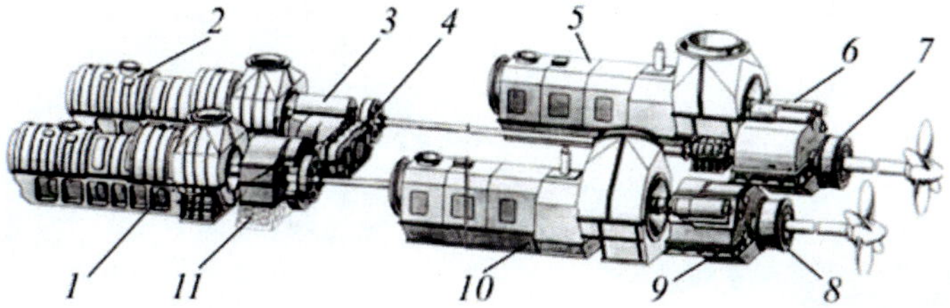

Fig. 6.47 The main gas turbine unit M7: *1,2*—sustainer M60 GTEs; *3, 11*—sustainer reducers; *4*—inter-reducer attachment; *5, 10*—full-stroke DO54R GTE; *6, 9*—full-stroke reducers; *7, 8*—sound-insulating couplings

continuous movement and up to 6.4 m of ice thickness when moving in successive strokes.

Combined plants with afterburners have found a wider application on the naval ships.

Let us consider the schemes of combined plants installed on military ships of various purposes. Their gas turbine units have been produced by UDBMashproekt—STP Zorya (Ukraine).

The main gas turbine unitM7 (Fig. 6.47) has been developed for the large anti-submarine ship "Burevestnik" with the displacement of 3200 tons and the speed of 32.5 knots.

The unit structure is as follows [2]:

- two reverse sustainer gas turbine engines M60 with the capacity of 4400 HPC;
- two sustainer two-speed reducers;
- sustainer reducer attachment, which connects sustainer reducers to each other;
- two reverse full-stroke DO54P GTEs the with the power of 14,700 kW each;
- two full-stroke reducers with high-speed tire-pneumatic couplings.

The maximum power of the unit is 34200 kW.

This unit employs an interesting solution regarding power transmission. The output shaft of the afterburner reducer is connected to the shafting from the sustainer reducer by means of the tire-pneumatic coupling. The main thrust bearing is located on the shaft. The shafts of the sides of the ship are connected with the help of the sustainer inter-reducer attachment, which allows a single sustainer engine to serve two propellers.

The main gas turbine unitM5 was designed by the SPB Mashproekt for the large anti-submarine ship of the "Berkut" project. The ship displacement (8500 tons) is much greater than that of the "Burevestnik". Two units were installed on the ship.

Each unit consists of a sustainer system and a full-stroke plant that serve a common propeller shaft (Fig. 6.48). The sustainer system comprises a reverse engine M60 with the capacity of 4400 HPC and a two-speed sustainer reducer.

The full-stroke plant is designed to provide a high speed. It consists of two DO54 (M3A) main engines and a combining main reverse reducer; the engines have the power of 14,700 kW and are located on one frame.

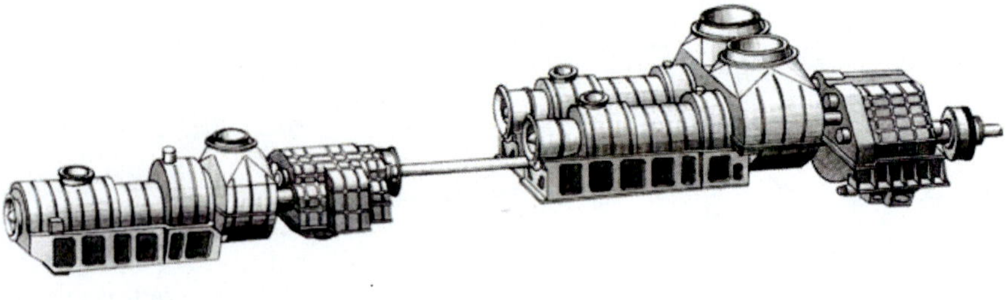

Fig. 6.48 Gas turbine unit M5

With the power of each main GTU being 33,800 kW, the total power of the marine power plant is 67600 kW. This is the most powerful marine power GTP.

The main gas turbine unit M15 was designed for the missile boat "Molniya" (displacement of 500 tons, speed of 43 knots).

The M15 unit consists of the sustainer plants and the full-stroke plants. The former include two DO76 sustainer reverse engines *1, 9* of the right and left ship boards, respectively, as well as two sustainer planetary two-speed reducers *2, 7*. The latter are two DO77 reverse GTEs *4, 5* of the right and left ship boards, as well as two angular reducers *3, 6* (Fig. 6.49) [88].

Sustainer reducers can work both simultaneously with the full-stroke plant and autonomously, since the plant is generally classified as a COGAG one.

Sustainer reducers of both ship boards are connected with each other by means of transferring shaft *8*, which allows transmitting the power of any sustainer engine to two propeller shafts, providing the economic stroke of the ship.

The full-stroke reducer is two-stage, with a split power and an integrated main thrust bearing which perceives the thrust from the propeller.

The maximum power of M15 is 24300 kW.

The main GTU allows reversing at any number of engines at a power not exceeding 40% of the nominal one, connecting and disconnecting any GTE without reducing the speed of the ship, and enabling the "propeller stop" mode when the turbine propellers have zero torque.

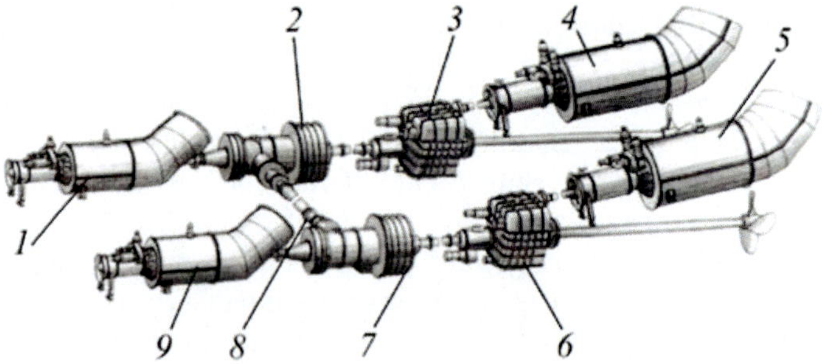

Fig. 6.49 Main gas turbine unit M15

The main gas turbine unitM27 is designed for use in the COGAG gas turbine power plant installed on the ship of the "Yastreb" project. Its design follows the scheme of the M7 unit. The unit is fitted with the third-generation GTE DS71 and DO90 of an increased resource [2].

The main characteristics of the M27 unit:

Power at full forward stroke, HPC (kW)	41,800
Full sustainer power, HPC (kW)	13,800
Full reverse power, HPC (kW)	8000

Resource, h, incl.:

Sustainer GTE (kW)	30,000
Afterburner GTE (kW)	20,000
Sustainer reducers (kW)	40,000
Afterburner reducers (kW)	20,000

The M27 unit was manufactured at the PA Zorya in 1988.

The main gas turbine unitM21 was designed for the "Atlant" ship (displacement of 11,500 tons, speed of 32.5 knots).

The main GTE M21 includes a sustainer plant and a full-stroke plant.

The full-stroke plant consists of two all-mode gas turbine engines DT59 and a reducer with a soundproof coupling.

The sustainer plant includes the all-mode reverse GTE DS71, a two-stage reducer with a soundproof coupling and a steam HRC, which uses the heat of the exhaust gases from the sustainer engine (Fig. 6.50).

The heat recovery circuit consists of a recovery steam boiler and an ST with a condenser. The steam boiler is located on the outlet pipe of the sustainer engine. Parameters of the superheated steam produced by the boiler are as follows: temperature of 310 °C, pressure of 1.1 MPa, steam capacity of 10.8 t/h.

The sustainer GTE and the steam turbine transmit rotation through springs to a common sustainer reducer, which is connected to the driven part of the soundproof coupling of the full-stroke reducer [2].

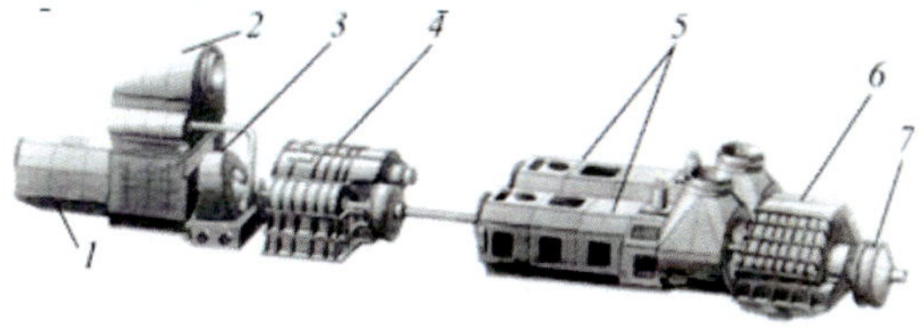

Fig. 6.50 The main gas turbine unit M21: *1*—sustainer reverse GTE DS71; *2*—recovery steam boiler; *3*—ST; *4*—sustainer reducer; *5*—full-stroke GTE DT59; *6*—full speed reducer; *7*—soundproof coupling

Both the sustainer and the full-stroke gas turbine plants operate in the full-stroke mode. The power plant of the ship belongs to the COGAG category. The total power of the main GTU in the full-stroke mode is 40000 kW, the specific fuel consumption of the sustainer plant is 238 g/(kW h), and the specific fuel consumption of the full-stroke plant is 300 g/(kW h). The resource of the sustainer GTE is 10000 h, while the resource of the full-stroke GTE is 5000 h.

The lead ship "Glory" was manufactured at the 61 Communards Shipyard in Mykolaiv. Equipment of the main GTU M21 was manufactured and delivered by the PA Zorya (the GTE and reducers), the SPE Mashproekt (steam turbine with a condenser), and the Black Sea Shipyard (recovery boiler) in June 1981. The sea trials were completed in June 1982.

Diesel-gas turbine units. The missile boat "Molniya-1" (speed of 41 knots, displacement of 470 tons) is fitted with the M15A unit, which includes sustainer DRUs with diesel engines with the power of 2950 kW (Fig. 6.51). The full stroke of the ship is provided by the operation of the DRU and the CODAG gas turbine engines DS77. The total power of the full stroke is 28500 kW [84].

The diesel-gas turbine unit M44 employs a reverse diesel with the power of 5500 kW as a sustainer diesel engine; it serves two propellers via an inter-reducer attachment and reducers (Fig. 6.52) [2].

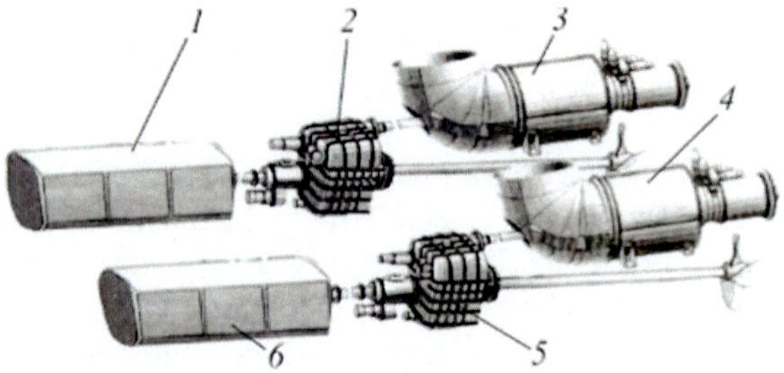

Fig. 6.51 Main gas turbine unit M15A: *1, 6*—sustainer diesels; *2, 5*—reducers; *3, 4*—full-stroke GTE DS77

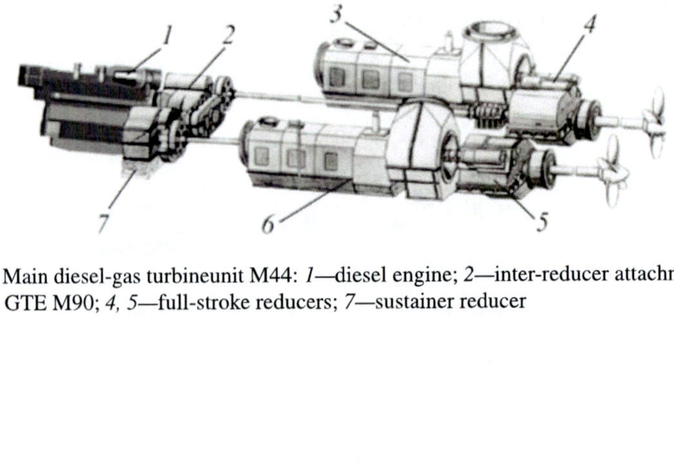

Fig. 6.52 Main diesel-gas turbineunit M44: *1*—diesel engine; *2*—inter-reducer attachment; *3, 6*—full-stroke GTE M90; *4, 5*—full-stroke reducers; *7*—sustainer reducer

The full stroke of the ship is provided by two reverse gas turbine engines DO90, each of which serves its propeller through a reducer. When the full-stroke engines are running, the diesel does not work (CODOG scheme). The total power of the full stroke is 35500 kW.

6.5 Characteristics of Marine Gas Turbine Engines of the World's Leading Manufacturers

Table 6.6 lits the main world manufacturers of marine gas turbine engines. The table data indicate that Ukraine certainly ranks in the top five, since the standard series of the gas turbine engines produced by Zorya-Mashproekt (Ukraine, Mykolaiv) fully covers the power range from 2.5 to 25 MW, there are developed GTEs with the power of 40 and 60 MW, and the company has even mastered the production of the power engineering engine of 110 MW. An exceptional advantage of the GTEs manufactured by this enterprise is that all of them have been created specifically for marine application.

As of 1990, 37% of the world's number of ship GTEs was manufactured in the Soviet Union. Taking into account the fact that the STP in Mykolaiv was the only enterprise in the USSR producing ship GTEs, we come to a conclusion that all these engines were produced in Mykolaiv.

Table 6.7 presents the characteristics of ship GTEs made by the main world manufacturers.

Table 6.6 Manufacturers of marine gas turbine engines [4]

Firm	Country	Power range of the GTE, kW
GE Marine	The USA	4420; 14,100; 24,300; 29,800; 42,200
Ishikawajima—HarimaHeavyIndustries	Japan	4100; 4150; 6380; 7730; 14,100; 24,300; 29,800; 42,200
KawasakiHeavyIndustries		3930; 12,600; 15,800; 17,780
MAN Turbo AG	Germany	11,050; 24,600
MitsubishiHeavyIndustries	Japan	24,460
MTU FriedrichshafenGmb	Germany	14,100; 24,300; 28,700
Pratt&WhitneyCanada	Canada	596; 724; 1940; 3990
Pratt&WhitneyPowerSystems	The USA	24,540; 27,100
Rolls-Royce	England	3900; 4440; 19,250; 24,900; 35,800
Turbomeca	France	1180
Vericor	The USA	2940; 3650; 3750
Zorya-Mashproekt	Ukraine	3120; 3310; 6260; 6700; 8320; 10,670; 14,700; 15,600; 16,000; 17,000; 20,200; 26,700

Table 6.7 Characteristics of marine gas turbine engines of the leading world manufacturers

GTE characteristics	GE Marine, the USA					Ishikawajima—Harima Heavy Industries, Japan				
	LM500	LM1600	LM2500	LM2500 +	LM6000 PC	501–KF5	601–KF9	601–KF11	LM500	160M500
Year of the serial production start	1980	1987	1969	1998	1997	1986	1997		1986	1987
Power accord. to ISO, kW:										
Maximum	–	14,700	–	–	–	–	–	–	–	–
Rated	4200	14,100	24,300	29,800	42,200	4150	6380	7730	4100	14,100
Specific fuel consumption, kg/(kW h):										
At maximum power	–	0.231	–	–	–	–	–	–	–	–
At rated power	0.285	0.236	0.231	0.220	0.204	0.312	0.257		0.280	0.236
Efficiency coefficient, %	29.5	35.7	36.5	38.3	41.3	27.0	32.8		30.1	35.7
Temperature of gases, °C:										
At the HPT inlet	–	1243		–	–	–	–	–	–	–
At the power turbine inlet	785	763	872	842	845	–	–	–	–	–

(continued)

Table 6.7 (continued)

GTE characteristics	GE Marine, the USA					Ishikawajima—Harima Heavy Industries, Japan				
	LM500	LM1600	LM2500	LM2500 +	LM6000 PC	501–KF5	601–KF9	601–KF11	LM500	160M500
Year of the serial production start	1980	1987	1969	1998	1997	1986	1997		1986	1987
At the engine outlet	565	514	570	522	459	585	529	504	565	504
Rate of pressure increase	14.4	21.4	19.3	22.2	28.5	10.1	15.0	19.8	14.4	21.4
Mass air consumption, kg/s	16.3	47.2	70.3	85.8	124.0	15.6	23.6	25.6	15.9	46.3
Power turbine rotation speed, rpm	7000		3600			14,200	11,500		7000	
Engine dry weight, kg	904	3720	4680	5240	8220	1134	1202	1600	610	3425
Dimensions, m:										
Length	2.9	4.2	6.7	6.7	7.3	2.7	2.0	2.4	2.1	4.6
Width	0.9	2.0	2.0	2.0	2.5	1.4	0.8	0.8	1.0	2.0
Height	0.9	2.0	2.0	2.0	2.5	1.4	1.0	1.0	1.0	2.1

(continued)

Table 6.7 (continued)

CGTE characteristics	Ishikawajima—Harima Heavy Industries, Japan			Kawasaki Heavy Industries, Japan				MAN Turbo AG, Germany		MHI, Japan
	LM2500	LM2500 +	LM6000 PC	Tyne RMIC	Olympus TM3B	Spey SM1A	Spey SM1C	THM 1304–11	FT 8 Marine	MFT 8
Year of the serial production start	1969	1997		1975	1973	1980	1987	1999	1990	1994
Power according to ISO, kW:										
Maximum	–	–	–	–	20,600	13,800	19,250	11,600	27,100	27,000
Rated	24,300	29,800	42,200	3930	15,800	12,600	17,770	11,050	24,600	25,480
Specific fuel consumption, kg/(kW h):										

(continued)

Table 6.7 (continued)

CGTE characteristics	Ishikawajima—Harima Heavy Industries, Japan			Kawasaki Heavy Industries, Japan				MAN Turbo AG, Germany		MHI, Japan
	LM2500	LM2500 +	LM6000 PC	Tyne RMIC	Olympus TM3B	Spey SM1A	Spey SM1C	THM 1304–11	FT 8 Marine	MFT 8
Year of the serial production start	1969	1997		1975	1973	1980	1987	1999	1990	1994
St maximum power	–	–	–	–	0.296	0.214	0.230	0.276	0.240	0.222
At rated power	0.231	0.220	0.204	0.291	0.318	0.244	0.232	0.278	0.224	0.223
Efficiency coefficient, %	36.5	38.3	41.3	28.9	26.5	34.5	36.3	30.3	37.6	37.8
Temperature of gases, °C:										
At the HPT inlet	–	–	–	–	–	–	–	995	–	–
At the power turbine inlet	–	–	–	891	855	880	908	–	–	–
At the engine outlet	557	518	456	442	433	408	461	509	465	470
Rate of pressure increase	19.3	22.2	28.5	12.5	10.5	18.5	21.9	11.0	18.0	20.4

(continued)

Table 6.7 (continued)

CGTE characteristics	Ishikawajima—Harima Heavy Industries, Japan			Kawasaki Heavy Industries, Japan				MAN Turbo AG, Germany		MHI, Japan
	LM2500	LM2500 +	LM6000 PC	Tyne RMIC	Olympus TM3B	Spey SM1A	Spey SM1C	THM 1304–11	FT 8 Marine	MFT 8
Year of the serial production start	1969	1997		1975	1973	1980	1987	1999	1990	1994
Mass air consumption, kg/s	69.9	85.8	124.0	21.0	108.5	58.3	66.7	49.0	83.3	84.8
Power turbine rotation speed, rpm	3600			13,970	5660	5220	5500	8600	3600	5000
Engine dry weight, kg	4760	5240	7303	14,070	30,870	25,480	25,660	9000	8600	6630
Dimensions, m:										
Length	6.7		7.3	5.6	9.2	7.5	7.5	5.6	8.2	7.6
Width	2.1		2.5	2.1	2.6	2.3	2.3	2.4	2.4	2.4
Height	2.1		2.5	2.1	2.6	3.4	3.4	3.1	2.4	2.4

(continued)

Table 6.7 (continued)

	MTU Friedrichshafen GntBH, Germany			Pratt & Whitney Canada, Canada				Pratt & Whitney Power Systems, the USA	
	LM1600	LM2500	LM2500+	STGL–794	STGL–812	ST–18A	ST–40	FT 8	FT 8–3
Year of the serial production start	1987	1969	1998	1976		1995	1999	1990	
Power according to ISO, kW:									
Maximum	–	–	–	666	830	2260	4840	27,190	29,250
Rated	14,100	24,300	28,700	596	724	1940	3990	24,540	27,106
Specific fuel consumption, kg/(kW h):									
At maximum power	–	–	–	0.348	0.329	0.274	0.258	0.221	0.220
At rated power	0.236	0.231	0.221	0.355	0.341	0.284	0.262	0.224	0.222
Efficiency coefficient, %	35.7	36.5	38.1	23.7	24.7	29.7	32.1	37.6	38.0
Temperature of gases, °C:									
At the HPT inlet	1243		–	–	–	–	–	–	–

(continued)

Table 6.7 (continued)

	MTU Friedrichshafen GntBH, Germany			Pratt & Whitney Canada, Canada				Pratt & Whitney Power Systems, the USA	
	LM1600	LM2500	LM2500+	STGL–794	STGL–812	ST–18A	ST–40	FT 8	FT 8–3
Year of the serial production start	1987	1969	1998	1976		1995	1999	1990	
At the power turbine inlet	763	872	833	–	–	–	–	–	–
At the engine outlet	509	560	526	565	546	536	548	465	485
Rate of pressure increase	21.4	19.3	22.2	7.0	8.2	14.0	16.9	18.8	19.7
Mass air consumption, kg/s	46.3	70.3	82.6	3.1	3.7	8.0	13.9	83.3	86.2
Power turbine rotation speed, rpm	7000	3600	360	33,000		18,900	14,875	3000 3600 5500	3000 3600 5500
Engine dry weight, kg	3430	4770	5080	104	136	350	525	8620	–
Dimensions, m:									
Length	4.6	6.7	7.02	1.25	1.30	1.53	1.71	5.7	–
Width	2.0	2.1	2.1	0.43	0.43	0.67	0.67	2.4	–
Height	2.1	2.1	2.1	0.55	0.55	0.82	0.98	2.4	–

(continued)

Table 6.7 (continued)

GTE characteristics	Rolls-Royce, England					Turbomeca, France	Vericor, the USA		
	AG9140	RR4500	Spey	WR21	MT30	MakilaT1	601-KF9	601-KF11	LM500
Year of the serial production start	1986	2004	1987	1997	2001	1988	1997		1986
Power according to ISO, kW:									
Maximum	–	–	–	–	–	1278	–	–	–
Rated	3900	4440	19,250	24,900	35,800	1180	6380	7730	4100
Specific fuel consumption, kg/(kW h):									
At maximum power	–	–	–	–	–	0.304	–	–	–
At rated power	0.292	0.266	0.230	0.204	0.216	0.317	0.257		0.280
Efficiency coefficient, %	28.5	31.7	36.7	41.3	39.0	26.5	32.8		30.1
Temperature of gases, °C:									

(continued)

Table 6.7 (continued)

GTE characteristics	Rolls-Royce, England					Turbomeca, France	Vericor, the USA		
	AG9140	RR4500	Spey	WR21	MT30	MakilaT1	601-KF9	601-KF11	LM500
Year of the serial production start	1986	2004	1987	1997	2001	1988	1997		1986
At the HPT inlet	–	–	–	–	–	–	–	–	–
At the power turbine inlet	–	–	–	851	–	–	–	–	–
At the engine outlet	556	519	461	358	478	515	529	504	565
Rate of pressure increase	11.3	14.3	21.9	16.2	24.0	9.6	15.0	19.8	14.4
Mass air consumption, kg/s	15.6	20.9	66.7	73.1	116.7	5.6	23.6	25.6	15.9
Power turbine rotation speed, rpm	14,340	14,600	5500	3600	3300 3600	6300	11,500		7000
Engine dry weight, kg	–	–	–	54,430	–	440	1202	1600	610
Dimensions, m:									
Length	–	–	–	7.96	–	1.9	2.0	2.4	2.1
Width	–	–	–	2.6	–	0.7	0.8	0.8	1.0
Height	–	–	–	4.5	–	0.7	1.0	1.0	1.0

(continued)

Table 6.7 (continued)

GTE characteristics	Zorya-Mashproekt Gas Turbine Research and Development Complex, Ukraine											
	UGT 3000	UGT 3000R	UGT 6000	UGT 6000R	UGT 6000 +	UGT 10,000	UGT 15,000	UGT 15000R	UGT 15,000 +	UGT 16,000	UGT 16000R	UGT 25,000
Year of the serial production start	1981		1978		1997	1998	1988			1982		1993
Power according to ISO, kW:												
Maximum	4120	3900	8390	7800	9570	–	18,700	17,700	22,100	20,400	19,100	31,200
Rated	3310	3120	6700	6260	8320	10,670	15,600	14,700	20,200	17,000	16,000	26,700
Specific fuel consumption, kg/(kW h):												

(continued)

Table 6.7 (continued)

GTE characteristics	Zorya-Mashproekt Gas Turbine Research and Development Complex, Ukraine											
	UGT 3000	UGT 3000R	UGT 6000	UGT 6000R	UGT 6000 +	UGT 10,000	UGT 15,000	UGT 15000R	UGT 15,000 +	UGT 16,000	UGT 16000R	UGT 25,000
Year of the serial production start	1981		1978		1997	1998	1988			1982		1993
At maximum power	0.260	0.287	0.266	0.281	0.245	–	0.238	0.251	0.235	0.257	0.273	0.225
At rated power	0.273	0.300	0.269	0.289	0.255	0.238	0.245	0.258	0.251	0.266	0.282	0.235
Efficiency coefficient, %	31.0	28.5	31.5	29.4	33.0	36.0	34.6	32.9	36.0	31.9	29.2	36.4
Temperature of gases, °C:												
At the HPT inlet	1020		1015		1092	1184	1035		1160	865		1245
At the power turbine inlet	–	–	–	–	–	–	–	–	–	–	–	–
At the engine outlet	420	428	420	428	442	458	405	410	470	345	355	465
Rate of pressure increase	13.7		13.9		15.7	19.6	18.6		19.4	12.8		21.0

(continued)

Table 6.7 (continued)

GTE characteristics	Zorya-Mashproekt Gas Turbine Research and Development Complex, Ukraine											
	UGT 3000	UGT 3000R	UGT 6000	UGT 6000R	UGT 6000 +	UGT 10,000	UGT 15,000	UGT 15000R	UGT 15,000 +	UGT 16,000	UGT 16000R	UGT 25,000
Year of the serial production start	1981		1978		1997	1998	1988			1982		1993
Mass air consumption, kg/s	16.5		31.0		33.6	37.4	69.0		74.0	99.0		87.6
Power turbine rotation speed, rpm	700	8800	3000 5300 7000 9300	4750 7200	7700	4900	5100	4400	3500	3900	3500	3300
Engine dry weight, kg	2500	2800	3500	3800	3500	4200	9000	9800	9000	16,000	16,900	14,000
Dimensions, m:												
Length	2.5	2.7	3.2	3.4	3.2	3.6	5.0	5.2	5.0	5.9	6.2	6.4
Width	1.3	1.8	1.6	1.8	1.6	1.7	2.6	2.7	2.6	2.7	2.8	2.5
Height	1.3	1.8	1.8	1.9	1.8	1.9	2.8	3.0	2.8	3.1	3.2	2.7

References

1. Artemov G. A., Gorbov V. M., Romanovskiy G. F. *Sudovyeustanovki s gazoturbinnymidvigatelyami: ucheb. posobie* [Ship plants with gas turbine engines: textbook]. Nikolaev, UGMTU Publ., 1997. 233 p.
2. *Nikolaevskiegazoturbinnyedvigateliiustanovki. Istoriyasozdaniya/GP NPKG "Zorya"–"Mashproekt", Tsentr NIOKR "Mashproekt"* [Nikolaev gas turbine engines and plants. History of establishment/The State Enterprise of Scientific and Production Complex of Gas Turbine Building Industry Zorya-Mashproekt, Research and Development Center Mashproekt]. Nikolaev, Yug—Inform Publ., 2005. 304 p.
3. Romanovskyi H. F., Ipatenko O. Ya., Patlaichuk V. M. *Teoriia ta rozrakhunokparovykh ta hazovykhturbin :navch. posib.* [Theory and calculation of steam and gas turbines: textbook]. Mykolaiv, UDMTU Publ., 2002. 292 p.
4. Horbov V. M. *Entsyklopediiasudnovoienerhetyky: pidruchnyk* [Encyclopedia of marine power engineering: textbook]. Mykolaiv, NUK Publ., 2010. 624 p.
5. Horbov V. M. *Enerhetychnipalyva :navch. posib.* [Energy fuels: textbook]. Mykolaiv, UDMTU Publ., 2003. 328 p.
6. Pakhomov Yu. A. *Sudovyeenergeticheskieustanovki s dvigatelyamivnut-rennegosgoraniya: uchebnik* [Marine power plants with internal combustion engines: textbook]. Moscow, TransLit Publ., 2007. 528 p.
7. Belkind L. O., Konfederatov I. Ya., ShneybergYa. A. *Istoriyatekhniki* [History of technology]. Moscow, Gosenergoizdat Publ., 1956. 490 p.
8. Spitsyn V. Ye., Kharchenko V. I. *Budovahazoturbinnohodvyhunatypu UGT 2500E :navch. posib.: u 2 ch. Ch. 1. Konstruktsiiadvyhuna* [The structure of the gas turbine engine UGT 2500E: textbook: in 2 parts. Part 1. Engine construction]. Mykolaiv, NUK Publ., 2007. 52 p.

Appendix

Characteristics of Modern Ship Internal Combustion Engines

Z. Yang et al., *Marine Power Plant*,
https://doi.org/10.1007/978-981-33-4935-3

Enginemodel	Engine cycle	Number and arrangement of cylinders L—row; V—V-like	Cylinder diameter	Pistonstroke	Circulation frequency, rpm	Power, kW		*Specific fuel consumption, g /(kWh)*
			mm			Cylinder	Enging	
Caterpillar Marine Power Systems								
Cat 3056	4	6 L	100	127	2100–2600	15.5–25.5	93–153	*222.3–280.7*
Cat 3508	4	8 L	170	190	1200–1800	65.8–107.1	526–857	*216.8–217.4*
Cat 3512	4	12 V	170	190	1200–1800	75.0–108.8	900–1305	*211.8–225.9*
Cat 3516	4	16 V	170	190	1200–1800	74.7–102.5	1195–1640	*201.2–220.3*
Cat C280-6	4	6 L	280	300	900–1000	287.5–338.7	2300–2710	–
Cat C280-16	4	16 V	280	300	900–1000	281.5–338.7	–	–
MaK M20C	4	6, 8, 9 L	200	300	900–1000	170–190	1020–1710	*186–190*
MaK M25C	4	6, 8, 9 L	255	400	720/750	290–330	1800–2970	*183–184*
MaK M32C	4	6, 8, 9 L 12, 16 V	320	480	600	480–500	2880–4500	*177–179*
MaK M 43C	4	6, 7, 8, 9 L 12, 16 V	430	610	500/514	900–1000	5400–9000 18,800–16,000	*175–178*
Cummins Engine Company Ltd								
KTA 19-M4	4	6 L	159	159	2100	87	522	*220.0*
QSK 19-M	4	6 L	159	159	1800–2100	82.0–99.5	492–597	*218.8–230.5*

(continued)

(continued)

Enginemodel	Engine cycle	Number and arrangement of cylinders L—row; V—V-like	Cylinder diameter	Pistonstroke	Circulation frequency, rpm	Power, kW		*Specific fuel consumption, g /(kWh)*
			mm			Cylinder	Enging	
KTA 38-M1	4	12 V	159	159	1600–1800	55.9–68.4	671–821	*204.7–212.5*
KTA 50-M2	4	16 V	159	159	1600–1950	65.3–87.4	1044–1398	*204.6–213.0*
QSK 60-M	4	16 V	159	190	1600–1900	93.3–116.5	1492–1884	*204.1–207.9*
Daihatsu Diesel MFG Co Ltd								
M2G	4	6 L	120	150	1200–1800	9.3–15.2	56–91	*269–272*
M3SG	4	6 L	140	160	1200–1800	33.0–44.0	198–264	*238*
M5SG	4	6 L	145	160	1200–1800	36.1–51.3	220–308	*231*
5DK-20	4	5 L	200	300	720–900	122–160	610–800	*192*
6DK-20	4	6 L	200	300	720–900	133–173	800–1046	*192–193*
8DK-20	4	8 L	200	300	720–900	133–170	1065–1360	*192*
6DK-28	4	6 L	280	390	720–750	332	1990	*190*
16DK-28	4	16 L	280	390	720–750	313	5000	*189*
5DC-17	4	5 L	170	270	900–1000	97,0	486	*190–197*

(continued)

(continued)

Enginemodel	Engine cycle	Number and arrangement of cylinders L—row; V—V-like	Cylinder diameter	Pistonstroke	Circulation frequency, rpm	Power, kW		*Specific fuel consumption, g /(kWh)*
			mm			Cylinder	Enging	
8DC-32	4	8 L	320	400	720–750	481	3844	*183*
Hanshin Diesel Works Ltd								
6L42GSH	4	6 L	240	400	400	80	480	*201*
6LUS40	4	6 L	400	640	340	466	2794	*190*
6LF54A	4	6 L	540	850	235	674	4045	*184*
LH28	4	6 L	280	460	395	172	1029	*193*
LH30	4	6 L	300	600	300	221	1323	*192*
LH34L	4	6 L	340	640	300	270	1618	*186*
LH36L	4	6 L	360	670	250	294	1765	*184*
LH38L	4	6 L	380	760	250	368	2206	*184*
LH41L	4	6 L	410	800	225	405	2427	*179*
Makita Corporation								
M30M	4	6 L	300	480	315	184.2	1105	*196.0*
LS38L	4	6 L	330	640	290	245	1470	*189.0*
LS42L	4	6 L	420	840	227	416.7	2500	*183.0*

(continued)

(continued)

Enginemodel	Engine cycle	Number and arrangement of cylinders L—row; V—V-like	Cylinder diameter	Pistonstroke	Circulation frequency, rpm	Power, kW		*Specific fuel consumption, g /(kWh)*
			mm			Cylinder	Enging	
S26MC	2	4–12 L	260	980	250	400	1600–4800	*179.0*
L35MC	2	4–12 L	350	1050	210	650	2600–7800	*177.0*
S35MC	2	4–12 L	350	1400	173	740	2960–8800	*181.5*
L42MC	2	4–12 L	420	1360	176	995	3980–11,940	*177.0*
S42MC	2	4–12 L	420	1764	136	1080	4320–12,960	*180.4*
S46MC-C	2	4–8 L	460	1932	129	1310	5340–10,480	*174.3*
S50MC	2	4–8 L	500	2000	127	1580	6320–12,640	*174.3*
S50MC-C	2	4–8 L	500	2000	127	1580	6320–12,640	*174.3*
MAN B&W Turbo Diesel								
K98ME7-C7	2	6–14 L	980	2660	97	4510–6230	27,780–87,220	*168–174*
G95ME-C9	2	5–12 L	950	3480	70–80	4520–6870	22,800–82,440	*159–166*
S90ME-C10	2	5–12 L	900	3260	72–84	4180–6100	20,900–73,200	*160–166*
G80ME-C9	2	6–9 L	800	3720	58–72	2860–4710	17,160–43,390	*159–166*
S80ME-C9	2	6–9 L	800	3450	72–78	3330–4510	19,880–40,690	*160–167*
S70MC-C8	2	5–8 L	700	2800	73–91	2100–3270	10,050–26,160	*163–169*

(continued)

(continued)

Enginemodel	Engine cycle	Number and arrangement of cylinders L—row; V—V-like	Cylinder diameter	Pistonstroke	Circulation frequency, rpm	Power, kW		*Specific fuel consumption, g /(kWh)*
			mm			Cylinder	Enging	
L70ME-C8	2	5–8 L	700	2360	91–108	2300–3270	11,000–26,160	*164–170*
S60ME-C8	2	5–8 L	600	2400	84–108	1520–2380	7600–19,040	*163–169*
G60ME-C9	2	5–8 L	600	2790	72–87	1500–2680	7500–21,440	*160–167*
S50ME-C8	2	5–9L	500	2000	108–127	1130–1660	5650–14,940	*164–170*
G50ME-C9	2	5–9 L	500	2500	79–100	1080–1720	5100–15,480	*161–167*
S46MC-C8	2	5–8 L	460	1932	110–129	800–1380	4500–11,040	*170–174*
S40ME-C9	2	5–8L	400	1770	124–146	770–1136	3850–9080	*173–177*
S35ME-B9	2	5–8L	350	1580	127–167	530–670	2650–6980	*171–175*
S30ME-B9	2	5–8 L	300	1326	148–195	300–640	1950–5120	*173–176*
L/V51/60DF	4	6–9 L 12–18 V	510	600	500–514	975–1000	5850–18,000	*180,5–183*
L/V48/60B	4	6–9 L 12–18 V	480	600	500–514	1180	6800–20,710	*182–186*
L/V32/44CR	4	6–10L 12–20 V	280	320	720–750	600	3800–12,000	*172–179*
L21/31	4	6–9 L	210	310	1000	215	1290–1935	*192*
Mitsubishi Heavy Industries Ltd								
UEC37LA	2	4–8 L	370	880	158–210	280–520	1120–4160	*168–175*

(continued)

(continued)

Enginemodel	Engine cycle	Number and arrangement of cylinders L—row; V—V-like	Cylinder diameter	Pistonstroke	Circulation frequency, rpm	Power, kW		*Specific fuel consumption, g /(kWh)*
			mm			Cylinder	Enging	
UEC45LA	2	4–8L	450	1350	119–158	480–890	1920–7120	*163–170*
UEC52LA	2	4–8 L	520	1600	100–133	640–1180	2560–9440	*160–167*
UEC60LA	2	4–8 L	600	1900	83–110	840–1550	3360–12,400	*159–166*
UEC60LS	2	4–8 L	600	2200	75–100	950–1770	3800–14,160	*159–166*
UEC37LS11	2	5–8 L	370	1290	140–186	419–773	2095–6180	*169–175*
UEC50LS11/Eco	2	4–8 L	500	1950	95–127	780–1445	3120–11,560	*160–167*
UEC50LSE/Eco	2	5–8 L	500	2050	99–124	1060–1660	5300–13,280	*166–171*
UEC60LS11/Eco	2	4–8 L	600	2300	79–105	1110–2045	4440–16,360	*159–166*
UEC75LS11	2	4–9 L	750	2800	63–84	1595–2940	6380–26,460	*158–165*
UEC85LS11/Eco	2	5–9 L	850	3150	54–76	1980–3860	9900–34,740	*156–163*
MTU Friedrichshafen GmbH								
700	4	4, 6 L	94	100	3800	24–39	126–235	–
2000	4	8, 10, 12, 16 V	135	156	2250–2450	90–112	720–1790	–
4000	4	8, 12, 16 V	165	190	1500–2100	87–170	700–2720	–
595	4	12, 16 V	190	210	1750–1800	225–270	3240–4320	–

(continued)

(continued)

Enginemodel	Engine cycle	Number and arrangement of cylinders L—row; V—V-like	Cylinder diameter	Pistonstroke	Circulation frequency, rpm	Power, kW		*Specific fuel consumption, g /(kWh)*
			mm			Cylinder	Enging	
1163	4	12, 16, 20 V	230	280	1200–1300	300–370	4440–7400	–
8000	4	20 V	265	315	1150	410–450	8200–9000	–
Rolls—Royce Marine AS, Engines—Bergen								
C25:33P	4	6, 8, 9 L	250	330	720–750 900–1000	250 330	1440–2250 1920–3000	–
B32:40A	4	6, 8, 9 L 12, 16 V	320	400	720–750	500	2880–8000	–
B35:40PG	4	12, 16, 20 V	350	400	750	437	5250–8750	–
KRGS-G4	4	6, 8, 9 L	250	300	900–1000	220	1190–1990	–
KVGS-G4	4	12, 16, 18 V	250	300	900–1000	220	2380–3970	–
B35:40AG	4	12, 16, 20 V	350	400	720–750	437	5050–8750	–
SEMT Pielstick								
PA4-185VG	4	6, 8 L 6, 8, 12, 16, 18 V	185	210	1500	123	740–2215	–
PA5	4	4, 5, 6, 8 L 12, 16, 18 V	255	270	1000	220	880–3960	–
PA6	4	6, 8, 9 L 12, 16, 18, 20 V	280	290	1050	325	1950–6500	–
PA6BSTC	4	12, 16, 18, 20 V	280	330	1050	405	4860–8100	–

(continued)

(continued)

Enginemodel	Engine cycle	Number and arrangement of cylinders L—row; V—V-like	Cylinder diameter	Pistonstroke	Circulation frequency, rpm	Power, kW		*Specific fuel consumption, g /(kWh)*
			mm			Cylinder	Enging	
PC2.6B	4	12, 14, 16, 18, 20 V	400	500	600	750	9000–15,000	–
PC4.2	4	10, 12, 14, 16, 18 V	570	620	430	1215	12,150–21,870	–
PC4.2B	4	10, 12, 14, 16, 18, 20 V	570	660	430	1400	14,000–28,000	–
Volvo Penta AB								
D3	4	5 L	81	93,2	3000–4000	16.2–28.0	81–140	–
D4	4	4 L	103	110	2800–3500	33.00–55.25	132–221	–
D5A	4	4 L	108	130	2300	23.75–29.50	118	–
D6	4	6 L	103	110	3500	38.0–53.3	228–320	–
D9	4	6 L	120	138	2200–2600	52.2–70.5	313–423	–
D12	4	6 L	131	150	1800–2300	36.0–95.0	216–570	–
D25A	4	6 L	170	180	1650	80.8–86.7	485–520	–
D30A	4	6 L	170	220	1400	81.7–88.3	490–530	–
D34A	4	12 V	150	160	2000	58.4–64.7	701–776	–
D49A	4	12 V	170	180	1650–1800	80.8–88.7	970–1040	–
D65A	4	16 V	170	180	1650–1800	80.6–86.3	1290–1380	–
Wärtsilä Corporation								

(continued)

(continued)

Enginemodel	Engine cycle	Number and arrangement of cylinders L—row; V—V-like	Cylinder diameter	Pistonstroke	Circulation frequency, rpm	Power, kW		*Specific fuel consumption, g /(kWh)*
			mm			Cylinder	Enging	
Wärtsilä 20	4	4, 6, 8, 9 L	200	280	1000	180–200	720–1620	*184–193*
Wärtsilä 26	4	6, 8, 9 L 12, 16 V	260	320	900–1000	325–340	1950–2925 3900–5200	*182–184*
Wärtsilä 32	4	6, 7, 8, 9 L 12, 16, 18 V	320	400	750	500	3000–4500 6000–9000	*173–180*
Wärtsilä 32DF	4	6 L 12, 18 V	320	350	720–750	335–350	2010 4020–6300	–
Wärtsilä 38	4	6, 8, 9 L 12, 16 V	380	475	600	675–725	4350–6525 8700–11,600	*173–175*
Wärtsilä 46F	4	6, 7, 8, 9 L 12, 16 V	460	580	600	1250	7500–11,250 15,000–2000	*170–173*
Wärtsilä 46	4	6, 8, 9 L 12, 16 V	460	580	500–514	975–1050	5850–8775 11,700–15,600	*170–177*
Wärtsilä 50F	4	6, 8, 9 L 12, 16, 18 V	500	580	500–514	950	5700–8550 11,400–17,100	–
Wärtsilä 64	4	6, 7, 8 L	640	900	327,3	2150	12,900–17,200	*164*
Wärtsilä RT-Flex50	2	5, 6, 7, 8 L	500	2050	99–124	1130–1760	5650–12,960	*165–171*
Wärtsilä RT-Flex60C	2	5, 6, 7, 8, 9 L	600	2250	91–114	1690–2420	8450–21,780	*164–170*
Wärtsilä RT-Flex96C	2	6–14 L	960	2500	92–102	4000–5720	24,000–80,080	*163–171*

(continued)

(continued)

Enginemodel	Engine cycle	Number and arrangement of cylinders L—row; V—V-like	Cylinder diameter	Pistonstroke	Circulation frequency, rpm	Power, kW		*Specific fuel consumption, g /(kWh)*
			mm			Cylinder	Enging	
Wärtsilä RT-Flex82	2	6–12L	820	2696	87–102	3620–4520	21,720–54,240	*167–173*
Wärtsilä X35	2	5–8L	350	1550	142–167	695–870	3475–6960	*170–176*
Wärtsilä X40	2	5–8L	400	1770	124–146	910–1135	4550–9060	*169–177*
Wärtsilä X52DF	2	5–8L	520	2315	82–105	970–1490	4850–11,820	*182.3*
Wärtsilä X62	2	5–8L	620	2658	80–100	1540–2680	6380–21,820	*160–167*
Wärtsilä X92	*2*	*6–12L*	*920*	*3456*	*70–80*	*4310–5850*	*25,860–70,200*	*160–166*

Uncited References

1. Veselovskiy O. N., ShneybergYa. A. *Energeticheskayatekhnikaieerazvitie :ucheb. posobie*[Power engineering and its development: textbook]. Moscow, Vysshayashkola Publ., 1976. 298 p.
2. Yepifanov O. A. *Konstruktsiiasudnovykhkotliv: navch. posibnyk* [Construction of ship boilers: textbook]. Mykolaiv, NUK Publ., 2016. 198 p.
3. Kornilov E. V., Afanashchenko V. N., Boyko P. V. *Vspomogatelnyeiutilizatsionnyekotly-morskikhsudov*[Auxiliary and recovery boilers of sea vessels]. Odessa, Feniks Publ., 2004. 170 p.
4. *Morskoyentsiklopedicheskiyslovar : v 3 t.* [Marine Encyclopedic Dictionary: in 3 volumes]. Leningrad, Sudostroenie Publ., 1991, Vol. 2, 584 p.
5. Shapiro L. S. *Serdtsekorablya* [Heart of the ship]. Leningrad, Sudostroenie Publ., 1990. 144 p.
6. Dokhum Van Klaus. Ship Knowledge. Modern encyclopedia. Meppel, DOKMAR Publ., 2003. 3 p.
7. MARPOL 73/78. Consolidated Edition. London, IMO Publ., 2002. 511 p.

Z. Yang et al., *Marine Power Plant*,
https://doi.org/10.1007/978-981-33-4935-3